ADVANCES IN FOOD PROCESS ENGINEERING

*Novel Processing, Preservation,
and Decontamination of Foods*

Innovations in Agricultural and Biological Engineering

ADVANCES IN FOOD PROCESS ENGINEERING

Novel Processing, Preservation, and Decontamination of Foods

Edited by
Megh R. Goyal, PhD, PE
N. Veena, PhD
Ritesh B. Watharkar, PhD

First edition published 2023

Apple Academic Press Inc.
1265 Goldenrod Circle, NE,
Palm Bay, FL 32905 USA

760 Laurentian Drive, Unit 19,
Burlington, ON L7N 0A4, CANADA

CRC Press
6000 Broken Sound Parkway NW,
Suite 300, Boca Raton, FL 33487-2742 USA

4 Park Square, Milton Park,
Abingdon, Oxon, OX14 4RN UK

© 2023 by Apple Academic Press, Inc.

Apple Academic Press exclusively co-publishes with CRC Press, an imprint of Taylor & Francis Group, LLC

Reasonable efforts have been made to publish reliable data and information, but the authors, editors, and publisher cannot assume responsibility for the validity of all materials or the consequences of their use. The authors, editors, and publishers have attempted to trace the copyright holders of all material reproduced in this publication and apologize to copyright holders if permission to publish in this form has not been obtained. If any copyright material has not been acknowledged, please write and let us know so we may rectify in any future reprint.

Except as permitted under U.S. Copyright Law, no part of this book may be reprinted, reproduced, transmitted, or utilized in any form by any electronic, mechanical, or other means, now known or hereafter invented, including photocopying, microfilming, and recording, or in any information storage or retrieval system, without written permission from the publishers.

For permission to photocopy or use material electronically from this work, access www.copyright.com or contact the Copyright Clearance Center, Inc. (CCC), 222 Rosewood Drive, Danvers, MA 01923, 978-750-8400. For works that are not available on CCC please contact mpkbookspermissions@tandf.co.uk

Trademark notice: Product or corporate names may be trademarks or registered trademarks and are used only for identification and explanation without intent to infringe.

Library and Archives Canada Cataloguing in Publication

Title: Advances in food process engineering : novel processing, preservation, and decontamination of foods / edited by Megh R. Goyal, PhD, PE, N. Veena, PhD, Ritesh B. Watharkar, PhD.

Names: Goyal, Megh R., editor. | Veena, N. (Nagaraj), editor. | Watharkar, Ritesh B., editor.

Series: Innovations in agricultural and biological engineering.

Description: First edition. | Series statement: Innovations in agricultural and biological engineering | Includes bibliographical references and index.

Identifiers: Canadiana (print) 2022044384X | Canadiana (ebook) 20220443890 | ISBN 9781774911143 (hardcover) | ISBN 9781774911150 (softcover) | ISBN 9781003303848 (ebook)

Subjects: LCSH: Food industry and trade. | LCSH: Food—Preservation. | LCSH: Food contamination—Prevention.

Classification: LCC TP370 .A38 2023 | DDC 664—dc23

Library of Congress Cataloging-in-Publication Data

....

CIP data on file with US Library of Congress

....

ISBN: 978-1-77491-114-3 (hbk)
ISBN: 978-1-77491-115-0 (pbk)
ISBN: 978-1-00330-384-8 (ebk)

ABOUT THE BOOK SERIES: INNOVATIONS IN AGRICULTURAL AND BIOLOGICAL ENGINEERING

Under this book series, Apple Academic Press Inc. is publishing book volumes over a span of 8–10 years in the specialty areas defined by the American Society of Agricultural and Biological Engineers (<asabe.org>). Apple Academic Press Inc. aims to be a principal source of books in agricultural and biological engineering. We welcome book proposals from readers in areas of their expertise.

The mission of this series is to provide knowledge and techniques for agricultural and biological engineers (ABEs). The book series offers high-quality reference and academic content on agricultural and biological engineering (ABE) that is accessible to academicians, researchers, scientists, university faculty and university-level students, and professionals around the world.

Agricultural and biological engineers ensure that the world has the necessities of life, including safe and plentiful food, clean air and water, renewable fuel and energy, safe working conditions, and a healthy environment by employing knowledge and expertise of the sciences, both pure and applied, and engineering principles. Biological engineering applies engineering practices to problems and opportunities presented by living things and the natural environment in agriculture.

ABE embraces a variety of the following specialty areas (<asabe.org>): aquaculture engineering, biological engineering, energy, farm machinery and power engineering, food, and process engineering, forest engineering, information, and electrical technologies, soil, and water conservation engineering, natural resources engineering, nursery, and greenhouse engineering, safety, and health, and structures and environment.

For this book series, we welcome chapters on the following specialty areas (but not limited to):

- Academia to industry to end-user loop in agricultural engineering;
- Agricultural mechanization;
- Aquaculture engineering;
- Biological engineering in agriculture;

- Biotechnology applications in agricultural engineering;
- Energy source engineering;
- Farm to fork technologies in agriculture;
- Food and bioprocess engineering;
- Forest engineering;
- GPS and remote sensing potential in agricultural engineering;
- Hill land agriculture;
- Human factors in engineering;
- Impact of global warming and climatic change on agriculture economy;
- Information and electrical technologies;
- Irrigation and drainage engineering;
- Nanotechnology applications in agricultural engineering;
- Natural resources engineering;
- Nursery and greenhouse engineering;
- Potential of phytochemicals from agricultural and wild plants for human health;
- Power systems and machinery design;
- Robot engineering and drones in agriculture;
- Rural electrification;
- Sanitary engineering;
- Simulation and computer modeling;
- Smart engineering applications in agriculture;
- Soil and water engineering;
- Micro-irrigation engineering;
- Structures and environment engineering;
- Waste management and recycling;
- Any other focus areas.

For more information on this series, readers may contact:

Megh R. Goyal, PhD, PE
Book Series Senior Editor-in-Chief: Innovations in Agricultural and Biological Engineering
E-mail: goyalmegh@gmail.com

OTHER BOOKS ON AGRICULTURAL AND BIOLOGICAL ENGINEERING BY APPLE ACADEMIC PRESS, INC.

Management of Drip/Trickle or Micro Irrigation
Megh R. Goyal, PhD, PE, Senior Editor-in-Chief
Evapotranspiration: Principles and Applications for Water Management
Megh R. Goyal, PhD, PE and Eric W. Harmsen, PhD Editors

Book Series: Research Advances in Sustainable Micro Irrigation
Senior Editor-in-Chief: Megh R. Goyal, PhD, PE

- **Volume 1:** Sustainable Micro Irrigation: Principles and Practices
- **Volume 2:** Sustainable Practices in Surface and Subsurface Micro Irrigation
- **Volume 3:** Sustainable Micro Irrigation Management for Trees and Vines
- **Volume 4:** Management, Performance, and Applications of Micro Irrigation Systems
- **Volume 5:** Applications of Furrow and Micro Irrigation in Arid and Semi-Arid Regions
- **Volume 6:** Best Management Practices for Drip Irrigated Crops
- **Volume 7:** Closed Circuit Micro Irrigation Design: Theory and Applications
- **Volume 8:** Wastewater Management for Irrigation: Principles and Practices
- **Volume 9:** Water and Fertigation Management in Micro Irrigation
- **Volume 10:** Innovation in Micro Irrigation Technology

Book Series: Innovations and Challenges in Micro Irrigation
Senior Editor-in-Chief: Megh R. Goyal, PhD, PE

- Engineering Interventions in Sustainable Trickle Irrigation: Water Requirements, Uniformity, Fertigation, and Crop Performance
- Management Strategies for Water Use Efficiency and Micro Irrigated Crops: Principles, Practices, and Performance

viii — Other Books on Agricultural and Biological Engineering

- Micro-Irrigation Engineering for Horticultural Crops: Policy Options, Scheduling, and Design
- Micro-Irrigation Management: Technological Advances and Their Applications
- Micro-Irrigation Scheduling and Practices
- Performance Evaluation of Micro-Irrigation Management: Principles and Practices
- Potential of Solar Energy and Emerging Technologies in Sustainable Micro-Irrigation
- Principles and Management of Clogging in Micro-Irrigation
- Sustainable Micro-Irrigation Design Systems for Agricultural Crops: Methods and Practices

Book Series: Innovations in Agricultural and Biological Engineering
Senior Editor-in-Chief: Megh R. Goyal, PhD, PE

- Advanced Research Methods in Food Processing Technologies
- Advances in Food Process Engineering: Novel Processing, Preservation and Decontamination of Foods
- Advances in Green and Sustainable Nanomaterials: Applications in Energy, Biomedicine, Agriculture, and Environmental Science
- Advances in Sustainable Food Packaging Technology
- Analytical Methods for Milk and Milk Products, 2-volume set:
  - o Volume 1: Sampling Methods, Chemical and Compositional Analysis
  - o Volume 2: Physicochemical Analysis of Concentrated, Coagulated and Fermented Products
- Biological and Chemical Hazards in Food and Food Products: Prevention, Practices, and Management
- Bioremediation and Phytoremediation Technologies in Sustainable Soil Management, 4-volume set:
  - o Volume 1: Fundamental Aspects and Contaminated Sites
  - o Volume 2: Microbial Approaches and Recent Trends
  - o Volume 3: Inventive Techniques, Research Methods, and Case Studies
  - o Volume 4: Degradation of Pesticides and Polychlorinated Biphenyls
- Dairy Engineering: Advanced Technologies and Their Applications
- Developing Technologies in Food Science: Status, Applications, and Challenges

Other Books on Agricultural and Biological Engineering

- Emerging Technologies in Agricultural Engineering
- Engineering Interventions in Agricultural Processing
- Engineering Interventions in Foods and Plants
- Engineering Practices for Agricultural Production and Water Conservation: An Interdisciplinary Approach
- Engineering Practices for Management of Soil Salinity: Agricultural, Physiological, and Adaptive Approaches
- Engineering Practices for Milk Products: Dairyceuticals, Novel Technologies, and Quality
- Enzyme Inactivation in Food Processing: Technologies, Materials, and Applications
- Field Practices for Wastewater Use in Agriculture: Future Trends and Use of Biological Systems
- Flood Assessment: Modeling and Parameterization
- Food Engineering: Emerging Issues, Modeling, and Applications
- Food Process Engineering: Emerging Trends in Research and Their Applications
- Food Processing and Preservation Technology: Advances, Methods, and Applications
- Food Technology: Applied Research and Production Techniques
- Functional Dairy Ingredients and Nutraceuticals: Physicochemical, Technological, and Therapeutic Aspects
- Handbook of Research on Food Processing and Preservation Technologies, 5-volume set:
  - o Volume 1: Nonthermal and Innovative Food Processing Methods
  - o Volume 2: Nonthermal Food Preservation and Novel Processing Strategies
  - o Volume 3: Computer-Aided Food Processing and Quality Evaluation Techniques
  - o Volume 4: Design and Development of Specific Foods, Packaging Systems, and Food Safety
  - o Volume 5: Emerging Techniques for Food Processing, Quality, and Safety Assurance
- Modeling Methods and Practices in Soil and Water Engineering
- Nanotechnology and Nanomaterial Applications in Food, Health, and Biomedical Sciences
- Nanotechnology Applications in Agricultural and Bioprocess Engineering: Farm to Table

Other Books on Agricultural and Biological Engineering

- Nanotechnology Applications in Dairy Science: Packaging, Processing, and Preservation
- Nanotechnology Horizons in Food Process Engineering, 3-volume set:
  - o Volume 1: Food Preservation, Food Packaging and Sustainable Agriculture
  - o Volume 2: Scope, Biomaterials, and Human Health
  - o Volume 3: Trends, Nanomaterials, and Food Delivery
- Novel and Alternative Methods in Food Processing: Biotechnological, Physicochemical, and Mathematical Approaches
- Novel Dairy Processing Technologies: Techniques, Management, and Energy Conservation
- Novel Processing Methods for Plant-Based Health Foods: Extraction, Encapsulation and Health Benefits of Bioactive Compounds
- Novel Strategies to Improve Shelf-Life and Quality of Foods: Quality, Safety, and Health Aspects
- Phytochemicals and Medicinal Plants in Food Design: Strategies and Technologies for Improved Healthcare
- Processing of Fruits and Vegetables: From Farm to Fork
- Processing Technologies for Milk and Milk Products: Methods, Applications, and Energy Usage
- Quality Control in Fruit and Vegetable Processing: Methods and Strategies
- Scientific and Technical Terms in Bioengineering and Biological Engineering
- Soil and Water Engineering: Principles and Applications of Modeling
- Soil Salinity Management in Agriculture: Technological Advances and Applications
- State-of-the-Art Technologies in Food Science: Human Health, Emerging Issues and Specialty Topics
- Sustainable and Functional Foods from Plants
- Sustainable Biological Systems for Agriculture: Emerging Issues in Nanotechnology, Biofertilizers, Wastewater, and Farm Machines
- Sustainable Nanomaterials for Biomedical Engineering: Impacts, Challenges, and Future Prospects
- Sustainable Nanomaterials for Biosystems Engineering: Trends in Renewable Energy, Environment, and Agriculture
- Technological Interventions in Dairy Science: Innovative Approaches in Processing, Preservation, and Analysis of Milk Products

Other Books on Agricultural and Biological Engineering

- Technological Interventions in Management of Irrigated Agriculture
- Technological Interventions in the Processing of Fruits and Vegetables
- Technological Processes for Marine Foods, From Water to Fork: Bioactive Compounds, Industrial Applications, and Genomics
- The Chemistry of Milk and Milk Products: Physicochemical Properties, Therapeutic Characteristics, and Processing Methods

ABOUT SENIOR-EDITOR-IN-CHIEF

Megh R. Goyal, PhD, PE

Megh R. Goyal, PhD, PE, is currently a retired professor of agricultural and biomedical engineering from the General Engineering Department at the College of Engineering at the University of Puerto Rico–Mayaguez Campus (UPRM); and Senior Acquisitions Editor and Senior Technical Editor-in-Chief for Agricultural and Biomedical Engineering for Apple Academic Press Inc.

During his long career, Dr. Megh R. Goyal has worked as a Soil Conservation Inspector; Research Assistant at Haryana Agricultural University and Ohio State University; Research Agricultural Engineer/Professor at the Department of Agricultural Engineering of UPRM; and Professor of Agricultural and Biomedical Engineering in the General Engineering Department of UPRM. He spent a one-year sabbatical leave in 2002–2003 at the Biomedical Engineering Department of Florida International University, Miami, USA.

Dr. Goyal was the first agricultural engineer to receive the professional license in agricultural engineering from the College of Engineers and Surveyors of Puerto Rico. In 2005, he was proclaimed the "Father of Irrigation Engineering in Puerto Rico for the Twentieth Century" by the American Society of Agricultural and Biological Engineers, Puerto Rico Section, for his pioneering work on micro irrigation, evapotranspiration, agroclimatology, and soil and water engineering.

During his professional career of 52 years, he has received many awards, including Scientist of the Year, Membership Grand Prize for the American Society of Agricultural Engineers Campaign, Felix Castro Rodriguez Academic Excellence Award, Man of Drip Irrigation by the Mayor of Municipalities of Mayaguez/Caguas/Ponce and Senate/Secretary of Agriculture of ELA, Puerto Rico, and many others. He has been recognized as one of the experts "who rendered meritorious service for the development of [the] irrigation sector in India" by the Water Technology Centre of Tamil Nadu Agricultural University in Coimbatore, India, and ASABE who bestowed on him the 2018 Netafim Microirrigation Award.

Dr. Goyal has authored more than 200 journal articles and edited more than 100 books. AAP has published many of his books, including *Management of Drip/Trickle or Micro Irrigation; Evapotranspiration: Principles and Applications for Water Management;* ten-volume set on *Research Advances in Sustainable Micro Irrigation.* Dr. Goyal has also developed several book series with AAP, including Innovations in Agricultural & Biological Engineering (with over 60 titles in the series to date), Innovations and Challenges in Micro Irrigation; and Innovations in Plant Science for Better Health: From Soil to Fork.

Dr. Goyal received his BSc degree in Engineering from Punjab Agricultural University, Ludhiana, India, and his MSc and PhD degrees from the Ohio State University, Columbus, Ohio, USA. He also earned a Master of Divinity degree from the Puerto Rico Evangelical Seminary, Hato Rey, Puerto Rico, USA.

ABOUT THE EDITORS

N. Veena, PhD
Assistant Professor,
Department of Dairy Chemistry,
College of Dairy Science and Technology,
Guru Angad Dev Veterinary and
Animal Sciences University,
Ludhiana–141004, Punjab, India

N. Veena, PhD, is working as an Assistant Professor in the Department of Dairy Chemistry, College of Dairy Science and Technology, Guru Angad Dev Veterinary and Animal Sciences University, Ludhiana, Punjab, India.

She received her BTech degree in Dairy Technology from Karnataka Veterinary, Animal, and Fisheries Sciences University, Bidar, Karnataka, India, and her MTech and PhD degrees in Dairy Chemistry from ICAR-National Dairy Research Institute, Karnal, Haryana, India.

Her area of expertise includes fortification of milk, development and characterization of functional dairy foods, and nanoencapsulation of nutraceuticals and their delivery. She is the recipient of junior research and institutional fellowships during her master's and doctoral programs from the ICAR-National Dairy Research Institute, Karnal, Haryana, India. She has been involved in different institutional and externally funded research projects such as Milkfed (Punjab), ICAR, DST, and RKVY as PI and Co-PI.

She has published 17 research papers in peer-reviewed journals, one edited book, 11 book chapters, 15 popular articles, and seven practical manuals. She has developed the e-learning content for post-graduate dairy chemistry courses under an NAHEP project. She has presented her research work at various national and international seminars and conferences and received several awards for best oral and poster presentations. She holds membership in various professional bodies such as the Indian Dairy Association, Association of Food Scientists and Technologists (India) and Dairy Technology Society of India, etc.

Ritesh B. Watharkar, PhD
Assistant Professor, College of Agriculture and Food Business Management,
MIT-ADT University, Rajbaugh,
Solapur–Pune Highway, Near Bharat Petrol Pump,
Loni Kalbhor 412201, Maharashtra–India

Ritesh B. Watharkar, PhD, is currently working as an Assistant Professor in the College of Agriculture and Food Business Management at MIT-ADT University, Maharashtra, India.

He received his BTech degree in Food Technology from Dr. Balasaheb Sawant Konkan Agriculture University, Dapoli, Maharashtra, India; his MTech in Food Processing and Engineering degree is from the Karunya University, India; his PhD in Food Engineering and Technology degree from Tezpur University (A Central University), Assam, India.

He was formerly a Senior Research Fellow under a funded project by the Ministry of Food Processing Industries, New Delhi (2013–2015), and Assistant Professor at K.K. Wagh Food Tech College (2011–2013); at Shramshakti College of Food Technology, affiliated by Mahatma Phule Agricultural University (2019–2020).

During his professional career, he received awards such as senior research fellow under the UNESCO/China Great Wall Fellowship Program and worked for one year in the School of Biosystem Engineering and Food Science at Zhejiang University, Hangzhou, China. He was also awarded for best popular article from *Krishidut Magazine,* Maharashtra, India. Currently, he is a life member of the Association of Food Scientists and Technologists, India. He has published more than 10 journal articles and attended more than 10 national and international conferences. He also published more than 10 popular articles in the *Marathi Literacy Magazine.*

CONTENTS

Contributors .... *xix*

Abbreviations .... *xxiii*

Preface .... *xxix*

PART I: Novel Food Processing Methods 1

1. **Ultrasound Processing of Foods: An Overview** 3
    Sandip Prasad, Shruthy Ramesh, and Preetha Radhakrishnan

2. **Supercritical Fluid Technology: Recent Trends in Food Processing** 25
    Mayuri A. Patil, Aniket P. Sarkate, Bhagwan K. Sakhale, and Rahul C. Ranveer

3. **Extrusion Technology of Food Products: Types and Operation** 45
    Anjali Sudhakar, Kanupriya Choudhary, Subir Kumar Chakraborty, and Cinu Varghese

4. **Fundamentals of Advanced Drying Methods of Agricultural Products** 69
    Shikha Pandhi and Arvind Kumar

5. **Encapsulation Methods in the Food Industry: Mechanisms and Applications** 93
    Rekha Rani, Payal Karmakar, Neha Singh, and Ronit Mandal

PART II: Membrane Technology in Food Processing 141

6. **Membrane Processing in the Food Industry** 143
    S. Abdullah, Rama C. Pradhan, and Sabyasachi Mishra

7. **Potential of Membrane Technology in Food Processing Systems** 177
    Olviya S. Gonsalves, Rahul S. Zambare, and Parag R. Nemade

PART III: Mathematical Modelling in Food Processing 215

8. **Mathematical Modeling in Fermentation Processes** 217
    Ankit Paliwal and Himjyoti Dutta

9. Modeling of Hydration Kinetics in Grains 237

Ankit Paliwal, Ashish M. Mohite, and Neha Sharma

PART IV: Decontamination Methods for Foods 253

10. Cold Plasma Processing Methods: Impact of Decontamination on Food Quality 255

Nikhil K. Mahnot, Sayantan Chakraborty, Kuldeep Gupta, and Sangeeta Saikia

11. Microbial Decontamination of Foods with Cold Plasma: A Non-Invasive Approach 277

Irfan Khan, Nazia Tabassum, and Abdul Haque

12. Food Irradiation: Concepts, Applications, and Future Prospects..........299

Subhajit Ray

Index..*321*

CONTRIBUTORS

S. Abdullah
PhD Research Scholar, Department of Food Process Engineering, National Institute of Technology, Rourkela–769001, Odisha, India

Sayantan Chakraborty
PhD Research Fellow, Department of Bio-Engineering and Technology, Gauhati University Institute of Science and Technology (GUIST), Gauhati–781014, Assam, India

Subir Kumar Chakraborty
Senior Scientist, Agro-Produce Processing Department, ICAR–Central Institute of Agricultural Engineering, Nabibagh, Bhopal–462038, Madhya Pradesh, India

Kanupriya Choudhary
PhD Research Scholar, Department of Agricultural Processing and Structures, ICAR–Central Institute of Agricultural Engineering, Nabibagh, Bhopal–462038, Madhya Pradesh, India

Himjyoti Dutta
Assistant Professor, Department of Food Technology, Mizoram University, Aizawl–796004, Mizoram, India

Olviya S. Gonsalves
PhD Research Fellow, Department of Chemical Engineering, Institute of Chemical Technology, Nathalal Parekh Marg, Matunga–400019, Mumbai, Maharashtra, India

Megh R. Goyal
Retired Faculty in Agricultural and Biomedical Engineering from College of Engineering at the University of Puerto Rico–Mayaguez Campus; and Senior Technical Editor-in-Chief in Agricultural and Biomedical Engineering for Apple Academic Press Inc.; P.O. Box 86, Rincon–PR–00677-0086, USA

Kuldeep Gupta
Research Associate, Department of Molecular Biology and Biotechnology, Tezpur University, Tezpur–784028, Assam, India

Abdul Haque
PhD Research Scholar, Department of Post-Harvest Engineering and Technology, Aligarh Muslim University, Aligarh–202002, Uttar Pradesh, India

Payal Karmakar
PhD Research Scholar, Dairy Chemistry Division, ICAR–National Dairy Research Institute, Karnal–132001, Haryana, India

Irfan Khan
Guest Faculty, Department of Food Processing and Technology, Gautam Buddha University, Greater Noida–201308, Uttar Pradesh, India

Arvind Kumar
Assistant Professor, Department of Dairy Science and Food Technology, Institute of Agricultural Sciences, Banaras Hindu University (BHU), Varanasi–221005, Uttar Pradesh, India

Contributors

Nikhil K. Mahnot
Assistant Professor, Department of Bio-Engineering and Technology, Gauhati University Institute of Science and Technology (GUIST), Gauhati–781014, Assam, India

Ronit Mandal
PhD Research Scholar, Department of Food, Nutrition, and Health, Faculty of Land and Food Systems, 2205 East Mall, University of British Columbia, Vancouver V6T 1Z4, BC, Canada

Sabyasachi Mishra
Assistant Professor, Department of Food Process Engineering, National Institute of Technology, Rourkela–769001, Odisha, India

Ashish M. Mohite
Assistant Professor, Amity Institute of Food Technology, Amity University, Noida–201313, Uttar Pradesh, India

Parag R. Nemade
UGC Assistant Professor, Department of Chemical Engineering and Department of Oils, Oleochemicals, and Surfactant Technology, Institute of Chemical Technology, Nathalal Parekh Marg, Matunga–400019, Mumbai, Maharashtra, India; and Deputy Director, ICT, Institute of Chemical Technology, Marathwada Campus, BT-6/7, Biotechnology Park, Additional MIDC, Jalna–431203, Maharashtra, India

Ankit Paliwal
Lecturer (Food Engineering), Department of Applied Sciences, Fiji National University, Suva, Nadi - Fiji

Shikha Pandhi
PhD Research Scholar, Department of Dairy Science and Food Technology, Institute of Agricultural Sciences, Banaras Hindu University (BHU), Varanasi–221005, Uttar Pradesh, India

Mayuri A. Patil
Assistant Professor, Yash Institute of Pharmacy, South City, Waluj, Aurangabad–431136, Maharashtra, India

Rama C. Pradhan
Associate Professor, Department of Food Process Engineering, National Institute of Technology, Rourkela–769001, Odisha, India

Sandip Prasad
PhD Scholar, Department of Food Process Engineering, SRM Institute of Science and Technology, Kattankulathur, Chengalpattu–603203, Tamil Nadu, India

Preetha Radhakrishnan
Associate Professor, Department of Food Process Engineering, SRM Institute of Science and Technology, Kattankulathur, Chengalpattu–603203, Tamil Nadu, India

Shruthy Ramesh
PhD Scholar, Department of Food Process Engineering, School of Bioengineering, SRM Institute of Science and Technology, Kattankulathur, Chengalpattu–603203, Tamil Nadu, India

Rekha Rani
Assistant Professor, Warner College of Dairy Technology, SHUATS, Prayagraj–211007, Uttar Pradesh, India

Rahul C. Ranveer
Department of Post-Harvest Management of Meat, Poultry, and Fish, PG Institute of Post-Harvest Management, Killa-Roha, Raigad–402116, Maharashtra, India

Contributors

Subhajit Ray
Associate Professor and Head, Department of Food Engineering and Technology,
Central Institute of Technology, Deemed to be University (Under MHRD, Government of India),
Kokrajhar–783370, Assam, India

Sangeeta Saikia
Visiting Faculty, Department of Bio-Engineering and Technology, Gauhati University Institute of
Science and Technology (GUIST), Gauhati–781014, Assam, India

Bhagwan K. Sakhale
Professor, Department of Chemical Technology, Dr. Babasaheb Ambedkar Marathwada University,
Near Soneri Mahal, Jaisingpura, Aurangabad–431004, Maharashtra, India

Aniket P. Sarkate
Assistant Professor, Department of Chemical Technology,
Dr. Babasaheb Ambedkar Marathwada University, Near Soneri Mahal, Jaisingpura, Aurangabad–431004,
Maharashtra, India

Neha Sharma
Associate Professor, Amity Institute of Food Technology, Amity University, Noida–201313,
Uttar Pradesh, India

Neha Singh
PhD Research Scholar, Food Science and Technology Section, SHUATS, Prayagraj–211007,
Uttar Pradesh, India

Anjali Sudhakar
PhD Research Scholar, Department of Agricultural Processing and Structures,
ICAR–Central Institute of Agricultural Engineering, Nabibagh, Bhopal–462038, Madhya Pradesh, India

Nazia Tabassum
Assistant Professor, Department of Post-Harvest Engineering and Technology,
Aligarh Muslim University, Aligarh – 202002, Uttar Pradesh, India

Cinu Varghese
PhD Research Scholar, Rural Development Center, Indian Institute of Agricultural Engineering,
Kharagpur–721302, West Bengal, India

N. Veena
Assistant Professor, Department of Dairy Chemistry, College of Dairy Science and Technology,
Guru Angad Dev Veterinary and Animal Sciences University, Ludhiana–141004, Punjab, India

Ritesh B. Watharkar
Assistant Professor, College of Management, Food and Agriculture Management Department,
MIT Art, Design, and Technology University, Loni Kalbhor, Pune – 412201, Maharashtra, India

Rahul S. Zambare
Post-Doctoral Researcher, Center for Advanced 2D Materials, National University of Singapore,
Singapore–117546

ABBREVIATIONS

ABE	acetone–butanol–ethanol
ADP	adenosine diphosphate
AEM	anion exchange membrane
AFM	atomic force microscopy
Al_2O_3	aluminum oxide
Ar	argon
ATP	adenosine triphosphate
AU	arbitrary units
a_w	water activity
BOD	biological oxygen demand
BPM	bi-polar membrane
Bq	becquerel
BU	biological units
C_2H_4	ethylene
CA	cellulose acetate
CAGR	compound annual growth rate
CAS	controlled atmosphere storage
CCFRA	Campden and Chorleywood Food Research Association
CDM	cell dry mass
CEM	cation exchange membrane
CFD	computational fluid dynamics
CFU	colony-forming unit
CFV	crossflow velocity
CMC	carboxymethylcellulose
CNF	cellulose nanofiber
Co	cobalt
CO_2	carbon dioxide
CoA	coenzyme A

COD	chemical oxygen demand
CP	cold plasma
CP	concentration polarization
CPT	cold plasma technology
Cs	Cesium
DBD	dielectric barrier discharge
DCMD	direct contact membrane distillation
DCS	downward concave shape
DF	diafiltration
DHA	docosahexaenoic acid
DI	dialysis
DLS	dynamic light scattering
DNA	deoxyribonucleic acid
DnaK	chaperone protein DnaK gene
DW	dry weight
ED	electrodialysis
EFSA	European Food Safety Authority
EHD	electrohydrodynamic
EMC	equilibrium moisture content
EPA	eicosapentaenoic acid
EVOH	ethylene-vinyl alcohol
FDA	Food and Drug Administration
FO	forward osmosis
GC	gas chromatography
GD	gas diffusion
GMP	good manufacturing practice
GRAS	generally recognized as safe
Gy	gray
H_2O_2	hydrogen peroxide
HDPE	high-density polyethylene
He	helium
HEF	high electric field
HFM	hollow-fiber modules

Abbreviations xxv

HNO$_2$	nitrous acid
HPLC	high-performance liquid chromatography
HTST	high-temperature short-time
HuNoV	human norovirus
IAEA	International Atomic Energy Agency
ICRP	International Commission on Radiation Protection
IG	immunoglobulin
IR	infrared
kDa	kilodalton
kGy	kilogray
kHz	kilohertz
kRad	kilorad
kV	kilovolt
L/D	length to diameter
LAB	lactic acid bacteria
LDPE	low-density polyethylene
LED	light-emitting diode
m/min	meter per minute
MAP	modified atmospheric packaging
MBR	membrane bioreactor
MD	membrane distillation
MD	microwave discharge
MeV	million electron volts
MF	microfiltration
MHz	megahertz
min	minutes
MP	membrane processing
MPa	megapascal
MPT	membrane process technology
Mrad	megarad
MS	manosonication
MTS	manothermosonication
MWCO	molecular weight cut off

N	nitrogen
NAD	nicotinamide adenine dinucleotide
NADH	nicotinamide adenine dinucleotide hydrogen
Ne	neon
NF	nanofiltration
nm	nanometer
NO_2	nitrite
NTP	non-thermal plasma
O/W	oil-in-water
O/W/O	oil-in-water-in-oil
O_2	oxygen
OD	osmotic distillation
OH•	hydroxyl radicals
OMD	osmotic membrane distillation
OxyR	oxygen-regulated gene
PA	polyamide
PAW	plasma-activated water
P_c	critical pressure
PDI	polydispersity index
PE	polyethylene
PES	polyethersulfone
PET	polyethylene terephthalate
PFM	plate and frame modules
PP	polypropylene
PPO	polyphenol oxidase
PS	polysulfone
PSEO	*Pimpinella saxifraga* essential oil
PTFE	polytetrafluoroethylene
PV	pervaporation
PVDF	polyvinylidene fluoride
PVP	polyvinylpyrrolidone
RNS	reactive nitrogen species
RO	reverse osmosis

Abbreviations

ROS	reactive oxygen species
RW	refractance window
SCF	supercritical fluid
SDBD	surface dielectric barrier discharge
SEC	size exclusion chromatography
SEM	scanning electron microscopy
SFC	supercritical fluid chromatography
SFD	spray-freeze-drying
SFE	supercritical fluid extraction
SiO_x	oxide of silicon
SLPEs	solid lipid particle emulsions
SWM	spiral wound modules
TEM	transmission electron microscopy
TiO_2	titanium oxide
TMP	transmembrane pressure
TS	thermosonication
TSS	total soluble solids
TVP	texturized vegetable protein
UAE	ultrasound-assisted extraction
UF	ultrafiltration
UNRA	United Nations risk assessment
US	ultrasound
UV	ultraviolet
W/O	water-in-oil
WHO	World Health Organization
WPC	whey protein concentrate
WPI	whey protein isolate
XRD	X-ray diffraction
ZrO_3	zirconium oxide

PREFACE

In the current era, the food industry sector is greatly motivated by the needs and demands of consumers. Present-day consumers demand healthy and safe foods that are processed in such a way so as to retain most of their natural properties. However, some level of processing is a must for any food product before it reaches from 'Farm to a consumer's fork.' However, food safety and quality aspects are major concerns in food processing. Novel food processing and preservation technologies are known for their product quality, safety, and efficiency. Along with product development, these novel technologies have exciting opportunities and interesting challenges in the area of food processing.

This book illustrates various applications of novel food processing, preservation, and decontamination methods. The book consists of four main sections. The first part of this book, "Novel Food Processing Methods," explores the principles, benefits, and techniques used, recent developments and applications of ultrasonication, supercritical fluid (SCF) extraction and supercritical fluid chromatography (SFC), extrusion technology, advanced drying and dehydration technologies, and encapsulation methods as an aid to processing of food. The second part of this book, "Membrane Technology in Food Processing," provides an overview of basic membrane processing (MP) technologies, their advantages and disadvantages, membrane modules, types of membranes, and membrane technologies, as well as various applications of membrane process in dairy processing, starch processing, beverage processing, sugar manufacturing, oil processing, and treatment of industrial food processing waste.

The third part of this book, "Mathematical Modelling in Food Processing," explores the application and use of mathematical models for measuring and regulating fermentation procedures, as well as understanding the hydration kinetics of grains that can help in the optimization and scaling of process on a large industrial scale. The last part of this book, "Decontamination Methods for Foods," gives an overview of various aspects, including concepts, basic principles, potential applications, and future prospects. The limitations of cold plasma (CP) technology and irradiation in the food processing sector have been summarized.

This book volume, *Advances in Food Process Engineering: Novel Processing, Preservation, and Decontamination of Foods*, is published in the book series on 'Innovations in Agricultural and Biological Engineering'. It is a treasure house of information and excellent reference material for researchers, scientists, students, growers, traders, processors, industries, and others on novel food processing and preservation techniques. This book is important because it consolidates recent literature evidence on the innovation in advanced food processing (ultrasonication, supercritical fluid extraction (SFE)/chromatography, membrane processing, extrusion technology) and preservation (cold plasma and irradiation) methods for the development of safe and quality foods. Apart from new food processing and preservation technologies, concepts of mathematical modeling and their application in food processing have been summarized.

This book has exceeded our anticipation due to the support of all contributing authors to this book, who have been most valuable in this compilation. The names of contributing authors are mentioned in respective chapters. The editorial team appreciates the authors for their proficiency, pledge, and perseverance. The effort of contributors and the editorial team will make this volume more informative for readers from different processing industries and academic institutions.

We would like to thank the staff at Apple Academic Press, Inc. for their valuable assistance, great support, and guidance in the right direction all the way through this project.

We appeal to the reader to suggest your feedback that may benefit from improving the subsequent edition of this book.

We also wish to thank our friends and families for their unlimited support, encouragement, love, and affection during the course of editing this book volume. Finally, and most importantly, we would like to commend our spouses, Subhadra, Mohan Kumar, and Aishwarya, for their understanding and patience throughout this project.

—Editors

PART I
NOVEL FOOD PROCESSING METHODS

CHAPTER 1

ULTRASOUND PROCESSING OF FOODS: AN OVERVIEW

SANDIP PRASAD, SHRUTHY RAMESH, and
PREETHA RADHAKRISHNAN

ABSTRACT

Nowadays, ultrasound (US) technology provides a promising role in the food industry for cutting, extraction, filtration, crystallization, deforming, emulsification, and so on. Food preservation by the US has a remarkable advantage. The US is also used for cooking, inactivation of microorganisms, drying, and freezing. The separation of fat in milk is one of its major applications in the dairy industry. In the juice processing industry, US-assisted filtration helps to increase the life of membrane filters. Moreover, it helps in functionality improvement in protein foods and also avoids the adhesion of the food residues during extrusion/demolding. The US is used for meat tenderization and fermented food preparation, including pickles.

1.1 INTRODUCTION

The usage and development of the US started at the beginning of 1790, inspired by bats. Bats and dolphins can identify US waves at different intensities for capturing prey. Bats can identify the low intensity of US waves, whereas dolphins can identify the higher intensity US waves [58]. The purpose of using the US in the field of electronics and the biomedical field is quite common. But recently, the US has been used in the food industry for

---

Advances in Food Process Engineering: Novel Processing, Preservation, and Decontamination of Foods.
Megh R. Goyal, N. Veena, & Ritesh B. Watharkar (Eds.)
© 2023 Apple Academic Press, Inc. Co-published with CRC Press (Taylor & Francis)

food preservation, drying of food, inactivation of microbes/enzymes in food, etc. [58]. US processing or ultrasonication uses high-frequency sound waves (above 20 kHz) commonly applied in the food industry for a wide range of applications. The US uses the food industry as a non-thermal mechanical method of processing where the nutritional value of the food is least affected [12]. The US is classified based on the amount of energy and its frequency. They are low energy-high frequency (intensity < 1 Wcm^2 and frequency > 100 kHz) and high energy-low frequency (intensity > 1 Wcm^2 and frequency ranging between 20 kHz and 500 kHz) waves of US [57]. The high-pressure US has become one of the major alternatives during large-scale food processing, which is used in extraction, degassing, extrusion, crystallization, and so on.

The US-assisted method has several advantages over conventional methods. One such advantage is the reduction of reaction time apart from improving the hydrolysis rate and mass transfer [2, 39]. The US and its advantage have been exploited extensively in the food industry. The non-destructive property of the ultrasound-assisted extraction (UAE) enabled researchers to identify and study the functional properties of complex phyto-chemicals since the US did not affect the organoleptic properties. UAE helps in reducing the extraction time; instrumentation adopted energy consumption and manpower utilization. UAE can operate at room temperature and can accommodate any solvent depending upon the desired extract. Vegetables, fish, meat, dairy products, and various other food qualities are evaluated using the US. The maturity of the cheddar cheese was also evaluated using the US technique in a non-destructive manner [22].

This chapter describes the concept of the US and the principle behind ultrasonication, methods of US processing, and advantages and limitations of ultrasonication. This chapter also discusses in detail its application in different food product processing, including milk, meat, fruits, vegetables, carbonated drinks, fish, etc.

1.2 PRINCIPLE OF OPERATION OF ULTRASOUND (US)

The cavities or bubbles are formed when the sound enters a medium, creating compression and rarefaction waves of particles in the medium. The cavities grow and become unstable with continuous cycles of the US, and they finally collapse, which results in chemical, thermal, and mechanical effects, including the development of high temperature and pressure. The

Ultrasound Processing of Foods: An Overview

temperature generated can reach up to 5,000 K and pressure up to 2,000 atm [67]. Even though these collapsing bubbles may reach extremely high temperatures, they are restricted to the collapsing bubble's core and surface areas [19]. The chemical effect includes the generation of free radicals, and mechanical effects include turbulence, shear stress, and collapse pressure. The principle of cavitation is shown in Figure 1.1.

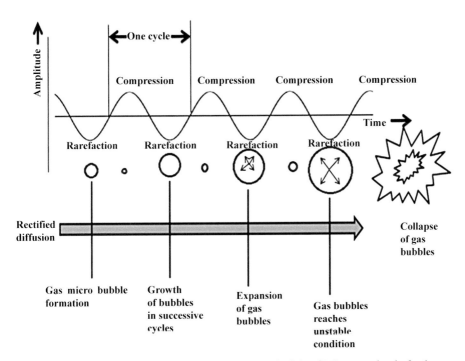

FIGURE 1.1 Schematic representation of working principle of US processing in foods.

1.3 METHODS OF ULTRASONICATION

Ultrasonication can be applied in two ways; either applying directly to the sample or indirectly through the container walls of the sample. Based on this, ultrasonication can be classified into two types; direct ultrasonication and indirect ultrasonication. In addition to that, the US can be applied in combination with temperature and pressure. Accordingly, it can be again classified into manosonication (MS), thermosonication (TS), and manothermosonication (MTS) (Figure 1.2) [58].

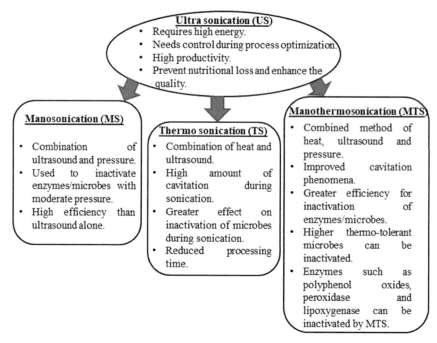

FIGURE 1.2 Types of sonication and its advantages.

1.3.1 DIRECT ULTRASONICATION

US waves were generated and applied with the help of ultrasonic probes in the direct method. The ultrasonic probes stay immersed in the solution throughout the ultrasonication process, and there are no barriers between the probe and the sample except the solution itself. Probes are nowadays made of glass material because of their non-reactive nature. Earlier, probes were made of pure titanium, which caused the metal to be detached from the probe and contaminate the sample. Also, the pure titanium-based probe gets easily contaminated by metals such as chromium and aluminum [61]. Moreover, volatile compounds can be lost due to the ultrasonic probe's use in an open condition.

1.3.2 INDIRECT ULTRASONICATION

In this method, the US waves have two barriers to cross. One is the solution, and the other is the sample container. The US waves must surpass these barriers to cause their effect on the sample, and this phenomenon eventually

Ultrasound Processing of Foods: An Overview

causes reduced US intensity reaching the sample. Indirect ultrasonication is done with a US bath or slightly powerful sonoreactor [61].

1.4 ULTRASONICATION TECHNOLOGY IN FOOD PROCESSING

1.4.1 DEFOAMING

Gallego et al. [20] reported that high-intensity ultrasonic waves could be used in a packed environment under sterile conditions to break form without high airflow, making it an energy-efficient process in the food industry. A focused ultrasonic generator was successfully utilized in high-speed bottling and canning lines in the food industry (carbonic beverages) to control the excess foam. The mechanism of the US-assisted defoaming is shown in Figure 1.3 [59].

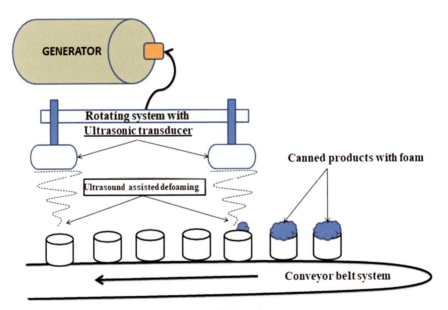

FIGURE 1.3 Mechanism of the US-assisted defoaming.

1.4.2 EXTRACTION

UAE has been reported widely in the food sector. Even though higher frequencies are utilized for the UAE process, commonly used frequency ranges from 20 kHz to 100 kHz. The cavitation phenomena generate a high shear force on the surface of the product and which leads to micro jetting. It causes erosion, surface peeling, and particle breakdown. Moreover, macro-turbulence and macro-mixing are caused by the implosion of cavitation bubbles [13]. Zhong and Wang [76] reported that UAE using solvents is an effective method to extract bioactive compounds from plants, crops, and herbs. Ultrasonication was used along with fermentation to improve the efficiency of extraction [56].

Hu et al. [31] have proved that a combination of ultrasonication with ethyl acetate as a solvent can replace an amine-based complex extraction system. Uju et al. [72] used ultrasonication at 40 kHz with varying process times to reduce phycoerythrin extraction time (red photosynthetic pigment) from *Kappaphycus alwarezii* seaweed and also enhanced its concentration.

1.4.3 FILTRATION

In the food industry, filtration is a major operation which mainly used to separate solid contents from liquids. During the filtration process, clogging/caking at the membrane surface is a major issue. It can be managed by ultra-sonically assisted filtration methods [3, 39]. Generally, US-assisted filtration uses in fruit juice industries for extraction. The study of Muthukumaran et al. [50] describes the successful membrane filtration using the US for apple juice extraction and infers an efficient flow rate. Similarly, in another study of whey filtration, US-assisted irradiation provides an improved flow rate [39]. The main advantage of the US-assisted filtration method is that it extends the life of filters and enhances the flow rate [28].

1.4.4 CRYSTALLIZATION/FREEZING

Preservation of food materials by freezing for processing is a common procedure in food preservation. During freezing, the water content inside the food items is converted into ice crystals and which leads to the deformation or structural loss or loss of quality of food items. The above-said drawback can be overcome with the help of US-assisted conventional cooling. It can be carried out within a shorter time. Moreover, it produces smaller ice crystals inside the food, hence reducing the damage and thereby improving the quality of food items [17].

Ultrasound Processing of Foods: An Overview

1.4.5 DRYING

Drying by conventional methods such as dehydration or desiccation takes considerable time to eliminate moisture content; and needs high temperature, which leads to the damage of food items [66]. The above-said problems will decrease the quality of the food items. Moreover, freeze-drying is an expensive procedure, and spray drying is preferred only for liquids. Due to increased mass transfer and accelerated diffusion, the drying of fruits, vegetables, fish, and meat is possible by the ultrasonic process (Figure 1.4). The main advantage of US-assisted drying is its reduced time, reduced energy consumption, and improved organoleptic qualities [18, 75].

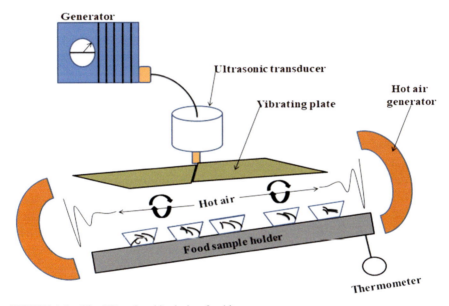

FIGURE 1.4 The US assisted in drying food items.

1.4.6 COOKING

During conventional cooking, the food item may be exposed to various temperature conditions, and such temperature variations can lead to undercooked or overcooked conditions. These issues tend to reduce the quality of food. In this context, US-based cooking can overcome the above-said drawbacks and improve its quality by providing proper heat transfer rates.

The study by Park [53] described that US-based cooking for beef was more efficient than conventional cooking. The main advantages of US-based cooking include reduced energy consumption, improved/efficient cooking speed, improved moisture retention rate, and improved textural property [12].

1.4.7 MICROBIAL INACTIVATION

There are various types of ultrasonication to achieve microbial inactivation, such as thermosonication (uses heat), MS (uses pressure), and MTS (uses both pressure and heat). The main factors that can cause microbial inactivation are its wave amplitude, treatment temperature, and time for ultrasonication. Combined effects of high temperature and amplitude play a significant role in microbial inactivation and which is usually done with the help of an ultrasonic probe, especially for longer treatment time. A reduction in the total count after US treatment of milk led to the inactivation of mesophilic bacteria [30]. Sugumar et al. [68] found that orange oil nanoemulsion developed with the US treatment was effective in the increased shelf life of food and beverages. Due to the anti-yeast activity of natural orange oil nanoemulsion, they could replace synthetic and semi-synthetic preservatives [30, 63].

Acoustic cavitation is the principle behind the US's microbial inactivation as it generates free radicals and hydrogen peroxide (H_2O_2). Microbial inactivation was possible at both the log and stationary phases of bacterial growth. *Enterobacter aerogenes, Bacillus subtilis,* and *Staphylococcus epidermis* in wastewater were inactivated with high-frequency ultrasonication [25]. Lv et al. [46] reported the sporicidal efficacy of ultrasonication by removing exosporium and destruction of the cortex in the spore structure. The combined effect of ultrasonication and acidic electrolyzed water treatment decreased the resistance of spores to individual treatments [55].

1.4.8 CUTTING

Cutting is one of the major operations during food processing, and conventional cutting methods are time-consuming and affect the quality of the food items [6]. The introduction of the US-based cutting of food items provides an overall efficient method for food processing. It is a novel trend in slicing/cutting of food items, which provides uniform size and shape for the food products. US-based cutting is affordable for both the small- and large-scale food industries. The main advantage of US-based cutting is less food waste

Ultrasound Processing of Foods: An Overview

production, and it also provides a continuous production process. The pictorial representation of the US-based cutting is shown in Figure 1.5.

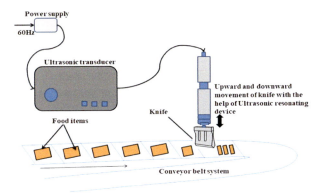

FIGURE 1.5 Mechanism of the US-based cutter/slicer in the food industry.

Ultrasonic resonance devices of different capacities are used to cut/slice food items. The cutting velocity depends upon the type of food items that are used to cut. In general, fragile food items such as cheese, cake, paneer, tofu, and other baked products are cut by US-based cutting. The vibration due to the US eliminates the tendency of food products to adhere to the blade, which prevents microbial growth, hence providing safe and quality food products [62].

1.4.9 NANOEMULSION PREPARATION

Nanoemulsions are oil-in-water (O/W) or water-in-oil (W/O) dispersions with a droplet size of 50–500 nm. Droplet sizes up to 200 nm are found to be transparent [48]. Also, there are double emulsions like oil-in-water-in-oil (O/W/O) and water-in-oil-in-water (W/O/W) available for various purposes [43, 73]. Both processing parameters (sonication time, power, and amplitude) and formulation parameters (oil type and oil volume fraction) control the emulsion droplet size and its polydispersity index (PDI). The impact of these parameters was tested with curcumin nanoemulsion and found that they had a strong influence on the droplet size, PDI, and zeta potential of the nanoemulsion [52]. In a study, sonication-assisted nanoemulsion was used for the probiotic storage along with prebiotic, which improved the viability/stability of the probiotic up to 60 days [40].

Akbas et al. [4] reported that nanoemulsions enhance the bioavailability of lipophilic compounds. The ultrasonication technique was used for the development of nanoemulsion for its improved color, particle size, and stability during storage. Leong et al. [44] used ultrasonication for skim milk emulsified with canola oil to develop cheddar cheese analogs in the form of single (O/W) or double (W/O/W) emulsions. Over a period of seven months of aging, the double emulsion cheese analogs showed a peculiar microstructure with skim milk droplets dispersed in the fat phase. They had greater free fatty acid content but, at the same time physically harder and melted less. The single emulsion cheese analogs had greater coalescence and highly spherical fat droplets yielding the same quantity of cheese, whey, and fat content compared to the control. They were soft but low in free fatty acids.

Syed et al. [70] reported that plant-based antimicrobial compounds could be used as an O/W emulsion system to prolong the shelf life of food products. Geraniol and carvacrol were stabilized with gum Arabic in O/W emulsion through the process of ultrasonication. Virgin coconut oil nanoemulsion was incorporated in surimi and had no impact on its structure and protein pattern. Textural properties were retained as original and simultaneously improved the whiteness and acceptance of surimi gel. Surimi gel has high elasticity and rigidity that helps in the binding of food materials [24]. High energy ultrasonication (20 kHz) was sufficient to provide high physiochemical stability to food, and it was reported for orange oil nanoemulsion [68]. Ultrasonication could enhance the anti-bacterial activity of eugenol-loaded chitosan nanoparticles. It was effective against both gram-negative (*Pseudomonas aeruginosa, Salmonella* species, and *E.coli* 0157:H7) and gram-positive (*Staphylococcus aureus*) bacterial strains. The antimicrobial activity achieved due to the enhanced solubility of eugenol through ultrasonication [63].

1.5 APPLICATIONS OF ULTRASONICATION IN FOOD PRODUCTS

The US-assisted technology plays a major role in nanotechnology, the biomedical sector, the manufacturing sector, and the food industry. In the food industry, the US is used for food processing, including meat tenderization, marination, pickling, extrusion, and fermentation, and for pretreatment processes such as crystallization, extraction, drying, and sterilization. In the case of the processing of fruits and vegetables, the US plays an important role. The US-assisted washing of fruits and vegetables proves its unchanged fresh quality and safety against microbial spoilage [1]. For example, in the study of Cheng et al. [15], the efficient retention of freshness and nutritional value

Ultrasound Processing of Foods: An Overview

of juices from tomato, guava, and orange was reported. During conventional processing, the loss of nutritional value results in the decreased shelf life and quality of fruits and vegetables [32].

Sonication can also use to extract plant compounds such as essential oils, flavors, cellulose nanofiber (CNF), color, and so on [33, 60, 65]. The early studies proved that sonication-assisted extraction enhances the final product [42]. The analysis of adulteration and aggregation of honey can be determined by applying the US method [21]. The US method is considered one of the major eco-friendly, non-toxic, and energy-saving processes in the food industry for enhancing food quality and safety [5]. Some of the other major applications of the US in the food industry are given in Table 1.1.

1.5.1 FAT SEPARATION IN MILK

The application of ultrasonication is remarkable in the dairy industry. Enzymatic pre-hydrolysis by Lipase enzyme and then combining with the US was effective in the separation of fat from the milk and dairy wastewater [2]. US frequencies ranging from 400 kHz to 1.6 MHz resulted in improved creaming of raw milk and also prepared recombined milk emulsion by dissolving skim milk powder in water (at 50°C) and milk fat which is insoluble. Sonication factors greatly influenced raw milk and coarse emulsion, and its influence became visible with the development of flocculated particles and clustering. The high-frequency US can initiate clustering phenomena, thereby helping the separation of milk fat [36].

1.5.2 FERMENTED PRODUCTS

Ultrasonication is found to have a positive impact on fermentation. Hashemi et al. [29] reported that US (100 W, 30 kHz, and 15 min) applied fermentation of milk enhanced the bioactive properties and growth of Lactobacillus strains at 60% amplitude. But ultrasonication did not produce any positive outcome at 90% amplitude. Fermentation of Korean six-row barley was aided with an ultrasonication bath at 40 kHz for higher ethanol production (13.18%) at 160 W [16]. The fermentation of apple pomace combined with ultrasonic pretreatment has increased the total phenolic content and antioxidant activity. Such fermented and ultrasonic apple pomace can enhance the texture, antioxidant, and functional properties in cereal formulate [45].

TABLE 1.1 Applications of the US for the Processing of Different Food Products

Ultrasound Methods and Principle	Applications	Products	Advantages	References
Airborne pressure/cavitation effect causes bubble collapse	Degassing and defoaming	Carbonated drinks	• Reduce wastage of raw materials • Reduce the usage of anti-foam chemicals • Reduce oxidation of nutrients in food commodities	[64]
High pressure and hydrodynamic shear force lead to bubble collapse by cavitation method	Emulsification	Mayonnaise, ketchup, and jam	• Efficient heat transfer • Reduced cooking time • Improved the quality and safety of food items	[38]
Superficial tissue disruption	Extraction	Essential oils and other natural components.	• Increased yield and rate of extraction	[41]
Vibration	Filtration	Juices	• Improves the life of filtration membranes • Requires less time	[77]
Increases the mass transfer and accelerates diffusion	Drying	Dehydrated vegetables, fruits, fish, and meat	• Reduce energy consumption • Improved organoleptic quality • Reduced drying time	[12]
The US-assisted cooking by uniform heat transfer	Cooking, curing, and marinating	Beef, pork, and hen breast	• Improved cooking time/processing time • Improved moisture retention capacity, energy-efficient and cooking yield • Enhanced safety and quality.	[49]
Reduce the microbial count; localized heating and production of free radicles	Sterilization	Fruits, vegetables, pulses, cereals, and honey	• Improves the quality and safety of the food at lower temperatures	[35]
Nucleation of the ice crystals due to uniform heat transfer	Crystallization and freezing	Fruits, vegetables, milk powder, and meat	• Decreased freezing time • Improves the quality of the frozen foods	[7]

Sonication in combination with temperature (thermosonication) can enhance bioethanol fermentation. Sulaiman et al. [69] proved that US-assisted fermentation of lactose with *Kluyeromyces marxianus* (ATCC 46537) increased bioethanol production. Gammoh et al. [23] confirmed that the combined effect of sonication and fermentation had several benefits. Through such a combination *Lb. delbrueckii* enhanced the biological and functional properties of camel milk whey, caused a structural modification in the casein protein, and improved antioxidant and antihypertensive potential. They also caused bioactive peptide liberation during fermentation.

1.5.3 TENDERIZATION OF MEAT

Tenderness, aroma, flavor, appearance, and juiciness are the quality determining factors of meat. The tenderness of the meat was considered the primary palatability factor preferred by the consumer in the determination of its quality [34]. Ultrasonication was able to break the muscular cell integrity through a biochemical effect due to enhanced enzymatic reactions [8]. Ohja et al. [51] reported that ultrasonication and encapsulation were able to enhance the lipid profile of pork meat. The application of ultrasonication in pork meat proved to be successful with an increase in healthy fatty acids. An increase in ultrasonication time caused an increase in docosahexaenoic acid (DHA) and eicosapentaenoic acid (EPA) fatty acids, irrespective of the formulation adopted. The improved physical property of meat is the main advantage of US-assisted tenderization. It can improve the cohesiveness and tenderness of meat without affecting the quality and water-binding capacity [12].

1.5.4 EXTRUDED PRODUCTS

Extrusion is one of the important techniques followed in the food industry during food processing. While extrusion, the adhesion of food items inside the extruder affects the operation condition of the machinery and affects the quality of the extruded products. The US-assisted extrusion provides a solution for the above-said issues. The same sticking/adherence issue was reported while large-scale cooking and during the demolding of food products from the processing vessel. The US can create vibration in the processing vessels and the extrusion tubes and thereby prevent adhering/ sticking of food products to vessels and extrusion tubes [54].

1.5.5   MARINATED AND PICKLED PRODUCTS

Marination/pickling/brining are common processes used in the food industry for food preservation and processing of food. Marination is done for food items such as meat, fish, fruits, vegetables, and cheese. The high concentration of salt and sugar absorption during the brining/pickling is a major issue. Other drawbacks of conventional marination (pickling or brining) are structural damage, bloating, and enzymatic softening [10]. Thus, US-assisted marination during food processing and preservation provides improved efficiency compared with conventional processes. Carcel et al. [10] reported that US-based processing during pickling allows absorption of low sodium chloride levels compared with conventional methods. Hence, further desalting of the final food products is not needed. In the cheese industry also, the effective usage of US-assisted brining was reported. The advantages include an increased rate of water removal and absorption of proper concentration of salt content in the cheese. It will also efficiently improve the internal or external mass transfer rate during processing [9].

1.5.6   FUNCTIONALITY ENHANCEMENT IN PROTEIN FOODS

Ultrasonication can alter the physical and functional properties of the protein, such as gelation, foamability, solubility, and emulsification. Thus, US is used for protein modification can enhance the chemical reaction of protein, producing protein conjugates and also improving the enzymatic protein hydrolysis [14]. The mechanism of cavitation during ultrasonication causes protein structure modification. Ultrasonication has an influence on improving the solubility of the protein. Protein denaturation and aggregation are dependent on its solubility, and this dependency is a critical parameter for the functionality of protein [27].

1.6   ADVANTAGES OF ULTRASONICATION

The US was found to be a non-destructive green processing technique with maximum retention of bioactive compounds, less energy consumption, and reduced risk and hazard to health and safety [37]. The US can be used for improving protein functionality and texture [22, 27]. Ultrasonication can

Ultrasound Processing of Foods: An Overview

develop emulsions with defined droplet size and structure. They can also enhance the efficacy of natural emulsifiers [42]. US pasteurization is better than conventional methods to retain the food quality parameters such as color, flavor, and other physiochemical properties due to the comparatively lower temperature used [71]. Ultrasonication helped in microbial inactivation when coupled with other non-thermal methods [46]. Ultrasonication can help in thermal sterilization, which reduces process time and prevents nutritional loss [47].

1.7 LIMITATIONS OF ULTRASONICATION

Ultrasonication is an antimicrobial tool that depends on the size, hydrophobicity, growth phase, and nature of the microorganism. Microorganisms have soft, thicker capsules that are mostly ultrasonic resistant [26]. When used on a commercial scale, ultrasonication consumes a large amount of energy [74]. The extraction of plant materials with ultrasonication was dependent upon their time, frequency, and temperature. They were also bound to plant material characteristics such as particle size, moisture content, and the carrier solvent [12]. Though ultrasonication (frequency ranging from 20 kHz to 100 kHz) can cause microbial inactivation, they are not effective when used alone [11]. The acoustic cavitation mechanism of high-frequency ultrasonication had no impact on yeast (*Aureobasilium pullulans*) [25].

1.8 SUMMARY

The ultrasonication method generally uses sound waves with a frequency greater than 20 kHz, and has a wide range of applications in the food industry, such as extraction of active components, sterilization, development of nanoparticles, emulsions, and improved texture of food, meat tenderization, and even causing a structural and functional modification of food or food constituents. Even though it is a high-energy utilization process, its non-thermal nature makes it a better option for food processing. Moreover, this technology prevents the loss of nutrients and is involved in texture improvement during processing. This non-destructive method can also replace most conventional food processing techniques to assist extraction and sterilization with the cavitation phenomenon.

KEYWORDS

- cavitation
- docosahexaenoic acid
- food processing
- microbial inactivation
- nanoemulsions
- ultrasonication
- ultrasound extraction

REFERENCES

1. Abid, M., Jabbar, S., Wu, T., Hashim, M. M., Hu, B., Lei, S., & Zhang, X., (2013). Effect of ultrasound on different quality parameters of apple juice. *Ultrasonics Sonochemistry, 20*(2), 1182–1187.
2. Adulkar, T. V., & Rathod, V. K., (2014). Ultrasound-assisted enzymatic pretreatment of high fat content dairy wastewater. *Ultrasonics Sonochemistry, 21*(3), 1083–1089.
3. Agi, A., Junin, R., Alqatta, A. Y. M., Gbadamosi, A., Yahya, A., & Abbas, A., (2018). Ultrasonic-assisted ultrafiltration process for emulsification of oil field produced water treatment. *Ultrasonics Sonochemistry, 8*(3), 148–166.
4. Akbas, E., Soyler, B., & Oztop, M. H., (2018). Formation of capsaicin loaded nanoemulsions with high pressure homogenization and ultrasonication. *LWT, 96*(5), 266–273.
5. Alarcon-Rojo, A. D., Janacua, H., Rodriguez, J. C., Paniwnyk, L., & Mason, T. J., (2015). Power ultrasound in meat processing. *Meat Science, 107*(1), 86–93.
6. Arnold, G., Leiteritz, L., Zahn, S., & Rohm, H., (2009). Ultrasonic cutting of cheese: Composition affects cutting work reduction and energy demand. *International Dairy Journal, 19*(1), 314–320.
7. Awad, T., Moharram, H., Shaltout, O., Asker, D., & Youssef, M., (2012). Applications of ultrasound in analysis; processing and quality control of food: A review. *Food Research International, 48*(3), 410–427.
8. Boistier-Marquis, E., Lagsir-Oulahal, N., & Callard, M., (1999). Applications of power ultrasound in the food industry. *Food and Agricultural Industries, 116*(3), 23–31.
9. Cameron, M., McMaster, L. D., & Britz, T. J., (2009). Impact of ultrasound on dairy spoilage microbes and milk components. *Dairy Science and Technology, 89*(4), 83–98.
10. Carcel, J. A., Benedito, J., Bon, J., & Mulet, A., (2007). High-intensity ultrasound effects on meat brining. *Meat Science, 76*(1), 611–619.
11. Chandrapala, J., Oliver, C., Kentish, S., & Ashokkumar, M., (2012). Ultrasonics in food processing. *Ultrasonics Sonochemistry, 19*(5), 975–983.
12. Chemat, F., & Khan, M. K., (2011). Applications of ultrasound in food technology: Processing, preservation, and extraction. *Ultrasonics Sonochemistry, 18*(4), 813–835.

Ultrasound Processing of Foods: An Overview

13. Chemat, F., Rombaut, N., Sicaire, A. G., Meullemiestre, A., Fabiano-Tixier, A. S., & Abert-Vian, M., (2017). Ultrasound-assisted extraction of food and natural products. Mechanisms, techniques, combinations, protocols, and applications: A review. *Ultrasonics Sonochemistry, 34*, 540–560.

14. Chen, L., Chen, J., Ren, J., & Zhao, M., (2011). Effects of ultrasound pretreatment on the enzymatic hydrolysis of soy protein isolates and on the emulsifying properties of hydrolysates. *Journal of Agricultural and Food Chemistry, 59*(6), 2600–2609.

15. Cheng, L. H., Soh, C. Y., Liew, S. C., & The, F. F., (2007). Effects of sonication and carbonation on guava juice quality. *Food Chemistry, 104*(1), 1396–1401.

16. Choi, E. J., Ahn, H., Kim, M., Han, H., & Kim, W. J., (2015). Effect of ultrasonication on fermentation kinetics of beer using six-row barley cultivated in Korea. *Journal of the Institute of Brewing, 121*(4), 510–517.

17. Delgado, A. E., Zheng, L., & Sun, D. W., (2008). Influence of ultrasound on freezing rate of immersion-frozen apples. *Food and Bioprocess Technology, 2*, 263–270.

18. Fernandes, F. A. N., Linhares, Jr. F. E., & Rodrigues, S., (2008). Ultrasound as pretreatment for drying of pineapple. *Ultrasonics Sonochemistry, 15*(6), 1049–1054.

19. Flannigan, D. J., & Suslick, K. S., (2005). Plasma formation and temperature measurement during single-bubble cavitation. *Nature, 434*(7029), 52–55.

20. Gallego, J. J. A., Rodríguez, C. G., Acosta, A. V. M., Andrés, G. E., Blanco, B. A., & Montoya, V. F., (2005). Ultrasonic Defoaming Procedure and System Using Emitters with Stepped Vibrating Plate. Patent No. ES 2 212 896 B1; p. 10.

21. Gallego, J. J. A., Rodriguez, G., Acosta, V., & Riera, E., (2010). Power Ultrasonic Transducers with Extensive Radiators for Industrial Processing. *Ultrasonics Sonochemistry, 17*(6), 953–964.

22. Gallo, M., Ferrara, L., & Naviglio, D., (2018). Application of ultrasound in food science and technology: A perspective. *Foods, 7*(10), 164.

23. Gammoh, S., Alu'datt, M. H., Tranchant, C. C., Al-U'datt, D. G., Alhamad, M. N., Rababah, T., Kubow, S., et al., (2020). Modification of the functional and bioactive properties of camel milk casein and whey proteins by ultrasonication and fermentation with *Lactobacillus delbrueckii* subsp. *lactis. LWT, 129*, 109501.

24. Gani, A., & Benjakul, S., (2018). Impact of virgin coconut oil nanoemulsion on properties of croaker surimi gel. *Food Hydrocolloids, 82*, 34–44.

25. Gao, S., Hemar, Y., Ashokkumar, M., Paturel, S., & Lewis, G. D., (2014). Inactivation of bacteria and yeast using high-frequency ultrasound treatment. *Water Research, 60*, 93–104.

26. Gao, S., Lewis, G. D., Ashokkumar, M., & Hemar, Y., (2014). Inactivation of microorganisms by low-frequency high-power ultrasound: 1. Effect of growth phase and capsule properties of the bacteria. *Ultrasonics Sonochemistry, 21*(1), 446–453.

27. Gharibzahedi, S. M. T., & Smith, B., (2020). The functional modification of legume proteins by ultrasonication: A review. *Trends in Food Science & Technology, 98*, 107–116.

28. Grossner, M. T., Belovich, J. M., & Feke, D. L., (2005). Transport analysis and model for the performance of an ultrasonically enhanced filtration process. *Chemical Engineering Science, 60*(12), 3233–3238.

29. Hashemi, S. M. B., & Gholamhosseinpour, A., (2020). Effect of ultrasonication treatment and fermentation by probiotic *Lactobacillus plantarum* strains on goat milk bioactivities. *International Journal of Food Science & Technology, 55*(6), 2642–2649.

30. Herceg, Z., Juraga, E., Sobota-Šalamon, B., & Režek-Jambrak, A., (2012). Inactivation of mesophilic bacteria in milk by means of high intensity ultrasound using response surface methodology. *Czech Journal of Food Sciences, 30*(2), 108–117.

31. Hu, Y., Kwan, T. H., Daoud, W. A., & Lin, C. S. K., (2017). Continuous ultrasonic-mediated solvent extraction of lactic acid from fermentation broths. *Journal of Cleaner Production, 145*(1), 142–150.

32. Jabbar, S., Abid, M., Hu, B., Wu, T., Hashim, M. A., Lei, S., & Zeng, X., (2014). Quality of carrot juice as influenced by blanching and sonication treatments. *LWT-Food Science and Technology, 55*(1), 16–21.

33. Jadhav, D., Rekha, B. N., Gogate, P. R., & Rathod, V. K., (2009). Extraction of vanillin from vanilla pods: A comparison study of conventional Soxhlet and ultrasound-assisted extraction. *Journal of Food Engineering, 93*(3), 421–426.

34. Jayasooriya, S. D., Bhandari, B. R., Torley, P., & D'Arcy, B. R., (2004). Effect of high power ultrasound waves on properties of meat: A review. *International Journal of Food Properties, 7*(2), 301–319.

35. Jovanovic-Malinovska, R., Kuzmanova, S., & Winkelhausen, E., (2015). Application of Ultrasound for enhanced extraction of prebiotic oligosaccharides from selected fruits and vegetables. *Ultrasonics Sonochemistry, 22*, 446–453.

36. Juliano, P., Kutter, A., Cheng, L. J., Swiergon, P., Mawson, R., & Augustin, M. A., (2011). Enhanced creaming of milk fat globules in milk emulsions by the application of ultrasound and detection by means of optical methods. *Ultrasonics Sonochemistry, 18*(5), 963–973.

37. Kehinde, B. A., Sharma, P., & Kaur, S., (2021). Recent nano-, micro-and macro technological applications of ultrasonication in food-based systems. *Critical Reviews in Food Science and Nutrition, 61*(4), 599–621.

38. Kentish, S., & Feng, H., (2014). Applications of power ultrasound in food processing. *Annual Review of Food Science and Technology, 5*, 263–284.

39. Kyllonen, H., Pirkonen, P., Nystrom, M., Nuortila-Jokinen, J., & Gronroos, A., (2006). Experimental aspects of ultrasonically enhanced cross-flow membrane filtration of industrial wastewater. *Ultrasonics Sonochemistry, 13*(6), 295–302.

40. Krithika, B., & Preetha, R., (2019). Formulation of protein-based inulin incorporated synbiotic nanoemulsion for enhanced stability of probiotic. *Materials Research Express, 6*(11), 114003.

41. Kumari, B., Tiwari, B. K., Hossain, M. B., Rai, D. K., & Brunton, N. P., (2017). Ultrasound-assisted extraction of polyphenols from potato peels: Profiling and kinetic modeling. *International Journal of Food Science & Technology, 52*(1), 1432–1439.

42. Leong, T. S. H., Manickam, S., Martin, G. J., Li, W., & Ashokkumar, M., (2018). *Ultrasonic Production of Nano-emulsions for Bioactive Delivery in Drug and Food Applications* (p. 36). Switzerland: Springer International Publishing.

43. Leong, T. S. H., Zhou, M., Kukan, N., Ashokkumar, M., & Martin, G. J. O., (2017). Preparation of water-in-oil-in-water emulsions by low frequency ultrasound using skim milk and sunflower oil. *Food Hydrocolloids, 63*, 685–695.

44. Leong, T. S. H., Ong, L., Gamlath, C. J., Gras, S. L., Ashokkumar, M., & Martin, G. J. O., (2020). Formation of cheddar cheese analogs using canola oil and ultrasonication: A comparison between single and double emulsion systems. *International Dairy Journal, 105*, 104683.

Ultrasound Processing of Foods: An Overview

45. Lohani, U. C., & Muthukumarappan, K., (2016). Effect of sequential treatments of fermentation and ultrasonication followed by extrusion on bioactive content of apple pomace and textural; functional properties of its extrudates. *International Journal of Food Science & Technology, 51*(8), 1811–1819.

46. Lv, R., Muhammad, A. I., Zou, M., Yu, Y., Fan, L., Zhou, J., Ding, T., et al., (2020). Hurdle enhancement of acidic electrolyzed water antimicrobial efficacy on *bacillus cereus* spores using ultrasonication. *Applied Microbiology and Biotechnology, 104*, 4505–4513.

47. Mason, T. J., Chemat, F., & Ashokkumar, M., (2015). Power ultrasonics for food processing; Chapter 27. In: Gallego-Juarez, J. A., & Graff, K. F., (eds.), *Power Ultrasonics: Applications of High-Intensity Ultrasound* (pp. 815–843). Cambridge–UK: Woodhead Publishing.

48. McClement, D. J., (2012). Advances in fabrication of emulsions with enhanced functionality using structural design principles. *Current Opinion in Colloid & Interface Science, 17*(5), 235–245.

49. McDonnell, C. K., Lyng, J. G., Arimi, J. M., & Allen, P., (2014). The acceleration of pork curing by power ultrasound: A pilot-scale production. *Innovative Food Science & Emerging Technologies, 26*, 191–198.

50. Muthukumaran, S., Kentish, S. E., Ashokkumar, M., & Stevens, G. W., (2005). Mechanisms for the ultrasonic enhancement of dairy whey ultrafiltration. *Journal of Membrane Science, 258*(1, 2), 106–114.

51. Ojha, K. S., Perussello, C. A., García, C. Á., Kerry, J. P., Pando, D., & Tiwari, B. K., (2017). Ultrasonic-assisted incorporation of nano-encapsulated omega-3 fatty acids to enhance the fatty acid profile of pork meat. *Meat Science, 132*(1), 99–106.

52. Páez-Hernández, G., Mondragón-Cortez, P., & Espinosa-Andrews, H., (2019). Developing curcumin nanoemulsions by high-intensity methods: Impact of ultrasonication and microfluidization parameters. *LWT, 111*, 291–300.

53. Park, S. H., & Roh, Y. R., (2001). *Cooker*. Int. Pat. WO 0113773.

54. Patist, A., & Bates, D., (2008). Ultrasonic innovations in the food industry: From the laboratory to commercial production. *Innovative Food Science & Emerging Technology, 9*(2), 147–154.

55. Patist, A., & Bates, D., (2011). Industrial applications of high power ultrasonics. In: Feng, H., Barbosa-Canovas, G., & Weiss, J., (eds.), *Ultrasound Technologies for Food and Bioprocessing* (pp. 599–616). New York: Springer.

56. Prakash, M. J., Manikandan, S., Vigna, N. C., & Dinesh, R., (2017). Ultrasound-assisted extraction of bioactive compounds from *Nephelium lappaceum* L. fruit peel using central composite face centered response surface design. *Arabian Journal of Chemistry, 10*(1), S1145–S1157.

57. Rana, A., Meena, & Shweta, (2017). Ultrasonic processing and its use in food industry: A review. *International Journal of Chemical Studies, 5*(6), 1961–1968.

58. Ravikumar, M., Suthar, H., Desai, C., & Gowda, S. A. J., (2017). Ultrasonication: An advanced technology for food preservation. *International Journal of Pure and Applied Bioscience, 5*(6), 363–371.

59. Riera, E., Gallego-Juárez, J. A., & Mason, T. J., (2006). Airborne ultrasound for the precipitation of smokes and powders and the destruction of foams. *Ultrasonics Sonochemistry, 13*(2), 107–116.

60. Rostagno, M. A., Palma, M., & Barroso, C. G., (2007). Ultrasound-assisted extraction of isoflavones from soy beverages blended with fruit juices. *Analytica Chimica Acta, 597*(2), 265–272.

61. Santos, H. M., Lodeiro, C., & Capelo-Martínez, J. L., (2009). The power of ultrasound; Chapter 1. In: Capelo-Martínez, J. L., (ed.), *Ultrasound in Chemistry: Analytical Applications* (pp. 1–16). Weinheim: John Wiley & Sons.

62. Schneider, Y., Zahn, S., Schindler, C., & Rohm, H., (2009). Ultrasonic excitation affects friction interactions between food materials and cutting tools. *Ultrasonics, 49*(6, 7), 588–593.

63. Shao, Y., Wu, C., Wu, T., Li, Y., Chen, S., Yuan, C., & Hu, Y., (2018). Eugenol-chitosan nanoemulsions by ultrasound-mediated emulsification: Formulation, characterization, and antimicrobial activity. *Carbohydrate Polymers, 193*, 144–152.

64. Shen, X., Shao, S., & Guo, M., (2017). Ultrasound-induced changes in physical and functional properties of whey proteins. *International Journal of Food Science & Technology, 52*(2), 381–388.

65. Shruthy, R., & Preetha, R., (2019). Cellulose nanoparticles from agro-industrial waste for the development of active packaging. *Applied Surface Science, 484*, 1274–1281.

66. Shwetha, & Preetha, R., (2017). Comparison of freeze-drying and spray drying for the production of anthrocyanin encapsulated powder from Jamun (*Syzygium cumini*). *Asian Journal of Chemistry, 29*(6), 1179–1181.

67. Soria, A. C., & Villamiel, M., (2010). Effect of ultrasound on the technological properties and bioactivity of food: A review. *Trends in Food Science & Technology, 21*(7), 323–331.

68. Sugumar, S., Singh, S., Mukherjee, A., & Chandrasekaran, N., (2016). Nanoemulsion of orange oil with non-ionic surfactant produced emulsion using ultrasonication technique: Evaluating against food spoilage yeast. *Applied Nanoscience, 6*, 113–120.

69. Sulaiman, A. Z., Ajit, A., Yunus, R. M., & Chisti, Y., (2011). Ultrasound-assisted fermentation enhances bioethanol productivity. *Biochemical Engineering Journal, 54*(3), 141–150.

70. Syed, I., & Sarkar, P., (2018). Ultrasonication-assisted formation and characterization of geraniol and carvacrol-loaded emulsions for enhanced antimicrobial activity against food-borne pathogens. *Chemical Papers, 72*, 2659–2672.

71. Tsukamoto, I., Yim, B., Stavarache, C. E., Furuta, M., Hashiba, K., & Maeda, Y., (2004). Inactivation of *Saccharomyces cerevisiae* by ultrasonic irradiation. *Ultrasonics Sonochemistry, 11*(2), 61–65.

72. Uju, Dewi, N. P. S. U. K., Santoso, J., Setyaningsih, I., & Hardingtyas, S. D., (2020). Extraction of phycoerythrin from *Kappaphycus alvarezii* seaweed using ultrasonication. In: *IOP Conference Series: Earth and Environmental Science* (Vol. 414, p. 012028). Bristol–UK: IOP Publishing Ltd.

73. Vaishanavi, S., & Preetha, R., (2020). Soy protein incorporated nanoemulsion for enhanced stability of probiotic (*Lactobacillus delbrueckii* subsp. *bulgaricus*) and its characterization. *Materials Today: Proceedings, 40*(1), S148–S153.

74. Yusaf, T., & Al-Juboori, R. A., (2014). Alternative methods of microorganism disruption for agricultural applications. *Applied Energy, 114*(3), 909–923.

75. Yuting, X., Lifen, Z., Jianjun, Z., Jie, S., Xingqian, Y., & Donghong, L., (2013). Power ultrasound for the preservation of postharvest fruits and vegetables. *International Journal of Agricultural and Biological Engineering, 6*(2)**,** 116–125.

76. Zhong, K., & Wang, Q., (2010). Optimization of ultrasonic extraction of polysaccharides from dried longan pulp using response surface methodology. *Carbohydrate Polymers, 80*(1), 19–25.
77. Zou, Y., Wang, W., Li, Q., Chen, Y., Zheng, D., Zou, Y., Zhang, M., et al., (2016). Physicochemical, functional properties and antioxidant activities of porcine cerebral hydrolysate peptides produced by ultrasound processing. *Process Biochemistry, 51*(3), 431–443.

CHAPTER 2

SUPERCRITICAL FLUID TECHNOLOGY: RECENT TRENDS IN FOOD PROCESSING

MAYURI A. PATIL, ANIKET P. SARKATE, BHAGWAN K. SAKHALE, and RAHUL C. RANVEER

ABSTRACT

In the supercritical fluid (SCF) technique, carbon dioxide (CO_2) has found a lot of applications in food production and processing. The use of organic solvent in food processing has a high risk of chemical contamination, and as there is increased awareness of health and safety hazards, it is beneficial to use SCF techniques as this technique is non-flammable, non-toxic, inert, eco-friendly, and easily recoverable. The food industry always prioritizes obtaining high purity and quality natural compounds. Considering its applications like low environmental impact, short time of separation, selectivity, low consumption of organic solvents, shorter retention time, and improved peak tailing, the food industry has increased its approach toward supercritical fluid chromatography (SFC). Mostly, the use of CO_2 as a mobile phase is preferred because of its properties like low viscosity and high molecular weight diffusiveness. Even though a food mixture is complex, SFC shows its best results, and hence it is considered a powerful tool in food analysis. The supercritical fluid extraction (SFE) technique has great potential in order to separate specific bioactive compounds, phytochemicals, and phytonutrients present in different food materials that can be commercially explored for the development of novel foods, functional foods, and nutraceuticals in the growth of food processing industry. In general, overall, greater applications

---

Advances in Food Process Engineering: Novel Processing, Preservation, and Decontamination of Foods.
Megh R. Goyal, N. Veena, & Ritesh B. Watharkar (Eds.)
© 2023 Apple Academic Press, Inc. Co-published with CRC Press (Taylor & Francis)

of SFC to other conventional techniques, including organic solvents, lead to the best quality, safety, and high purity of products in the food industry.

2.1 INTRODUCTION

If the temperature and pressure of a substance are greater than its critical values (T_c and P_c, respectively), then a pure substance is in the supercritical state [6, 7]; for example, carbon dioxide (CO_2) has T_c of 31°C and P_c of 7.48 MPa. In its supercritical state, a fluid exhibits a physicochemical characteristic intermediate between liquid and gas. The high density of the supercritical fluid (SCF) gives it a good solvating power, whereas excellent penetrating and mixing with SCF is possible because of its low viscosity, lower surface tension, and greater diffusivity [53]. Boron Gagniard de la Tour initially reported the supercritical state in 1822. Demonstration of all SCFs containing solvating power was done by Hannay and Hogarth in 1879 [61]. There are several applications of SCF. It is used as a solvent for active substances. Moreover, it is also used as an anti-solvent which leads to the precipitation of active substances; also used in SC-CO_2-based spray-drying methods [62].

Figure 2.1 shows two points as the triple and critical points. A triple point indicates a path at which three states (solid, liquid, and gas) coexist. Whereas the critical point is a point that lies at the boiling point where two states (liquid and gas) exist. The region above the critical point is called a supercritical point or supercritical region. All SCFs in this region possess physicochemical properties of both liquid and gas; hence these fluids have a wide capacity to dissolve poorly soluble compounds or insoluble compounds in separate gas or liquid [36]. Food technology is currently developing with a lot of variants in applied research and industry. Production efficiency and environmental preservation could be achieved by using alternatives to conventional processing, preservation, and extraction procedure [8].

In the food industry, SCFs have many applications. For modifying purposes, SCF is used. Solvent characteristics of fluid can be modified by adding modifiers, and the most used modifier is CO_2. It is also used in SFE and SFC (Figure 2.2). Less polarity is a limitation of CO_2; this problem could be overcome by modification of CO_2 with polar organic solvents like methanol [12]. SFC permits the use of higher flow rate and low pressure from columns; it improves efficiency in time analysis and decreases the use of organic solvents. SFC improves resolution and gets sharper peaks as compared to conventional methods. It is a fast method because of the less time required for column equilibration. Analysis of thermolabile and polar compounds is also possible with SFC. Considering these benefits, SFC is widely used [5, 19].

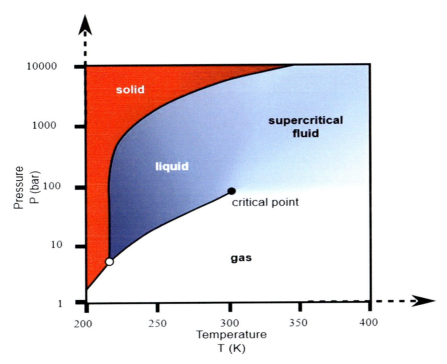

FIGURE 2.1 Phase diagram for supercritical fluids.
Source: Rifleman 82. Public domain.

Best separation technologies are always looked forward by food industries for pure and best quality products. SFE and SFC have attracted many platforms as an alternative to conventional food processing methods [56].

Successful use of supercritical CO_2 (SC-CO_2) for the extraction of foodstuffs is well accepted by the food industry. Supercritical water is also used with SC-CO_2 for food processing, and usage of organic solvents is avoided nowadays in the food industry [29]. SC-CO_2 is also used as a method to lower the viscosity of plasticizers [4]. It is preferred by the food industry because it isolates and removes molecules with better precision, separation of chemically close molecules or mixtures occurs very efficiently, it enriches the liquid, it can be preceded at low temperatures with better integrity of molecules, and it also reduces post-processing steps.

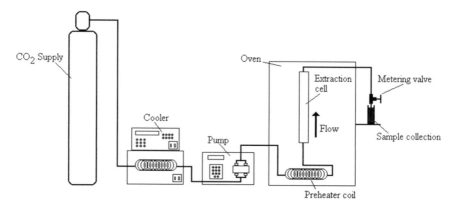

FIGURE 2.2 Supercritical extraction system with a CO_2 phase diagram.
Source: Stainless316. Public domain.

This chapter explores various advances and advantages of SC-CO_2 in the food industry as well as applications of SFE and SFC in food processing systems.

2.2 SUPERCRITICAL FLUID EXTRACTION (SFE) TECHNOLOGY

Supercritical fluid extraction (SFE) technology is an efficient and effective technique for the extraction of bioactive compounds as compared to current chemical analysis methods. Its application is also applied for qualitative and quantitative identification of molecules of naturally occurring and thermolabile compounds [22, 23].

Various operations and processes could be possible and employed by extraction with supercritical fluids (SCFs) for applications like extraction as well as fractionation of edible fats or oils [6, 9], purification of the solid matrices, separation of tocopherols or other antioxidants, the concentration of fermentation broth, fruit juices among others [43], extracting β-carotene as well as lycopene from tomato processing [11, 36], decaffeination of tea, coffee [25, 45], extraction of pesticides from soluble solid matrices [1], in production of natural flavors and fragrances, extraction of mycotoxins [3, 33], microbial reduction and food pasteurization, fat removal from food to produce high-protein low-fat dietary food [2, 10].

SFE is also used in food safety and as a sterilizing agent in food processing [34, 45], as an anti-solvent for encapsulation of food in the food industry [41], extraction of lipid from food products such as butter and fish oil [30],

Supercritical Fluid Technology: Recent Trends

extraction of nutraceuticals [38, 52], extraction of food coloring agent [54], inactivation of food-related bacteria in the treatment of food waste [28, 65], recovery of aroma from food [17, 55], extraction of turmeric [57], extraction of bioactive compounds [60], extraction of grapes byproduct which is a source of polyphenolic compounds, used in the food, pharmaceutical, and cosmetic industry [16].

2.2.1 APPLICATIONS OF SUPERCRITICAL FLUID (SCF) TECHNOLOGY IN THE FOOD INDUSTRY

The demand for chemical-free natural food ingredients is increasing day by day, and it is an important trend now. Most of the SFE process is carried out batch-wise because of limited solubilization. Application of high solvent is required to feed ratios. Supply of fresh solvent in a high flow rate, SCF should be continuously recycled. Continuous contact SCF with solid substrate carries out the extraction process. In SFE, a solid sample is dissolved in SCF (CO_2) in the extraction vessel. After extraction, precipitation of the entire material occurs due to depressurization of extract in a fraction chamber, and hence CO_2 loses its solvating power. Thus, precipitated material is extracted, and CO_2 is recycled [24, 46].

2.2.1.1 EXTRACTION AND FRACTIONATION OF EDIBLE FATS AND OILS

In a conventional method, the production of sesame oil is done by roasting the sesame seeds at a temperature higher than 200°C and then compressing them at 180°C to extract the oil [42]. After this process, sesame oil goes under oxidation, and the flavor is lost. However, the use of SC-CO_2 as a solvent will prevent the loss of flavor from sesame oil. SFE is also used to separate or extract lipid, cholesterol, and fat from food products. SFE and SFC are also used to assay fat-soluble vitamins in the hydrophobic ointment as well as essential oil [39, 49].

Rozi and Singh [54] have reviewed the lipid extraction method which is used for the extraction of various oils from brown seaweed, butter oil, cotton seed, corn bran, ground beef, rapeseed, milk fat, pistachio, pork, *Pythium irregulare*, oat bran, pecan, rice bran, safflower, soybean, etc. According to them, this study has occurred using various instruments at different pressures [31]. Some parameters under study for the extraction of oil are shown in Table 2.1.

TABLE 2.1 Conditions of Extraction for Recovery of Oil [31, 54]

Sample	Temperature (°C)	Pressure (MPa)	Oil Recovered
Almond fruits	40	35	54%
Borago seed	10–55	20–30	31%
Corn germ oil	42	30	45–55%
Flaked seed	60	40	7.09% (w/w)
Orange peel	55	15	2.22% (w/w)
Walnut	25–70	20–40	74%
Wheat germ oil	40	30	9%

Lipid is extracted from various nuts and oilseeds like pecan and rice bran [31, 54]. The temperature required for the extraction of oil from pecan is 75°C, and the pressure required is 66.8 MPa. Breakage of pecan kernel can happen during the SFE process due to an increase in moisture.

A byproduct of rice milling is rice bran which contains about 15 to 20% of oil. This oil is extracted by SFE. The temperature and pressure required for oil extraction are in the range of 0–60°C and 17–31 MPa, respectively. This study is further compared by using solvent hexane to extract the oil. The quality of oil extracted by SCF and oil extracted from hexane at 80°C is compared. This study reveals that the quality of the oil obtained by SCF technology is far better than oil extracted by hexane. SFE of wheat germ oil was carried out by Shao et al. [58]. The temperature varied from 40, 50, and 60°C, whereas pressure varied from 20, 27.5, and 35 MPa. The flow rate is also changed from 15, 20, and 25 L/h. SFE yields 10.15% of wheat germ oil at a temperature of 50°C, a pressure of 35 MPa, and a solvent flow rate ranges 22.5 to 25 L/h [58].

2.2.1.2 EXTRACTION FROM DAIRY PRODUCTS

Dairy products contain large amounts of fats in it. Removal of vitamin A and vitamin E from 0.5 gm sample of powdered milk and liquid milk is carried out by SFE. The pressure required to carry out this extraction process is 37 MPa, and the temperature was 80°C. However, the time required for this process is shorter than conventional methods. In another study, fractionation of milk fat by SCF technology was carried out by Rizvi et al. [53].

Supercritical Fluid Technology: Recent Trends

2.2.1.3 USE OF SFE IN NUTRACEUTICALS

SFE is also used in the production of nutraceuticals as well as functional foods. Omega-3 fatty acids were extracted from seaweed and fungi. The yield obtained after the SFE process is greater than the Soxhlet extraction process. γ-Linolenic acid is also extracted with SFE from *Cunnighamella echinulata*. Extraction conditions like the temperature of 50°C and pressure of 30 MPa are used to yield more than 110% of γ-linolenic acid [34, 59]. β-carotene and lycopene contain natural antioxidant activity. It is extracted from tomato waste which is in paste form at a temperature of 65°C. The yield obtained by using SFE was 54% higher than the traditional method.

2.2.1.4 DECAFFEINATION OF COFFEE

Decaffeination of coffee can also be done with the SFE technique, with 97% of caffeine can be extracted from coffee beans. After extraction, the bitter taste and strong smell of coffee get reduced. The extraction condition used for decaffeination is the temperature of 93°C and a pressure of 25 MPa. It does not affect protein and carbohydrates in coffee beans because of the selectivity property of CO_2 [25].

2.2.1.5 EXTRACTION OF NATURAL FLAVORS AND FRAGRANCES

Various flavoring agents are added to food products, and these are widely used in the food industry. There are many sources from which aromatic compounds can be obtained. Natural flavors (plant or animal raw material), artificial flavors (not natural, produced by chemical modification), and nature-identical (flavors synthesis through the chemical process) are available. There are so many traditional methods available to separate flavors and fragrances from natural origin. Today there is a need to minimize the use of synthetic materials and chemicals for this process, and hence a need for a cost-effective and best alternative extraction method. Therefore, the SFE method is used to separate out flavor and aroma from various natural sources like plants, fruits, vegetables, and cereals. Aroma compounds in spices like cinnamon, clove bud, basil, chamomile, cardamom, coriander, ginger, eucalyptus, lavender, oregano, caraway, hop, pepper, peppermint, turmeric, thyme, and vanillin.

SCFs are used to extract aromatic compounds from fruits, plants, vegetable oils, and processed seeds. This process is carried out from solids and liquids in batches of single-stage or multistage. There are two steps in single-stage extraction: Extraction and extract separation. The requirement of solvent at the time of the process is low in the case of $SC-CO_2$ as compared to traditional methods [17, 55].

Anklam et al. [3] observed that the time required for extraction and separation of aromatic compounds using SFE is less than any traditional method used before. The quality of oil obtained from basil after SFE followed by multistage separation is higher than conventional methods [51].

SFE technique is also used for the extraction of thermolabile components such as matricin from flower heads of chamomile by single-step separation [66]. It was observed that its aroma and composition did not change after SFE when compared to the conventional extraction method or distillation method. Thus, the SFE technique is used successfully for the extraction of essential oils without any thermal degradation.

SFE with gas chromatography (GC) is used to extract carvone and limonene from caraway (dried) for a rapid extraction process [27]. $SC-CO_2$ has been used for the extraction of cardamom oil. The quality and shelf life of the oil is compared with traditionally extracted commercial oils [20]. Cinnamon oil also can be extracted by SFE technology. Oil extracted is then compared with oil extracted by the conventional method by analyzing with the help of HPLC [14]. It has been shown that the solvent-assisted SCE is more superior to the conventional extraction method using the Soxhlet apparatus.

Flavor compounds like eugenol acetate and vanillin are extracted from the clove bud using SFE. The study has also shown that extracts obtained from conventional methods (such as steam distillation) undergo thermal degradation, and the quantity obtained after the process is also less [32]. Storage quality of extracted clove oil by SFE and by commercial distilled oils is also compared; the change in the composition of clove bud oil was comparable in both methods [20]. 6-Gingerol from the ginger rhizome is also extracted using $SC-CO_2$. The obtained oil from ginger using SFE was far more superior than conventional steam distillation and solvent extraction methods in terms of levels of major constituents.

Lavender flavor is used in high amounts in all food products which are extracted from the dried flowers of lavender. Extracts of lavender oil obtained from all three methods such as SFE, steam distillation, solvent extraction method, were then analyzed by GC and results shown that chemical profile of all these methods were comparable [48]. Oregano is one of the most

Supercritical Fluid Technology: Recent Trends 33

widely used spices generally used in meat and dairy products and as topping on fast foods. The study has shown that oregano has been extracted rapidly with good quantity using SFE technique.

Pepper is commonly used in food industries. Extract of pepper obtained from supercritical technology were observed to be the same as obtained from the conventional method like steam distillation and solvent extraction. After comparing both the methods, it is concluded that contained odor and flavor of oil obtained from SFE technology is better than any conventional method.

Menthol and menthone are also extracted from peppermint leaves by SFE. Peppermint oil is an important and widely used flavoring agent in many products. The problem associated with SFE of peppermint leaf oil was presence of cuticular waxes. This problem is overcome by using SFE using CO_2 followed by a two-stage fractional separation [50].

Turmeric is another very important and widely used spice as a flavoring and coloring agent. Extraction of turmeric rhizomes is carried out by using SFE. The main ingredient and component of turmeric is curcumin, can be recovered up to about 90% under optimized conditions [3].

Anklam et al. [3] reported the extraction of vanillin by SFE using CO_2. Extraction of vanillin was carried out from wood pulps by two methods, i.e., SFE, and conventional method using Soxhlet. Both the results were compared for its concentration of p-hydroxybenzaldehyde, vanillic acid, vanillin, and para-hydroxy benzoic acid. In ethanolic extract, the ratio of acids to the corresponding aldehydes was found to be a little higher than SFE method [13, 29, 47].

Various flavors extracted by SFE method have its own application for the food industry [55]. List of samples from which flavor is extracted by SFE technique and application is given in Table 2.2.

SFE is also used to extract aroma compounds from various fruits, vegetables, and cereals. SC-CO_2 is also used to extract lemon peel oil. Extracted oil is then compared with oil which is obtained from the conventional method that is a cold-pressed method. Allicin is a major constituent of garlic which is used in the food industry. Allicin was extracted very well by SFE than conventional methods. It is observed that above 35°C, garlic undergoes thermal degradation. Extraction of fresh, dried, and stir-fried garlic can also be carried out in which extract contains 1% of frying oil [3, 33].

2.2.1.6 *EXTRACTION OF MYCOTOXINS*

Mycotoxins are toxic compounds present in food and animal feeding stuffs. Traditional extraction methods have some drawbacks, like poor recoveries

34   Advances in Food Process Engineering

of compounds and many interfering compounds during the process. Kalinoski et al. [26] proposed the use of the SFE technique to overcome this problem with chemical ionization mass spectrometry for the extraction of trichothecene mycotoxins. For this study, wheat samples were doped with different levels of three trichothecenes (T-2 toxin, deoxynivalenol, and diacetoxyscirpenol) one year before the analysis. The solvent used for SFE was CO_2 at 61°C and a pressure of 0.5 MPa. According to Kalinoski et al. [26], it is difficult to get quantitative extraction of trichothecenes and also interference from sample matrices during extraction of deoxynivalenol. Therefore, extraction of diacetoxyscirpenol and T-2 toxin was possible from wheat.

TABLE 2.2 Sample Matrix for Flavor and its Applications in SFE [55]

Sample Matrix	Solvent Used in SFE	Application of SFE
Apple	Supercritical CO_2	Used to study the effect of temperature, solvent, and pressure and to feed extraction ratio
Green tea		Different types of solvents are used like ethyl acetate, ethanol, and ethyl lactate with static extraction and SC-CO_2 for dynamic extraction
Citrus		Ethanol is used as co-solvent to optimize extracted yield
Wine		Different CO_2 and wine ratios are used. Aroma of rose wine is recovered in two steps
Grape spirit		For extraction, number of solvents to feed ratio are used
Sugar cane spirit		On extraction process effect of temperature as well as pressure is studied
Brandy		On extracted yield, effect of temperature as well as pressure is observed

Aflatoxin is also one of the most toxic substances because it contains carcinogenic substances. Therefore, the presence of aflatoxin in food needs to be analyzed. In traditional methods of extraction, organic solvents like acetone, chloroform, or ethanol were used to extract aflatoxins. A combination of various temperatures and pressures are used to study the extraction of aflatoxin with the help of SFE. SFE using CO_2 and organic modifiers has been used to extract alfotoxin B_1 from field-inoculated corn samples [64]. The optimum conditions were 34.47 MPa and 80°C to extract 94.6% aflatoxin B_1 by using SFE.

2.2.1.7 APPLICATIONS IN DIETARY FOODS

SCFs and SFE are also used in food pasteurization techniques. For denaturation of proteins, the pH of the solution should get changed. Process is done with the help of SC-CO$_2$ because it can influence pH at a varied temperature and pressure of solution. Advantage of using SC-CO$_2$ for pasteurization of food is original flavor and color of solution doesn't get changed and the process occurs without any thermal degradation of substance. Pressure of CO$_2$ required for the process is about 20 MPa as compared to other SCF like nitrogen, nitrous oxide, etc., which required pressure of 300–600 MPa. Day-by-day people are becoming very aware about health and moving towards healthy and dietary food. SFE is also used in the preparation of diet food with very low content of fat.

Anggrianto et al. [2] presented a study for extraction of fat from dietary food. Study was performed on black-eyed pea (*Vigna unguiculata*) and peanut (*Arachis hypogaea*) to produce high protein-low fat diet products. The pressure of supercritical CO$_2$ varies from 25 to 35 MPa and temperature varies from 40 to 60°C. Maximum yield of 5.4% and 48.5%, respectively for black-eyed pea (with optimized conditions were 25 MPa, 60°C, 10 g/min) and peanut (with optimized conditions were 35 MPa, 60°C, 15 g/min).

2.2.1.8 APPLICATIONS IN FOOD SAFETY

SCFs are also used as sterilizing agents in the food industry. SFE method has its application in food safety also. Term food safety comes due to the adulteration and contamination in food products nowadays. Food pollutants were found in large amounts in food products, and this is an important issue in food safety for consumers and for regulatory authorities. Detection of food pollutants is then an essential process which has to be done before marketing of food products.

Cleanup and extraction process of food pollutants were done by Soxhlet or saponification method. The disadvantage of this traditional method is that it is a time-consuming process and large amounts of organic chemicals are required. To restrict the use of organic chemicals and to speed up the process, SFE technique is developed to remove food pollutants. By using SFE technique, analysis of various pesticides in tomato, potato, lettuce, cereals, vegetable canned soups, diet foods, and honey can be carried out with SC-CO$_2$ and 10% acetonitrile. Food pollutants in the form of pesticides are hazardous for the human body. These can be found in fruit, vegetables, cereals, and soil. Therefore, SFE and SFC method is used to analyze and extract these pesticides from food products [54].

SFE is used to extract pesticides from the fat-rich foods. In conventional methods of extraction of pesticides, generally organic solvents like hexane and dichloromethane were used. To avoid any use of organic solvents, SFE is the best alternative to extract pesticide from its sample matrix. The advantage of using SFE over traditional methods is that it requires less time to extract pesticides and less volume of solvent required [1]. SFE with CO_2 or methanol-modified CO_2 has been successfully used in 'on-line,' 'off-line' or combination modes (such as SFE-SFC-GC) to estimate the organochlorine and organophosphorus pesticides in various food products like milk and meat products.

2.2.1.9 ANTISOLVENT FOR ENCAPSULATION OF FOODS

To stabilize any substance or material to react with any chemical or environment and go under oxidation reaction or to control the release of drug, encapsulation method is useful. Core substance or material is physically coated with another material called an encapsulating agent in a capsule form. This encapsulating material isolates the active substance from the surrounding environment and protects it from light, humidity, oxygen, or pH. There are many solvents available as encapsulating agents but SCFs have a greater advantage of their properties of both the phases that are liquid and gas, and hence it is largely used for encapsulation of the substances [41]. Surface tension of SCFs is near to zero, densities are like liquid and viscosity is like gas. Hence, due to mix properties of liquid and gas SCFs are extensively used in this process [33].

2.2.1.10 EXTRACTION OF LIPIDS

SFE and SFC are used to separate out low and high fat molecular weight lipid molecules from food products. Most of the research work related to SFE mainly been concentrated on the extraction of various lipid fractions (such as triglycerides, free fatty acids, sterols, fat-soluble vitamins, etc.), rather than the total fat content. This technique has been successfully applied for the extraction of lipid fractions from different food products like mushrooms, beef, chicken fat, egg yolk, etc.

Supercritical Fluid Technology: Recent Trends

Extraction conditions used were pressure between 8 MPa and 15 MPa and temperature at 40°C. This condition was used to separate out 43% polyphenols from cocoa seed by SFE [30].

There are various samples from which the lipid is extracted by using the SFE technique [45]. Pressure and temperature conditions for the extraction are also summarized (Table 2.3).

TABLE 2.3 Conditions for Extraction of Lipids from Various Foods [30, 54]

Sample	Pressure (MPa)	Temperature (°C)	Analyte
Brown seaweed	24.1–37	40–50	Fatty acid composition
Butter oil	10–27.6	40–70	Cholesterol
Corn bran	13.8–69	40–80	Ferulate phytosterol esters
Cotton seed	51.7–62	100	Fatty acid composition
Milk fat	6.9–17.2	40–60	Lipid profile; vitamin; solid fat content
Pork	7.3–34	50–150	Cholesterol
Rice bran	17–31	0–60	Fatty acid composition
Rice bran	48.2–62	70–100	α-tocopherol; sterol
Soybean	51.7–62	100	Fatty acid composition
Sunflower	51.7–62	100	Fatty acid composition

2.2.1.11 EXTRACTION OF FOOD COLORING AGENTS

Food coloring agents can also be extracted by the use of SFE. *Bixaorellana* L. also called Bixin, is a pigment of seed of annatto from which a food color is obtained. SC-CO_2 is used as solvent for extraction. Extraction condition for this process is temperature range in between 40°C and 80°C and pressure range is 40.41 to 60.62 MPa. Co-solvents (4% acetonitrile and 0.05% trifluoroacetic acid) are also used in this method. The optimum temperature and pressure conditions were found to be 40°C and 60.62 MPa, respectively. Coloring pigments are also obtained from paprika extract [54].

2.2.1.12 INACTIVATION OF MICROORGANISMS

High pressure food product market is increasing day by day. First in 1992, pressure processed food without heat processing was used in Japan. After that, high pressure foods with or without heat were used on a higher scale by the food industry all around the world. Pressure range for food is 100 to 1,000 MPa. *E. coli* is mainly found in pressurized food products. Inactivation of *E. coli*, which is pressure-resistant, is observed in fruit extracts and pH buffers. Pressure treatment of 300–600 MPa and temperature of 20–50°C for 5–10 minutes is useful for reduction of microorganisms in vegetables. During the pasteurization of milk, inactivation of heat-resistant bacteria is also important. *Saccharomyces cerevisiae* is inactivated at a pressure of 120–300 MPa and a temperature range of 20–50°C by pseudo-first-order kinetics. Many types of viruses, bacteria, and spores can be inactivated by using SC-CO$_2$ [28].

2.2.1.13 EXTRACTION OF BIOACTIVE COMPOUNDS

Bioactive molecules are found in plants, and they can prevent the formation of hydroperoxide by interfering in the formation of free radicals. Some examples of bioactive compounds are ascorbic acid, citric acid, lecithin, polyphenols, tocopherol, flavonoid, and lycopene. There are so many traditional methods available for separation of bioactive compounds. Advantages of SFE over traditional methods (like precipitation, membrane filtration, crystallization, evaporation, etc.), are rapid separation, more stability, extract heat-labile compounds, and less product loss. Some compounds are thermo-labile; in that case, SFE is a choice for extraction to avoid degradation of substance [60, 63]. In one of the study, pressure range set for the extraction of lycopene with SFE was 20, 25 and 30 MPa, whereas temperature ranges from 35°C, 45°C, 55°C and 65°C and flow rate of 2, 4, and 8 kg/h. A suitable extraction method for lycopene was observed as temperature at 55°C, pressure 30 MPa and flow rate was 4 kg/h.

Turmeric (*Curcuma longa* L.) is used as an additive and coloring agent in the food industry. Its oleoresin curcumerin gives it a property of coloring agent. Extraction conditions for the experiment consist of temperature at 60°C, pressure of 25 MPa and flow rate of 2 mL/min. The yield obtained was the same as obtained from traditional petroleum ether extraction. SFC helps to remove low polarity compounds easily, which can interfere in the process [57].

2.2.2 SUPERCRITICAL FLUID CHROMATOGRAPHY (SFC)

Supercritical fluid chromatography (SFC) is a separation technique, which is very widely used nowadays [35]. The solvent used for the chromatographic

Supercritical Fluid Technology: Recent Trends

technique is SCFs. Generally, SC-CO_2 is used as a solvent in many chromatographic techniques [40]. SFC indicates usage of high flow rate with low pressure, which falls through the column and this leads to higher efficiency and very less analysis time.

The setup of analytical SFC is a little bit similar to that of HPLC. The temperature used in SFC is critical temperature, that is the difference and more pressure is used in SFC by the SCFs like CO_2. Therefore, temperature, and pressure are two important things that differ in SFC from any other classical method of separation.

Sharper peaks can be obtained by using SFC as compared to any traditional method of separation. In the food industry, SFC nowadays is used because of its advantages like less time for separation and it can also extract thermolabile compounds and polar compounds. SCFs are widely used as a substitute for solvent like hexane for separation of soybean oil [21].

Three major qualities of SFC are selectivity, efficiency, and sensitivity. Different critical pressure and temperature conditions of SCFs are given in Table 2.4.

TABLE 2.4 Supercritical Fluids with Critical Temperature and Pressures Values [37, 40, 41]

Supercritical Fluids	Critical Temperature (°C)	Critical Pressure (MPa)
Ammonia	132.5	10.98
Carbon dioxide	31.1	7.2
Dichlorodifluoromethane	111.7	10.98
Diethyl ether	193.6	6.38
Ethane	32.2	4.76
n-Butane	152	7.06
Nitrous oxide	36.5	7.06
Tetrahydrofuran	267	5.05

Analysis of glycoproteins, lipids, vitamins, carotenoids, and other food products can be separated and analyzed by SFC. Separation of non-polar food compounds is observed according to the number of carbon present [9, 18].

Food components that are separated by SFC are hydrocarbons, pesticides, pigments, terpenes, spices, flavors, oils, nutraceuticals, and carbohydrates. It is an excellent technique to observe reaction chemistry between lipid species. It is also applied in esterification and transesterification of lipid and randomization of fats. SFC is cost and time saving technique [31].

It is a technique to detect, purify, and separate chiral compounds. It is possible to separate out a large range of compounds by using columns. Columns should contain different functionality and the same mobile phase. It is widely used as an alternative for liquid chromatography [15]. SFC (300 bar and 313 K) was used to fractionate the phospholipids from egg yolk lecithin (78.9% phospholipid content) using silica bonded with aminophenyl stationary phase and SC-CO$_2$ + ethanol as mobile phase [44].

2.3   SUMMARY

Nowadays, SFE and SFC is a very beneficial technique for the food industry. Applications of these two technologies in the food industry have been described in this chapter. Various studies are available for the extraction of food products using these technologies. Removal of fat and oil from the food is a very important process in the food industry which is in demand and for this SFE method has been used. Various food colors and flavors are also extracted from food products rather than using chemicals as a flavor and coloring agents. SC-CO$_2$ is mostly used SCF as compared to other fluids like diethyl ether, n-butane, hexane, etc. The most important advantage of SCFs is due to its supercritical temperature and pressure, thermolabile components can also be extracted, which was difficult with traditional and classical extraction techniques. SCFs speed up the process and fast extraction is obtained at lower cost. Therefore, overall usage of SFE and SFC, these are ranking higher as compared to traditional techniques like distillation, evaporation, Soxhlet extraction, membrane filtration, and centrifugation.

KEYWORDS

- **carbon dioxide**
- **food applications**
- **food processing**
- **gas chromatography**
- **supercritical fluid chromatography**
- **supercritical fluid extraction**

REFERENCES

1. Abbas, K. A., Mohamed, A., Abdulamir, A. S., & Abas, H. A., (2008). A review on supercritical fluid extraction as a new analytical method. *American Journal of Biochemistry and Biotechnology, 4*(4), 345–353.
2. Anggrianto, K., Salea, R., Veriansyah, B., & Tjandrawinata, R. R., (2014). Application of supercritical fluid extraction on food processing: Black eyed pea (*Vigna unguiculata*) and peanut (*Arachis hypogaea*). *Procedia Chemistry, 9*, 265–272.
3. Anklam, E., Berg, H., Mathiasson, L., Sharman, M., & Ulberth, F., (1998). Critical fluid extraction (SFE) in food analysis: A review. *Food Additives and Contaminants, 15*(6), 729–750.
4. Balentic, J. P., Ackar, D., Jozinovic, A., Babic, J., Milicevic, B., Jokic, S., Pajin, B., & Subaric, D., (2017). Application of supercritical carbon dioxide extrusion in food processing technology. *Hemijska Industrija, 71*(2), 127–134.
5. Bernal, J. L., Martín, M. T., & Toribio, L., (2013). Supercritical fluid chromatography in food analysis. *Journal of Chromatography A, 1313*, 24–36.
6. Brunner, G., (2010). Applications of supercritical fluid. *Annual Review of Chemical and Biomolecular Engineering, 1*, 321–342.
7. Brunner, G., (2005). Supercritical fluids: Technology and applications to food processing. *Journal of Food Engineering, 67*(1–2), 21–33.
8. Chemat, F., Rombaut, N., Meullemiestre, A., Turk, M., Perino, S., Fabiano-Tixier, A. S., & Abert-Vian, M., (2017). Review of green food processing techniques. Preservation, transformation, and extraction. *Innovative Food Science & Emerging Technologies, 41*, 357–377.
9. Chester, T. L., Pinkston, J. D., & Raynie, D. E., (1998). Supercritical fluid chromatography and extraction. *Analytical Chemistry, 70*(12), 301R–320R.
10. Chester, T. L., Pinkston, J. D., & Raynie, D. E., (1996). Supercritical fluid chromatography and extraction. *Analytical Chemistry, 68*(12), 487R–514R.
11. Ciftci, O. N., (2012). Supercritical fluid technology: Application to food processing. *Journal of Food Processing and Technology, 3*(7), 1–2.
12. Clifford, A. A., & Williams, J. R., (2000). Introduction to supercritical fluid and its applications; Chapter 1. In: Williams, J. R., & Clifford, A. A., (eds.), *Supercritical Fluid Methods and Protocols* (Vol. 13, pp. 1–16). Totowa–New Jersey: Humana Press.
13. Ehlers, D., & Bartholomae, S., (1993). High pressure liquid chromatographic analysis of vanilla carbon dioxide high pressure extracts. Comparison with conventionally produced alcoholic vanilla extracts. *Journal of Food Investigation and Research, 197*, 550–557.
14. Ehlers, D., Hilmer, S., & Bartholomae, S., (1995). HPLC analysis of supercritical carbon dioxide extracts of cinnamon and cassia in comparison with cinnamon and cassia oils. *Journal of Food Investigation and Research, 200*, 282–288.
15. Erkey, C., & Akgerman, A., (1990). Chromatography theory: Application to supercritical fluid extraction. *AIChE Journal, 36*(11), 1715–1721.
16. Fiori, L., De Faveri, D., Casazza, A. A., & Perego, P., (2011). Grape byproducts: Extraction of polyphenolic compounds using supercritical CO_2 and liquid organic solvent: A preliminary investigation. *CyTA Journal of Food, 7*(3), 163–171.
17. Furia, T. E., & Bellanca, N., (1975). *Fenaroli's Handbook of Flavor Ingredients* (2nd edn., Vol. II, p. 551). Boca Raton: CRC Press, Inc.

18. García-Risco, M. R., Vicente, G., Reglero, G., & Fornari, T., (2011). Fraction of thyme (*Thymus vulgaris L.*) by supercritical fluid extraction and chromatography. *The Journal of Supercritical Fluids, 55*(3), 949–954.

19. Gere, D. R., (1983). Supercritical fluid chromatography. *Science, 222*(4621), 253–259.

20. Gopalakrishnan, N., (1994). Studies on the storage of carbon dioxide extracted cardamom and clove bud oils. *Journal of Agricultural and Food Chemistry, 42*(3), 796–798.

21. Gopaliya, P., Kamble, P. R., Kamble, R. K., Chauhan, C., & Nobel, B., (2014). A review article on supercritical fluid chromatography. *International Journal of Pharma Research & Review, 3*(5), 59–66.

22. Hawthorne, S. B., (1990). Analytical-scale supercritical fluid extraction. *Analytical Chemistry, 62*(11), 633A–642A.

23. Herrero, M., Mendiola, J. A., Cifuentes, A., & Ibanez, E., (2010). Supercritical fluid extraction: Recent advances and applications. *Journal of Chromatography A, 1217*(16), 2495–2511.

24. Hrncic, M. K., Cor, D., Verboten, M. T., & Knez, Z., (2018). Application of supercritical and subcritical fluids in food processing. *Food Quality and Safety, 2*(2), 59–67.

25. Ilgaz, S., Sat, I. G., & Polat, A., (2018). Effects of processing parameters on the caffeine extraction yield during decaffeination of black tea using pilot-scale supercritical carbon dioxide extraction technique. *Journal of Food Science and Technology, 55*(4), 1407–1415.

26. Kalinoski, H. T., Udseth, H. R., Wright, B. W., & Smith, R. D., (1986). Supercritical fluid extraction and direct fluid injection mass spectrometry for the determination of trichothecene mycotoxins in wheat samples. *Analytical Chemistry, 58*(12), 2421–2425.

27. Kallio, H., Kerrola, K., & Alhonmaki, P., (1994). Carvone and limone in caraway fruits (*Carum carvi L.*) analyzed by supercritical carbon dioxide extraction-gas chromatography. *Journal of Agricultural and Food Chemistry, 42*(11), 2478–2485.

28. Khosravi-Darani, K., (2010). Research activities on supercritical fluid science in food biotechnology. *Critical Reviews in Food Science and Nutrition, 50*(6), 479–488.

29. King, J. W., (2000). Advances in critical fluid technology for food processing. *Food Science and Technology Today, 14*(4), 186–191.

30. King, J. W., & Srinivas, K., (2015). Development of multiple unit-fluid process and bio-refineries using critical fluids. In: Fornari, T., & Stateva, R., (eds.), *High Pressure Fluid Technology for Green Food Processing* (pp. 455–478). Switzerland: Springer.

31. King, J. W., & Taylor, S. L., (2002). Preparative scale supercritical fluid extraction and supercritical fluid chromatography of corn bran. *Journal of American Oil Chemists' Society, 79*(11), 1133–1136.

32. Kollmannsberger, H., & Nitz, S., (1994). The flavor-composition of supercritical gas extracts: III. clove (*Syzygium aromaticum*). *Chemistry-Microbiology-Technology-of-Food, 16*, 112–123.

33. Krukonis, V. J., (1988). Processing with supercritical fluids: Overview and applications; Chapter 2. In: Charpentier, B. A., & Sevenants, M. R., (eds.), *Supercritical Fluid Extraction and Chromatography* (Vol. 366, pp. 26–43). Washington, DC: ACS Publications.

34. Liu, L. X., Zhang, Y., Zhou, Y., Li, G. H., Yang, G. J., & Feng, X. S., (2020). The application of supercritical fluid chromatography in food quality and food safety: An overview. *Critical Reviews in Analytical Chemistry, 50*(2), 136–160.

35. Lopez-Avila, V., Dodhiwala, N. S., & Beckert, W. F., (1990). Supercritical fluid extraction and its application to environmental analysis. *Journal of Chromatographic Science, 28*(9), 468–476.

Supercritical Fluid Technology: Recent Trends

36. Majid, A., Phull, A. R., Khaskheli, A. H., Abbasi, S., Sirohi, M. H., Ahmed, I., Ujjan, S. H., et al., (2019). Applications and opportunities of supercritical fluid extraction in food processing technologies: A review. *International Journal of Advanced and Applied Sciences, 6*(7), 99–103.
37. Mansoori, G. A., (2002). The use of supercritical fluid extraction chromatography in food processing. *Food Technology Magazine, 20*, 134–139.
38. Martinez, J. L., (2007). *Supercritical Fluid Extraction of Nutraceuticals and Bioactive Compounds* (1st edn., p. 424). Boca Raton: CRC Press.
39. Masuda, M., Koike, S., Handa, M., Sagara, K., & Mizutani, T., (1993). Applications of supercritical fluid extraction and chromatography to assay fat-soluble vitamins in hydrophobic ointment. *Analytical Sciences, 9*, 29–32.
40. McHugh, M. A., & Krokonie, V. J., (1993). *Supercritical Fluid Extraction: Principles and Practices* (p. 503). Stoncham, MA: Butterworth–Heinemann.
41. Meireles, M. A. A., & Silva, E. K., (2014). Encapsulation of food compounds using supercritical technologies: Application of supercritical carbon dioxide as an anti-solvent. *Food and Public Health, 4*(5), 247–258.
42. Mendes, R. L., (2007). Supercritical fluid extraction of active compounds from algae; Chapter 6. In: Martinez, J. L., (ed.), *Supercritical Fluid Extraction of Nutraceuticals and Bioactive Compounds* (1st edn., pp. 189–213). Boca Raton: CRC Press.
43. Mohamed, R. S., & Mansoori, G. A., (2002). The use of supercritical fluid extraction technology in food processing. *Food Technology Magazine, 20*, 134–139.
44. Montañés, F., & Tallon, S., (2018). Supercritical fluid technology as a technique to fractionate high-valued compounds from lipids. *Separations, 38*(5), 1–13.
45. Nautiyal, O. H., (2016). Food processing by supercritical carbon dioxide-review. *E. Chronicon Chemistry, 2*(1), 111–135.
46. Palmer, M. V., & Ting, S. S. T., (1995). Application for supercritical fluid technology in food processing. *Food Chemistry, 52*(4), 345–352.
47. Parhi, R., & Suresh, P., (2013). Supercritical fluid technology: A review. *Journal of Advanced Pharmaceutical Science and Technology, 1*(1), 13–36.
48. Pellerin, P., (1991). Supercritical fluid extraction of natural raw materials for the flavor and perfume industry. *Perfumer & Flavorist, 16*, 37–39.
49. Reverchon, E., (1997). Supercritical fluid extraction and fractionation of essential oil and related products. *The Journal of Supercritical Fluids, 10*(1), 1–37.
50. Reverchon, E., Ambruosi, A., & Senatore, F., (1994). Isolation of peppermint oil using supercritical carbon dioxide extraction. *Flavor and Fragrance Journal, 9*(1), 19–23.
51. Reverchon, E., Donsi, G., & Pota, F., (1992). Extraction of essential oils using supercritical carbon dioxide effect of some process and pre-process parameters. *Italian Journal of Food Science, 4*, 187–194.
52. Riaz, M. N., (2000). *Extruders in Food Applications* (p. 240). Boca Raton: CRC Press.
53. Rizvi, S. S. H., Mulvaney, S. J., & Sokhey, A. S., (1995). The combined application of supercritical fluid and extrusion technology. *Trends in Food Science & Technology, 6*(7), 232–240.
54. Rozzi, N. L., & Singh, R. K., (2002). Supercritical fluids and the food industry. *Comprehensive Reviews in Food Science and Food Safety, 1*(1), 33–44.
55. Saffarionpour, S., & Ottens, M., (2018). Recent advances in techniques for flavor recovery in liquid food processing. *Food Engineering Reviews, 10*, 81–94.

56. Sahena, F., Zaidul, I. S. M., Jinap, S., Karim, A. A., Abbas, K. A., Norulaini, N. A. N., & Omar, A. K. M., (2009). Application of supercritical CO_2 in lipid extraction: A review. *Journal of Food Engineering, 95*(2), 240–253.

57. Sanagi, M. M., Ahmad, U. K., & Smith, R. M., (1993). Application of supercritical fluid extraction and chromatography to the analysis of turmeric. *Journal of Chromatographic Science, 31*(1), 20–25.

58. Shao, P., Sun, P. L., & Ying, Y. J., (2008). Response surface optimization of wheat germ oil yield by supercritical carbon dioxide extraction. *Food and Bioproducts Processing, 86*(3), 227–231.

59. Sharif, K. M., Rahman, M. M., Azmir, J., Mohamed, A., Jahurul, M. H. A., Sahena, F., & Zaidul, I. S. M., (2014). Experimental designs of supercritical fluid extraction: A review. *Journal of Food Engineering, 124*, 105–116.

60. Shi, J., Kassama, L. S., & Kakuda, Y., (2007). Supercritical fluid technology for extraction of bioactive compounds; Chapter 1. In: Shi, J., (ed.), *Functional Food Ingredients and Nutraceuticals: Processing Technology* (pp. 3–44). Boca Raton: CRC Press.

61. Sihvonen, M., Jarvenpaa, E., Hietaniemi, V., & Huopalahti, R., (1999). Advances in supercritical carbon dioxide technologies. *Trends in Food Science & Technology, 10*(6, 7), 217–222.

62. Silva, E. K., & Meireles, M. A. A., (2014). Encapsulation of food compounds using supercritical technologies: Application of supercritical carbon dioxide as an antisolvent. *Food and Public Health, 4*(5), 247–258.

63. Sugiyama, K., Shiokawa, T., & Moriya, T., (1990). Application of supercritical fluid technology and supercritical fluid extraction to the measurement of hydroperoxides in food. *Journal of Chromatography, 515*, 555–562.

64. Taylor, S. L., King, J. W., Richard, J. L., & Greer, J. L., (1993). Analytical-scale supercritical fluid extraction of aflatoxin B_1 from field-inoculated corn. *Journal of Agricultural and Food Chemistry, 41*(6), 910–913.

65. Viganó, J., Machado, A. P. D. F., & Martínez, J., (2015). Sub-and supercritical fluid technology applied to food waste processing. *The Journal of Supercritical Fluids, 96*, 272–286.

66. Vuorela, H., Holm, Y., Hiltunen, R., Harvala, T., & Laitinen, A., (1990). Extraction of volatile oil in chamomile flowerheads using supercritical carbon dioxide. *Flavor and Fragrance Journal, 5*(2), 81–84.

CHAPTER 3

EXTRUSION TECHNOLOGY OF FOOD PRODUCTS: TYPES AND OPERATION

ANJALI SUDHAKAR, KANUPRIYA CHOUDHARY,
SUBIR KUMAR CHAKRABORTY, and CINU VARGHESE

ABSTRACT

Extrusion is a commercially viable technology due to its widespread application in food industries and consumers demand for extruded products. This chapter describes the history of extrusion technology, theories, and types of extrusion process, equipment used for extrusion, effect of extrusion on the physicochemical characteristics of food, and its application in the food industry.

3.1 INTRODUCTION

The functional food market is demanding processed food products with innovative designs and nutritional benefits. Extrusion technology is undoubtedly the best process to meet these demands. Extrusion is a continuous process in which a material is forced to pass through a restricted area or die under controlled conditions. The raw material is fed through a hopper into a barrel inside which the structural deformation occurs due to extreme temperature, pressure, and shear; then finally, the material comes out through die down the barrel. It is an eco-friendly technology that has been used for molding, forming, shaping, coating, etc., in the plastic and metallurgical industries.

The principle of extrusion was applied to the food materials as well. Now it has become one of the most successful technologies for food processing.

Advances in Food Process Engineering: Novel Processing, Preservation, and Decontamination of Foods.
Megh R. Goyal, N. Veena, & Ritesh B. Watharkar (Eds.)
© 2023 Apple Academic Press, Inc. Co-published with CRC Press (Taylor & Francis)

There are numerous reasons for the acceptance of extrusion technology for commercial benefit. It provides high productivity with low process costs and no process effluents. This process has become very popular with food industrialists due to the versatility of produce that can be handled, resulting in a wide variety of food products. Extruded products can have different shapes like tubes, rods, doughnut, spheres, rings, strips, etc. Extruded products may require subsequent processes like drying, frying, or packaging to improve its consumer acceptability and shelf life. Compared to other cooking techniques extruded products can be stored for a longer period due to low water activity (a_w 0.1–0.4) of products.

Consumers are using extruded products in different forms like breakfast cereals, baby food, ready-to-eat snacks, etc. It has been reported that the value of international food extrusion market was USD 58.75 billion in 2018 and by 2026 it is expected to reach $87.86 billion at a compound annual growth rate (CAGR) of 5.1% [23]. The upcoming demand of this technology requires a deep knowledge about its working, equipment, applications, and products.

This chapter focuses on various aspects of extrusion technology, such as history, types, and effects on the physicochemical characteristics of food.

3.2 HISTORY

The word extrusion is derived from the Latin word *extrudere*. '*Ex*' means out and '*trudere*' means thrust. Extrusion technology was first used in 1797 for the production of seamless lead pipe in a metallurgical industry. The research on extrusion continued to find its applications in food industries. The development of an extruder in 1870 for the preparation of sausage was the first achievement in the food sector [22]. In mid-1930s, the first twin-screw extruder for food material was developed, and in the same year, single-screw extruder became common for manufacturing spaghetti or pasta. Later it was used to pre-cook dough for making ready-to-eat cereal. In 1946, during World War II, the US developed a single-screw extruder for the development of expanded snack made from corn flour and rice flour. The application of extrusion technology in food industries accelerated after the mid-1950s, when the first patent on the application of twin-screw extrusion was filed. In 1970, the first continuous extrusion-cooking process was developed. From then to now, extrusion has been applied to several food materials to develop a wide variety of commercial products.

Extrusion Technology of Food Products

3.3 EXTRUSION TECHNOLOGY

In the extrusion process, two main energy inputs are involved: (i) the energy transferred due to the rotation of screws; (ii) heat energy transferred from the heater through the barrel walls. This results in the generation of enough heat in the barrel resulting in evaporation of moisture or melting of solid material being used for extrusion.

3.3.1 EXTRUSION-COOKING

Extrusion-cooking is a thermo-mechanical process that operates in dynamic steady-state equilibrium, where the input and the output variables are comprehensively balanced. Extrusion-cooking can perform various unit operations, including shaping, kneading, mixing, conveying, forming, shearing, sizing, cooling, heating, flavor generation, and partial drying or puffing depending upon the equipment and material used. It is also a high-temperature short-time (HTST) process and thus used for microbial decontamination and enzymes inactivation. In this process, food is cooked when its temperature is elevated to 100–180°C. Extrusion-cooking is a low moisture process (normally 10–40% on a wet basis). The pre-conditioned mix is fed to the extruder where the high temperature and high pressure (10–20 bar) and high shear is applied. The heat provided through the heater and frictional heat is used for cooking of food. The high temperature reduces the cooking time and allows the raw material to transform into its functional form within a period of approximately 30–120 s. Initially, steam is injected to start the cooking process. During extrusion-cooking, the food constituents undergo complex changes like gelatinization of starch, denaturation of proteins and melting of fats.

3.3.2 RAW MATERIALS FOR EXTRUSION-COOKING

Selection of raw materials for extrusion-cooking is a crucial factor that influences product characteristics [8]. The type of material, physical state, chemical composition, moisture content, and pH are the main characteristics that should be considered while deciding the raw materials depending on the product and type of extrusion process. Among different types of materials, starch, and protein are the most commonly used raw materials. Starch is mainly used as a base material for the development of expanded snacks, breakfast cereals, and biscuits, while protein is used to produce extruded

48   Advances in Food Process Engineering

product in the form of texturized vegetable protein (TVP) and pet foods [10]. The selection of raw materials is a confusing and complex task for extrusion-cooking technologists due to the wide options of raw materials. This problem can be overcome by Guy classification of ingredients published in 1994 at Campden and Chorleywood Food Research Association (CCFRA). This provides an easy understanding of the process of extrusion. The classification was based on the functional role of raw materials using a physicochemical approach. Initially, six groups were formed, but in later, more groups were included to describe the functional role of ingredients (Table 3.1).

TABLE 3.1   Classification of Raw Materials Based on Functional Properties

Group of Ingredients	Examples
Coloring substances	β-carotene, canthaxanthin, annatto
Dispersed phase filling materials	Albumin, amylomaize starches
Flavoring substances	Synthetic and natural flavoring
Nucleating substances	Bran, calcium carbonate, insoluble calcium phosphate
Plasticizers and lubricants	Water, oils, and fats
Soluble solids	Salt, sugar
Structure forming materials	Cereals, potato, cassava, maize, soya, wheat gluten, oilseed proteins

3.4   THEORY OF EXTRUSION

3.4.1   WORKING PRINCIPLE

An extruder consists of mainly three components: screw, barrel, and die (Figure 3.1). Screw rotates inside the barrel which helps to convey the material and the flow is known as drag flow. Die (single or multiple) is the restricted area available at the discharge end of the barrel. There are three zones inside barrel: feed zone, compression zone and metering zone. The pressure at this zone is comparatively low may be due to the constant root diameter of the screw. In the compression zone, the material gets compressed because flight of the barrel restricts the flow and thus volume increases. High compaction creates a positive pressure gradient between screws and feed which causes material to flow forward (pressure flow).

Further, the material moves into the metering zone where kneading takes place causing rheological changes in the material. These rheological changes occur due to frictional heat and external heat (in the case of hot extrusion), shearing, and pressure. Finally, the material is forced through die where expansion occurs due to the pressure difference between the atmosphere and inside the barrel. A cutting system is also equipped near die, which helps to cut the extrudate.

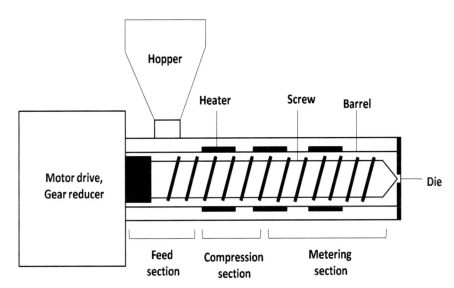

FIGURE 3.1 Components of a screw extruder.

3.4.2 ENGINEERING ASPECTS

In the extrusion process, several transport phenomena occur, including mass transfer, thermal energy transfer, flow of material in the barrel. The understanding of these phenomena helps in the selection of process conditions, extruder type and raw materials, designing extruder parts and rheological changes in feed. The basic numerical approaches to study these phenomena are energy balance, mass balance and power law. These approaches formed the basis to develop several mathematical models of extrusion technology.

3.4.3 MODEL TYPES FOR TWIN-SCREW EXTRUDER

Basically, for twin-screw extruders, the models were developed for three main topics as listed in Table 3.2. In the following sections, the most important models are discussed.

50　　　　　　　　　　　　　　　　　Advances in Food Process Engineering

TABLE 3.2　Types of Models for Twin-Screw Extruder

Topic	Model	References
Flow behavior	Tadmor and Klein	[27]
	Janssen	[13]
	Tayeb and co-workers	[28]
Rheological behavior	Harper and co-workers	[11]
	Fletcher and co-workers	[9]
	Vergnes and Villemaire	[31]
	Morgan and co-workers	[19]
Energy	Jepso	[14]
	Yacu	[32]
	Mohammad and Ofoli	[18]
	Van Zuilichem and co-workers	[30]

3.4.3.1　YACU MODEL

The temperature and pressure profiles for three main sections (i.e., conveying zone, pumping zone, and shearing zone) in twin-screw extruder were predicted by Yacu [33]. Some assumptions in this model were:

- The extruder operates under steady and uniform conditions;
- The flow of the melt is non-Newtonian and in the laminar flow regime;
- The viscosity of the melt is non-isothermal and considers the effect of moisture and fat;
- The effect of gravity is ignored;
- The degree of cooking is uniform over the cross-section;
- The screw is assumed to be adiabatic.

In a twin-screw extruder, the mode of heat transfer between barrel and feed material is through conduction in the conveying zone. Inside the barrel, heat transfer occurs due to convection during mixing of feed. Considering all the modes of heat transfer, the temperature at any boundary condition is determined as follows:

$$T = T_b - \left(T_b - T_f \right) e^{\frac{-FU_s Ax}{C_{ps} Q_m}} \tag{1}$$

where; T is the temperature at an unknown point (x); T_b, T_f is the temperature of the barrel and the feed, respectively; F is the degree of fill; Q_m is the feed

Extrusion Technology of Food Products   51

rate; C_{ps} is heat capacity of solid matter; and U_s is the pseudo heat transfer coefficient.

The value of heat transfer coefficient (U_s) is unreal due to phase discontinuity and additional resistance between solid particles. In the pumping zone, the melting of solid-powder food material occurs rapidly. According to Martelli [17], energy dissipates at four locations within the channel. The total energy converted per channel per screw turn (E_t) is expressed by Eqn. (2).

$$E_t = C_{1p}\mu_p N^2 \tag{2}$$

Using the heat balance, the rate of change of temperature in the pumping section is evaluated by Eqn. (3):

$$\frac{dT}{dx} = C_{2p}e^{-b_1T} + C_{3p}(T_b - T) \tag{3}$$

where; μ_p is the viscosity of the product; 'N' is the rotational speed of the screw; and C_{1p}, C_{2p}, C_{3p} are the geometry factor for pumping section screw.

3.4.3.2 VAN ZUILICHEM MODEL

This model elaborates on the total heat transfer along the axis in an extruder. This model calculates the heat in two parts: (a) heat transferred from barrel to the extrudate, and (b) heat generated due to viscous dissipation [20]. The input parameters of the model are: torque, feed rate and temperature profile imposed at the barrel. The torque is converted into phase transition energy, pressure energy, and temperature increase. This model was validated for twin screw extruder. A slight deviation (maximum 10°C) was observed between predicted and measured temperatures.

3.5 TYPES OF EXTRUSION

3.5.1 COLD EXTRUSION

Cold extrusion is also known as low temperature extrusion. In this type, the extrusion temperature of the food remains that of the ambient temperature [24]. It is often used to mix for more specific shape products, i.e., pasta, meat products, etc. In this process, no external heat is added and extrudate is pumped through the die. Cold extrusion system is a simple process like kneading steps in preparation of dough before baking [26]. In general, cold extrusion is defined as a process used for kneading, mixing, dispersing, dissolving, and forming a product ingredient. Some popular cold extrusion products are noodle, pasta, hot dog, candy, pastry dough, etc. [8]. It is also known as low pressure extrusion because in this type of extrusion process low shear is created resulting in a low pressure at the upstream end from the die.

3.5.2 HOT EXTRUSION

Hot extrusion is also known as high temperature extrusion or extrusion-cooking. High temperature extrusion is more widely applied than the low temperature extrusion [24]. As the name suggests, high temperature extrusion is operated at high temperatures and high pressures. Therefore, this process can be considered as HTST process and in addition, this process also helps in reducing microbial contamination and inactivating spoilage enzymes [8]. In this process, heat is added to the melt from an external source, this addition of energy occurs at the surface of the barrel and then the energy is transferred through the barrel wall and surfaces to the melt. If external heat is not added in this process, then heat is generated by friction between surfaces and ingredients within the barrel of the extruder and is dissipated as thermal energy into the melt [26]. This process is mainly used for starch-based extruded products.

3.6 TYPES OF EXTRUDERS

Extruders can be divided into two categories on the basis of barrel configuration, such as single screw extruder and twin-screw extruder.

3.6.1 SINGLE SCREW EXTRUDER

The main structural element of a single screw extruder is the barrel, which is a hollow cylindrical shield inside which screw is fitted. Barrel inside surface can be smooth or grooved, the section can be cylindrical or conical; but for food extruder only smooth or grooved barrel are used. This screw has thick root and shallow flights and rotates inside the barrel, while flights of the rotating screw push the raw material along a flow channel formed between

Extrusion Technology of Food Products

the screw roots and the barrel. The width of the flow channel depends on the screw pitch; it is significantly larger than its thickness. The gap between the barrel surface and screw tip is made lowest as possible. At the exit end of an extruder, a limited passageway is provided known as die. Its function is to serve as a pressure release valve and to give the desired shape to the extruder which depends on aperture cross-section. A rotating knife of variable speed is provided to cut extrudates into desired pieces.

Different kinds of elements are provided for cooling or heating the barrel. These elements are mainly divided into three sections (Figure 3.1), which help in keep up change in temperature according to the zones of the extruder. A gravity feeding type hopper or positive feeding type auger is provided for feeding. Ports are needed for pressure release and for the application of water, steam, and other fluids. A drive having ability to control torque and speed variation capability is provided. Different type of measuring devices, i.e., feed rate, temperature, and pressure are also provided.

High viscous raw material like dough or particulate solids is fed into the barrel through a hopper and move towards die end due to the rotation of the screw. The flight of screw and friction between wall of barrel and screw decides the rate of flow of material inside the flow channel. Normally, for the occurrence of internal shear, the frictional force at the surface of the barrel should be sturdy [2]. Screw configuration is made in such a way so that the flow channel constantly decreases to allow compression of material as it flows towards the die. By using different type of screw configuration, a continuous decrease in flow area is achieved, but the most common method is by decreasing the screw pitch constantly and by increasing the root diameter along the flow channel. In a single screw extruder, there are three zones:

1. **Feed Zone:** This zone act as a screw conveyor, its main function is to convey raw material from the entrance to the barrel. In this zone, no modification or compression of mass takes place.
2. **Transition Zone:** In this zone, heating and compression of material takes place.
3. **Metering Zone:** In this zone, melting, texturization, kneading, chemical reactions, etc., occurs through mixing and shear.

The power is required to turn the screw shaft and this power is later dissipated into the material as heat through friction. If additional heat is required, then it is transferred through the external source to the barrel surface and is also added by the direct injection of steam. High pressure is generated in the extruder due to the compression; therefore, the melt is heated above 100°C temperatures, sometimes this temperature reaches up to 200°C. When the material reaches the exit end of the extruder (i.e., at the die), flash evaporation takes place due to the presence of water in the product. As a result of flash evaporation of water, the product is puffed out as a result of atmospheric air finding place inside the extrudate matrix. The opening of the die at the last section of the extruder controls the degree of puffing [8]. The pressure gradient is built up in the opposite direction of the material movement due to compression. In extruder, two types of flow occurs in opposite direction, i.e., drag flow and backflow. Drag flow direction is from the feed end to the exit end, and it is a result of the mechanical thrust due to the screw flights, whereas backflow is due to pressure difference between the exit end and feed end of the extruder. Mixing in extruder takes place due to the flow of mass in both directions, i.e., in the direction of backflow and of drag flow. The extent of backflow depends on the resistance of the die and restriction in mass flow.

3.6.2 TWIN SCREW EXTRUDER

In twin screw extruder, parallel screw rotates in the barrel and forms 8-shaped cross-section [2]. Twin-screw extruder has two parallel screws (Figure 3.2), which can be co-rotating (both moving in the same direction) or counter rotating (both moving in the opposite direction).

FIGURE 3.2 Arrangement of screws in twin-screw extruder.
Source: Motorhead. https://creativecommons.org/licenses/by-sa/2.5/

Mostly in food industry co-rotating, self-wiping type extruders are used. There are different models present in self-wiping type extruders. In one of the model, two screws are intermeshed. In another model, the barrel is partitioned into two longitudinal halves, this helps in screw inspection and cleaning. In other models, the barrel contains short and detachable modules. Different screw configurations are used for enhancing the mixing and reducing pressure on a shaft screw elements for different compression ratios, mixing sections, reverse-pitch screw sections restrictions (for building pressure), etc. [2]. This type of flexibility was not present in earlier models with fixed screw configuration.

In a twin-screw extruder, the feed is transported in a C-shaped chamber formed between screws. The amount of power requirement for pumping is more for twin-screw extruder in comparison to a single screw extruder. Therefore, during extrusion-cooking, heat is supplied from an external source. Slippage or friction on the barrel surface is less significant. Large screw angles in conveying section leads to conveying the material at high forward axial velocity. This is also called a heated screw conveyor. As a result of high velocity, the filling ratio at conveying section is low. The heating in this section occurs due to heat transfer from the barrel surface and low hold-up causes rapid heating of the material. Temperature of the material reaches close to the barrel surface temperature. Conveying section does not have a significant contribution for pressure build up inside the barrel. Therefore, pressure is attained after the conveying section; this pressure build-up section is known as the metering section or melts pumping section. In this section, requisite die pressure is achieved by restricting mass flow. The restriction in mass flow is caused by the configuration of the screw. The filling ratio in this section gradually increases towards the exit end. The screw part facing towards the barrel helps in the translation motion of the melt while most of the mixing occurs in the intermeshing section between the parallel screws.

3.7 ADVANTAGES AND DISADVANTAGES OF TWIN SCREW COMPARED TO SINGLE SCREW EXTRUDERS

Advantages are:

1. **Pumping Efficiency:** It depends less on the flow properties of material and better than single screw extruder.
2. **Mixing:** It mixes material better than single screw extruder because mixing is done mostly in the metering zone of the screw.
3. **Heat Exchange Rate:** It is faster and uniform.
4. **Versatility:** Due to flexible configuration of screw and barrel.
5. **Feeding:** In twin-screw extruder, we can easily handle highly moist and sticky materials while it is not possible in a single screw extruder. Thus, difficulties of feeding cohesive material is solved easily by a twin-screw extruder.

6. **Residue Build-up:** Due to self-wiping features of twin-screw extruder, no residue build up takes place inside the barrel.

Disadvantages are:

1. **Higher Cost:** Capital and operation cost are higher than single screw extruder.
2. **Complexity:** It is mechanically more complex and less robust than single screw extruder. Therefore, twin-screw extruder is more profound to mechanical abuse, i.e., high torque.

3.8 PROCESS CONTROL OF AN EXTRUDER

The aim of process control technology is to provide improved and consistent quality of the final product, reduction in waste generation and energy usage, more precise control, and a stable operation, increase throughput and yield, reduced product gives away. The quality of an extruded product mainly depends on four variables of an extruder:

- Configuration of screw and the length to diameter (L/D) ratio of screw;
- Design of die, i.e., diameter, shape, length, and number of dies on die plate;
- Design of cutting knife;
- Process variables that can control operator during operation.

The configuration of screw-in an extruder affects the amount of mixing, shear force and heat generated by the friction and residence time distribution in different sections of the extruder. All these factors are responsible for the degree of cooking and product quality and consistency. The residence time for an extruder is depends on L/D ratio of screw, which is also accountable factor for degree of cooking. In extruder, most of the work takes place in barrel, but design of die has a notable effect on size and shape of the product, amount of generated energy, melt rheology and the pressure drop across die. All these factors affecting from design of the die are responsible for product quality [10]. The design of rotating cutting knife affects the size and shape of the product. The variables affecting product quality like configuration of screw, die design and cutting knife could not change during operation. However, some variables like screw speed, application rate of water, feed rate, rotating speed of cutting knife, and temperature profile in barrel can change during operating conditions of extruder.

Extrusion Technology of Food Products 57

In an extruder, residence time, the melt viscosity, the amount of shear action, heat generated due to friction and barrel fill are affected by screw speed. The application rate of water affects the amount of moisture in the barrel, which decides the amounts of heat generated and melt viscosity. The barrel temperature in an extruder is maintained by external energy source like direct injection of steam, hot oil and electrical conduction and induction and for cooling purpose cold water is circulated.

There are four key control variables in an extruder as listed below:

- Die temperature;
- Die pressure;
- Flow rate through die; and
- Specific mechanical energy.

Die temperature and die pressure are crucial variables, which can be controlled by the process parameters. The flow rate through the die is the measure of overall flow rate through the extruder to the cross-sectional area of the die. Specific mechanical energy is the measure of how much mechanical energy from the motor is consumed by the extrusion system per unit mass.

3.8.1 CONTROL VARIABLES OF AN EXTRUDER

The objective of controlling the key control variable is to develop the product with high and uniform quality. During the extrusion process, it is factual that for known screw configuration and die design, the main process variables (i.e., screw speed, feed rate, barrel temperature profile, rate of steam injection and speed of cutting knife) should be maintained properly, then only key-controlled variables will be maintained at the desired levels and extrudates with desired quality can be obtained. If we are able to maintain these key-controlled variables at the particular levels, then we will get desired quality products.

During extrusion process due to instability in an extruder, we get non-uniform products. The emptying of barrel is responsible for this instability. This takes place when screw speed is much higher than feed rate. Hence, it is essential to control the screw speed and feed rate in an extruder, so that barrel is not flooded or emptied during the process. Therefore, it is essential to use a precise feed rate controller and a pump to control the water injection, because a small fluctuation can affect the product quality [10].

3.9 EFFECTS OF PROCESS CONTROL ON PHYSICOCHEMICAL AND FUNCTIONAL PROPERTIES OF A FOOD PRODUCT

3.9.1 BULK DENSITY

The bulk density of the extrudate seeks to define the degree of puffing undergone by the melt while extruding out of the die of the extruder. It has positive co-relation with feed moisture content, if amount of feed moisture increases the bulk density also increase. This is because as the amount of moisture in feed increases during extrusion-cooking, this reduces the elasticity of the feed through plasticization of the melt, resulting in reduced specific mechanical energy and thus decreased gelatinization, reducing the expansion and increases the amount of bulk density of the extrudates. However, bulk density of the extrudates decreases with increase in die temperature and screw speed. The decrease in bulk density might be due to starch gelatinization at higher temperatures [17].

3.9.2 EXPANSION RATIO

The degree of puffing of the extrudates is characterized in term of expansion ratio. Expansion ratio is positively co-related with screw speed; this is because an increase in screw speed results in an increased structural breakdown [17]. The Expansion ratio and bulk density are dependent physical properties, i.e., an increase in expansion ratio results in the decrease in bulk density and vice-versa. It has been reported that extrudate product expansion ratio depends on moisture content and extrusion temperature [4].

3.9.3 WATER ABSORPTION INDEX

It is used as an index of gelatinization, because water absorption index measures the amount of water absorbed by starch [1]. It measures the water held by the starch after swelling in excess water which corresponds to the weight of the gel formed [1]. It decreases with the increase in feed moisture due to the fact that with the subsequent increase in moisture, tendency to absorb water decreases [5].

3.9.4 WATER SOLUBILITY INDEX

Water solubility index determines the number of polysaccharides released from granules after the addition of excess water. The degradation of

Extrusion Technology of Food Products

molecular components is indicated by water solubility index. As feed moisture content increases in extrusion-cooking water solubility index value decrease, this is due to a decrease in the level of protein denaturation. The increase in temperature might increase the degree of starch gelatinization that could increase the amount of soluble starch resulting in an increase in water solubility index [12].

3.9.5 CHEMICAL CHANGES

3.9.5.1 CARBOHYDRATES

In extrusion-cooking, carbohydrate is the major food component that undergoes substantial transformation. Humans cannot easily digest ungelatinized starch. The key effect of extrusion on starch is gelatinization. The starch suspension is heated with water for gelatinization. The moisture content of the raw material decides the temperature requirement for starch gelatinization. Gelatinization occurs at low moisture content (12–20%) during the process of extrusion-cooking. The rate of gelatinization increases with an increase in processing conditions, i.e., temperature, pressure, and shear. Dietary fiber, lipids, and sucrose are also responsible for rate for gelatinization. Thus, during extrusion-cooking starch is pre-digested. Hence, the loss of molecular weight of amylopectin and amylose is observed due to sheared off branches of amylopectin molecules during its travel through the barrel [10].

3.9.5.2 PROTEIN

Extrusion-cooking improves digestibility of protein by denaturing it. Due to denaturation of protein most of the enzymes and enzyme inhibitors lose their activity. The level of protein denaturation is measured by the solubility of protein in aqueous solution. These changes are more prominent under high shear actions and also influenced by temperature and moisture content [10]. Researchers reported that solubility of wheat protein used in pasta making is reduced at a relatively low temperature [29]. The nutritional evaluation of extruded products revealed that protein content is less in these products. High temperature and low moisture results in Maillard reaction during extrusion-cooking; therefore, reducing sugar (formed by shearing of sucrose and starch) react with lysine, which reduce the nutritional value of the protein.

3.9.5.3   FIBER

During extrusion-cooking, total fiber content remains unaffected but insoluble fiber reduces when the product is extruded at 25% moisture content. However, soluble fiber increases in extruded samples when processed at 30% moisture and 180°C.

3.9.5.4   VITAMINS AND MINERALS

Vitamins are important for a healthy immune system in humans. Being a HTST process, extrusion-cooking destroys vitamins by heat and oxygen. During the extrusion process, vitamin loss takes place due to temperature, but at the same time micronutrient fortification is also popular in extruded products. Therefore, in extrusion-cooking precursor of vitamins is added. But at high temperatures these are also unstable, so nowadays, manufacturers are adding vitamins in the form of spray at post extrusion.

Researchers reported that mineral content after extrusion increases in high fiber products; i.e., potato peel extruded under high temperature conditions result in 38% increase in iron content. However, extruded corn, i.e., low fiber content product, not showed a significant difference in soluble iron. Any resulting changes in mineral content due to extrusion-cooking did not affect utilization of zinc and iron from wheat and wheat bran in adult human volunteers [10].

3.9.5.5   ANTIOXIDANTS

Antioxidants (such as pigments like anthocyanin) provide attractive color and can protect vision and cardiovascular diseases. Researchers have reported that anthocyanin content in blueberry and ascorbic acid content in breakfast cereals is considerably reduced by the extrusion-cooking. Polymerization and browning are also responsible for the loss of antioxidants in extruded process.

3.10   APPLICATIONS OF EXTRUSION TECHNOLOGY

The application of extrusion is not only limited to some unit operations (already discussed above) in food processing. This technology has been used for stabilization of rice bran, precooking of soy flours, treatment of oilseeds

Extrusion Technology of Food Products 61

for oil extraction, flavor encapsulation, determination of spices, enzymatic liquefaction of starch for fermentation, destruction of aflatoxin in peanuts and many more [6]. Besides all these applications, extrusion technology provides various opportunities that have brought advancement and a new revolution in food engineering. Some of the important applications of extrusion technology are discussed in subsections.

3.10.1 AGRO-BASED BYPRODUCT UTILIZATION

An immense amount of byproduct is generated from food industries; there is a major concern for proper utilization and management of this waste in an environment friendly manner. However, most of the agro-produce byproducts like fruit pomace, peel, leaf trimmings, hulls, and bran, oilseed cake are rich in nutrients. These byproducts can be utilized to develop value-added products. Most of the extruded products are made from starch which makes its nutritionally inadequate. So, incorporation of these byproducts can add fiber and other nutrients to develop a healthy snack. Nevertheless, the nutritional loss in extrusion-cooking is comparatively smaller than the other cooking. Therefore, extrusion technology has a great potential to utilize the waste products. Several products have been developed (Table 3.3) by incorporating agro-based byproducts with a base material to improve nutritional value and sensory attributes [15].

TABLE 3.3 List of Extruded Products Developed from By-Products of Foods

By-Product	Extruded Product
Apple pomace	Puffed snack
Buckwheat bran	Pasta
Carrot pomace	Puffed snack
Cauliflower trimmings	Puffed snack
Linseed/rapeseed/soybean/sunflower cake	Fish pellet
Potato peel	Cookies
Soybean hull	Puffed snack

3.10.2 3D FOOD PRINTING

In the present era of digitalization, 3D food printing is grabbing the attention of food technologists because it has a great potential to produce customized food with complex designs and personalized nutrition. 3D printed food can be manufactured using extrusion-based printer, inkjet printer, binder-jet printer, laser printer and selective sintering with hot air [7]. Extrusion based printing is one of the most famous and widely adopted method. Extrusion-based 3D food printing is a digital and robotic process which aims to develop a complex 3D food by using several layers of raw materials [34]. The raw material is initially loaded in the printer and then forced out of the nozzle in a pre-defined and controlled manner, and eventually the deposited layer bond together to form a coherent solid structure (Figure 3.3). The extrusion-based food printer consists of a multi-axis stage and one or more extrusion units. Either hot extrusion or cold extrusion can be used for printing 3D foods. Extrusion-based 3D printers are relatively cheap, easy for customization of food and most importantly, a wide variety of raw materials can be used. This technology can be applied for making chocolate, confection, decorations made of sugar, candies, etc.

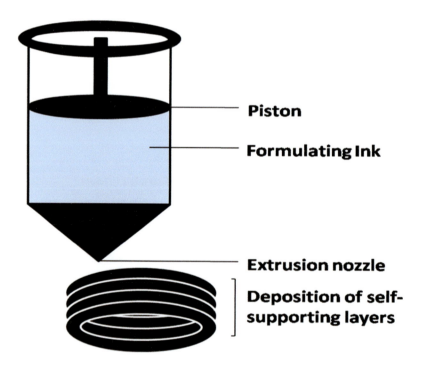

FIGURE 3.3 Hot-melt extrusion technology for 3D food printing.

Extrusion Technology of Food Products 63

3.10.3 DEVELOPMENT OF WIDE VARIETY OF EXTRUDED FOOD PRODUCTS

Extrusion technology serves as an option of controlling several parameters which is difficult in other cooking processes. Therefore, various types of extruded products can be produced by selecting different extrusion methods, changing the type of extruder, ingredients, operating conditions. The versatile range of extruded products is discussed in subsections.

3.10.3.1 BREAKFAST CEREALS

Nowadays, people like to have nutritious and easy-to-cook food items in their daily diet. Breakfast cereals like corn flakes, oat flakes, muesli, and expanded coated breakfast cereals have gained popularity due to such reasons. Flaked cereal is easy to cook because the cooking time gets reduced. Flaked oats contribute to more than 15% of the total revenue share, corn flakes market can grow with a CAGR more than 8% in future and rice flakes segment can share around 4% volume of the share by the end of 2025. Expanded coated breakfast cereal is globally popular due to their attractive shapes and coating materials like chocolate, sugar, honey, etc. The global breakfast cereal market size is projected to expand at a CAGR of 4.3% from 2017 to 2025 [21].

3.10.3.2 SNACK FOODS

Whenever we say the word "snack" then we think of a light meal that can be consumed at any time of a day. There are numerous kinds of snack food available in the market. Chips, popcorns, fried snacks, crisps are the major snack products but biscuits, nuts, and confectionaries are the other range of snack products. Most of the extruded snack products are made from starch based raw materials like rice flour, maize flour, potato. But fibers or proteins can be also incorporated into the starch based materials to enhance its nutritional quality or sometimes sensory characteristics [3, 4, 21]. The global extruded snacks market is estimated to account for USD 48.3 billion in 2019 and is projected to reach USD 65.2 billion by 2026, recording a CAGR of 4.4% during the forecast period.

3.10.3.3 BABY FOOD

Babies require easily digestible food, which most importantly has to be nutritious. Cereal-based foods that have soft, soupy, or creamy texture are used as baby food because it is easy to swallow and digest. The most common baby food is pre-cooked flour which attains perfect texture for swallowing when it is rehydrated with milk or water. Extrusion technology is an ideal process for making pre-cooked flour as it allows pre-gelatinization of starch granules without by far affecting the other nutritional parameters. The size and shape of the food does matter for small kids because large-sized foods may cause choking. Extrusion technology provides a customized solution for developing different shapes and sizes of food products.

3.10.3.4 MEAT ANALOGUE

Meat analog is a protein-rich food material having a similar flavor, texture, and appearance like animal meat. It is mainly prepared from soy-based or gluten-based material. It is very difficult to process the defatted soy flour as it requires a controlled thermal treatment. Extrusion process helps in the texturization of soy protein, which finally gives the fibrous and meat-like texture to the final product. It also reduces the degree of solubility of protein. Several physical and chemical changes occur during the preparation of meat analog or TVP like the combination of lipids (0.2–0.5 micron) to larger drops of fat, denaturation of protein, breakdown of soy cell wall, release of cell content, all this merge into a uniform mass. Dry extrusion and moist extrusion are the two models that are available in the market for the production of meat analogs.

3.10.3.5 INSTANT PASTA OR NOODLES

The consumption of pasta and noodles vary in the world depending upon the region; but in many countries it is consumed in as a part of daily diet. Instant pasta or noodle is very popular in the food market as it is pre-cooked and requires rehydration in boiling water before consumption. It is prepared by cold extrusion followed by drying. Refined wheat flour or semolina is the conventional raw material used for preparing pasta and noodles. But, the introduction of extrusion-cooking has extended the gambit of raw materials to rice flour, maize flour, legume flour and

Extrusion Technology of Food Products 65

vegetable fibers. According to Verified Market Research, the global pasta and noodles market was valued at USD 59.56 billion in 2018 and it is expected to reach USD 74.2 billion by 2026, growing at a CAGR of 2.94% from 2019 to 2026.

3.10.3.6 *PET FOODS*

Generally, most of the pet foods or animal feeds are in the pellet form. The formation of pellets requires high energy consumption (~0.1 kW/kg of the product) for the chemical transition of raw material using thermal treatment and agglomeration of high moisture feed material (> 30%). Due to high energy input extrusion-cooking came in demand to produce pet food. Also, some of the anti-nutritional components of the raw materials are inactivated during extrusion-cooking and changes occur in the functional quality which makes the food easily digestible and suitable for animals. Extrusion-cooking is an affordable technology for the production of animal feed from under-utilized and inconsistent grains which are otherwise not acceptable for human consumption.

3.11 SUMMARY

The changing lifestyle of the world is calling for food items, which requires less time for cooking or can be easily served at any time, all this while not compromising on the nutritional aspects. Food extrusion process is capable of developing easy-to-cook food items or ready-to-eat snacks. Besides, this extrusion technology is applied for the development of 3D-printed food. This process requires less manpower, low running cost, and provides versa-tile products. The food is cooked due to high temperature, high pressure, and high shear. Hot extrusion and cold extrusion are two types of extrusion process used to develop extrudates of different shapes and sensory attributes. Extruders can be divided into two categories on the basis of barrel configura-tion – single screw extruder and twin screw extruder. The physicochemical and functional properties of food are changes due to extrusion-cooking as well. Wider application of extrusion technology is coming up in the era of manufacture of analogs from various raw materials. There is a promising application of this technology which can lead consumers to savor vegetarian foods with the textural features of non-vegetarian foods.

KEYWORDS

- compound annual growth rate
- extruder
- extrusion-cooking
- functional property
- physicochemical property
- process parameters
- texturized vegetable protein

REFERENCES

1. Anderson, R. A., Conway, H., & Peplinski, A. J., (1970). Gelatinization of corn grits by roll cooking, extrusion cooking and steaming. *Starch, 22*(4), 130–135.
2. Berk, Z., (2009). *Food Process Engineering and Technology* (1st edn., p. 624). Burlington–USA: Academic Press.
3. Chakraborty, S. K., Kumbhar, B. K., Chakraborty, S., & Yadav, P., (2011). Influence of processing parameters on textural characteristics and overall acceptability of millet enriched biscuits using response surface methodology. *Journal of Food Science and Technology, 48*(2), 167–174.
4. Chakraborty, S. K., Singh, D. S., & Chakraborty, S., (2009). Extrusion: A novel technology for manufacture of nutritious snack foods. *Journal of Beverage and Food World, 42*, 23–26.
5. Chakraborty, S. K., Singh, D. S., Kumbhar, B. K., & Singh, D., (2009). Process parameter optimization for textural properties of ready-to-eat extruded snack food from millet and legume pieces blends. *Journal of Texture Studies, 40*(6), 710–726.
6. Cheftel, J. C., (1986). Nutritional effects of extrusion-cooking. *Food Chemistry, 20*(4), 263–283.
7. Dankar, I., Haddarah, A., Omar, F. E., Sepulcre, F., & Pujolà, M., (2018). 3D printing technology: The new era for food customization and elaboration. *Trends in Food Science & Technology, 75*, 231–242.
8. Fellows, P. J., (2000). *Food Processing Technology: Principles and Practice* (2rd edn., p. 608). Boca Raton–USA: CRC Press.
9. Fletcher, S. I., McMaster, T. J., Richmond, P., & Smith, A. C., (1985). Rheology and extrusion of maize grits. *Chemical Engineering Communications, 32*(1–5), 239–262.
10. Guy, R., (2001). *Extrusion Cooking: Technologies and Applications* (p. 216). Cambridge–United Kingdom: Woodhead Publishing.
11. Harper, J. M., Rhodes, T. P., & Wanninger, L. A., (1971). Viscosity model for cooked cereal doughs. *AIChE Symposium Series, 67*(108), 40–43.
12. Hernández-Díaz, J. R., Quintero-Remos, A., Barnard, J., & Balandrán-Quintana, R. R., (2007). Functional properties of extrudates prepared with blends of wheat flour/pinto

Extrusion Technology of Food Products

beans meal with added wheat bran. *Food Science and Technology International, 13*(4), 301–308.

13. Janssen, L. P. B. M., (1978). *Twin Screw Extrusion* (p. 172). North-Holland: Elsevier Scientific Publishing Company.

14. Jepson, C. H., (1953). Future extrusion studies. *Industrial & Engineering Chemistry, 45*(5), 992–993.

15. Leonard, W., Zhang, P., Ying, D., & Fang, Z., (2020). Application of extrusion technology in plant food processing byproducts: An overview. *Comprehensive Reviews in Food Science and Food Safety, 19*(1), 218–246.

16. Martelli, F. B., (1983). *Twin-Screw Extruders: A Basic Understanding* (p. 137). New York–US: Springer.

17. Mercier, C., & Feillet, P., (1975). Modification of carbohydrate component by extrusion cooking of cereal products. *Cereal Chemistry, 52*, 283–297.

18. Mohamed, I. O., & Ofoli, R. Y., (1990). Prediction of temperature profiles in twin-screw extruders. *Journal of Food Engineering, 12*(2), 145–164.

19. Morgan, R. G., Steffe, J. F., & Ofoli, R. Y., (1989). A generalized viscosity model for extrusion of protein doughs. *Journal of Food Process Engineering, 11*(1), 55–78.

20. Moscicki, L., (2011). *Extrusion-Cooking Techniques: Applications, Theory, and Sustainability* (p. 234). Weinheim–Germany: Wiley-VCH.

21. Őzer, E. A., İbanoğlu, Ş., Ainsworth, P., & Yağmur, C., (2004). Expansion characteristics of a nutritious extruded snack food using response surface methodology. *European Food Research and Technology, 218*(5), 474–479.

22. Ramachandra, H. G., & Thejaswini, M. L., (2015). Extrusion technology: A novel method of food processing. *International Journal of Innovative Science, Engineering & Technology, 2*(4), 358–369.

23. Reports and Data, (2019). *Food Extrusion Market to Reach USD 87.86 Billion by 2026.* https://www.globenewswire.com/news-release/2019/10/15/1930063/0/en/Food-Extrusion-Market-To-Reach-USD-87-86-Billion-By-2026-Reports-And-Data.html (accessed on 11 June 2022).

24. Saravacos, G. D., & Kostaropoulos, A. E., (2016). *Handbook of Food Processing Equipment* (2ⁿᵈ edn., p. 775). Switzerland: Springer International Publishing.

25. Shukla, R. M., Dwivedi, M., Deora, N. S., Mishra, H. N., & Rao, P. S., (2018). 3D-food printing using extrusion: An insight. *Journal of Nutritional Health & Food Engineering, 8*(6), 449–451.

26. Singh, R. P., & Heldman, D. R., (2014). *Introduction to Food Engineering* (5ᵗʰ edn., p. 900). London: Academic Press**.**

27. Tadmor, Z., & Klein, I., (1970). *Engineering Principles of Plasticating Extrusion* (p. 500). New York: Van Nostrand Reinhold Company.

28. Tayeb, J., Vergnes, B., & Valle, G. D., (1988). Theoretical computation of the isothermal flow through the reverse screw element of a twin-screw extrusion cooker. *Journal of Food Science, 53*(2), 616–625.

29. Ummadi, P., Chenoweth, W. L., & Ng, P. K. W., (1995). Changes in solubility and distribution of semolina proteins. *Cereal Chemistry, 72*(6), 564–567.

30. Van, Z. D., Van Der, L. E., & Kuiper, E., (1990). The development of a heat transfer model for twin-screw extruders. *Journal of Food Engineering, 11*(3), 187–207.

31. Vergnes, B., & Villemaire, J. P., (1987). Rheological behavior of low moisture molten maize starch. *Rheologica Acta, 26*(6), 570–576.

32. Yacu, W. A., (1985). Modeling a twin-screw co-rotating extruder. *Journal of Food Process Engineering, 8*(1), 1–21.
33. Yacu, W. A., (1984). Modeling of a two-screw co-rotating extruder. In: Zeuthen, P., Cheftel, J. C., Eriksson, C., Jul, M., Leniger, H., Linko, P., Varela, G., & Vos, G., (eds.), *Thermal Processing and Quality of Foods* (1st edn.). New York: Elsevier Applied Science Publishers.
34. Yang, F., Zhang, M., & Bhandari, B., (2017). Recent development in 3D food printing. *Critical Reviews in Food Science and Nutrition, 57*(14), 3145–3153.

CHAPTER 4

FUNDAMENTALS OF ADVANCED DRYING METHODS OF AGRICULTURAL PRODUCTS

SHIKHA PANDHI and ARVIND KUMAR

ABSTRACT

Agricultural products encounter significant losses after harvesting due to their extremely perishable nature. Drying and dehydration are amongst the most primeval and superlative physical methods to curtail postharvest losses and aid in food preservation. Drying and dehydration technologies are based on the principle of simultaneous heat and moisture removal to lower the moisture content of food, wherein heat generation facilitates the escape of free water molecules from the product causing a significant loss in product mass. Dried products offer numerous advantages such as extended shelf-life, reduced the bulk of the material, reduced transportation and packaging costs, and extend the availability of products even in the offseason. Conventional dehydration methods are based on conductive and convective heat transfer modes that are highly energy-intensive and may result in significant product quality losses. Advanced drying and dehydration technologies like microwave- or ultrasound (US)-assisted drying, high electric field (HEF) drying, heat-pump drying, spray freeze drying, refractance window (RW) drying and impinging stream drying have come up as new and improved drying technologies with better efficiency and low-energy consumption whilst simultaneously preserve the product quality. The probable application of these technologies has received great attention as an alternate, to at least partially replace the conventional technologies, as the industries are embarking towards more efficient but economically sustainable technologies. This chapter provides a

Advances in Food Process Engineering: Novel Processing, Preservation, and Decontamination of Foods.
Megh R. Goyal, N. Veena, & Ritesh B. Watharkar (Eds.)
© 2023 Apple Academic Press, Inc. Co-published with CRC Press (Taylor & Francis)

succinct overview of various advanced drying and dehydration technologies employed for sustainable drying of agricultural products and eatables along with its fundamental principles and promising applications in brief.

4.1 INTRODUCTION

Crops form an imperative part of the human diet. Some of the crops are highly seasonal and its plentiful availability in the market during the season leads to an unnecessary stock in the market which ultimately results in its spoilage. Considering its seasonal and perishable nature, the adoption of an effective preservation strategy stands obligatory for preventing the huge crop wastage and enhancing its availability during the off-seasons at a remunerative cost [25]. Because of this, drying and dehydration are employed as the most primordial and pre-eminent method of food preservation that aims at reducing the water content of food commodities to extend its shelf-stability, packaging, storage, and distribution cost along with extending the availability of a variety of offseason crops or products to consumers [40]. It is predominately employed for agricultural products such as fruits, vegetables, grains, spices, and products that are 'highly perishable' with moisture content (> 80%) [12, 48]. It serves as an essential operation for postharvest management of crops due to poorly established low-temperature distribution and handling facilities. Given this, over 20% of perishable crops are dried globally to extend its shelf life [55]. Agricultural products dried to remove the maximum moisture with active nutrients remained; it can be safely preserved for many days. It is also a process in which free water in the food is significantly reduced, thus leading to a concentration of dry matter without damaging the tissue, wholesomeness, and physical appearance of the food by slowing down the deterioration process [33].

Water forms an essential part of foods, especially in the case of fruits and vegetables that contains about 90–98% of moisture. It acts as an essential medium to carry out various chemical and biological reactions and stabilizes various biological components such as enzymes, proteins, DNA, cellular membranes, etc., due to its ability to form a hydrogen bond. Thus, to preserve the quality and wholesomeness of foods, moisture plays an important role as it controls the growth of microorganisms, browning reactions, lipid oxidation, etc., which renders the food unsafe or unacceptable to the consumers [15]. Drying or dehydration being a complex operation works on the principle of simultaneous heat and mass transfer in which there

Fundamentals of Advanced Drying Methods

is a continuous transfer of energy and mass in the form of heat and moisture, respectively. The process of dehydration reduces the amount of available water that encourages the growth of microorganisms, the rate of chemical and enzymatic reactions that decelerate the process of deterioration [53]. The phenomenon of heat and mass transfer depends invariably upon the shape, size, type, composition, and structure of the product which is to be dried [44]. Despite various advantages offered by the process of dehydration, there are certain nutritional, sensorial as well as functional parameters that are affected during the course of this unit operation that makes it important to vigilantly select the appropriate dehydration technique based on the product characteristics [47]. The engineering and technological aspects that play a significant role in the selection of a food drying technique are: (i) rate of drying; (ii) product quality; and (iii) energy consumption. Many conventional drying or dehydration methods are used as a postharvest technique for the preservation of agricultural products such as solar drying, osmotic dehydration, freeze-drying, hot-air drying (tunnel or cabinet drying), fluidized bed drying, vacuum drying, etc. [47]. These conventional dryers are coupled with certain limitations such as: (i) longer drying times; (ii) changes in organoleptic characteristics such as color, texture, etc.; (iii) higher operating cost; (iv) low drying performance; (v) non-uniform product quality; and (vi) high energy-consumption [4, 5, 19, 37]. Consequently, with the advancement of technological operations, various novel and advanced drying and dehydration technologies have been developed with improved technique, equipment, energy efficiency, and product quality [40].

Novel and emerging drying techniques such as ultrasound (US)- or microwave-assisted drying, superheated steam drying, refractance window (RW) drying, etc., may serve as a promising approach that can put forth a positive advantage to postharvest processing of crops in view of product final quality and process efficiency [40]. These advanced techniques offer inventive benefits of shorter processing time, drying of a variety of products, better process control, lower operating, and maintenance cost [42]. Drying systems can be subdivided into different generations based on historical advancement such as first, second, third, or fourth (the latest) generation. The novel and most advanced drying or dehydration techniques such as RW drying, infrared (IR) drying, microwave-assisted drying, etc., have been categorized under the fourth generation that primarily emphasizes on retaining maximum product quality possible [45]. This chapter tends to present a comprehensive overview of basic drying and dehydration process, its elementary mechanism in brief. It mainly put frontward details about various advanced drying and dehydration techniques, its prospects, and probable application in postharvest processing of agricultural produce.

4.2 DRYING AND DEHYDRATION OF AGRICULTURAL PRODUCE

Agricultural crops and produce are harvested to fulfill the need for food, fodder, and other uses. These crops mainly include grains, legumes, fruits, vegetables, oilseeds, and spices and can be broadly categorized into perishable and non-perishable crops [25]. Fresh produce such as fruits and vegetables are the most perishable amongst these due to its higher moisture content at the time of harvest that consequently results in their increased susceptibility to microbial deterioration resulting in poor quality and lower shelf-life of this agricultural produce [71]. Spices, grains, and legumes constitute the non-perishable category due to their comparatively lower moisture content that makes them stable for an extensive longer period as compared to perishable crops. However, instead of their low-moisture content, they are still subjected to postharvest processing for their adequate storage in the warehouse. Most of the food processing and preservation operations aim to prevent postharvest losses and value-addition to the product by enhancing its stability so that it can be stored for a longer duration of time. It enhances the seasonal accessibility, retains the nutritional attributes, and offers convenience to the handlers, producers, and consumers [25].

Drying and dehydration is an important preservation operation that reduces the amount of free water imparting stability to the product that can be stored for a longer period [70]. It prevents the crops or products from the microbial deterioration by retarding the rate of moisture mediated deteriorative reactions. The drying and dehydration process results in reduced postharvest losses during the storage period through the prevention of pre-mature germination of crops, the proliferation of insects and microbial attack [25]. Various drying and dehydration technologies are used for the treatment of agricultural produce. These technologies can be broadly categorized as: (i) natural method, which utilizes solar energy as a source of heat; and (ii) artificial method, which makes use of mechanical or electrical equipment for efficient removal of moisture under control condition. Natural drying using solar energy largely depends on the weather condition and is mainly practiced at a small scale whereas, artificial methods are commercially exploited for drying and dehydration of agricultural produce on a large-scale [38].

4.3 FUNDAMENTALS OF FOOD DRYING AND DEHYDRATION PROCESS

Water is the most profusely found constituent in foods. It forms an essential component of various biological and chemical reactions that brings about deteriorative changes in fresh produce. It serves as a medium wherein various reactive biological and chemical agents move and interact. Moreover, it significantly brings about changes in the various chemical and physical characteristics of foods such as heat conductivity, heat capacity, dielectric behavior, boiling point, freezing point, etc. Thus, necessitates the need to establish a correlation between various properties of food, associated reactions, and water for efficient dehydration operation [15]. Water in food is chiefly expressed in two distinctive ways such as "unbound or free" and "bound" water. The "unbound" fraction of water is held within the pores and interstitial spaces of a biological matrix through a physical force linked with surface tension. It is more likely associated with the physical configurations of the product and has a similar vapor pressure along with the similar latent heat of vaporization as that of water in pure form at the same temperature. The other fraction of water, i.e., "bound" water constitutes that fraction of water that is detained on the interior and the exterior surface of the biological matrix through interaction between a water molecule and solid matrix to form a water monolayer. It is majorly linked with the chemical arrangements in the product and has a lesser vapor pressure and higher vaporization heat than pure water under similar conditions. Free moisture is mostly associated with deterioration and is predominately removed during the dehydration process whereas bound water is removed only to a limited extent with definite difficulty [13].

The amount of free water available for biological reactions and promotes the growth of microorganisms is expressed as water activity (a_w). It is described as the ratio of vapor pressure of moisture in food (P) to the vapor pressure of pure water (P_o) at the same temperature. It is given by:

$$a_w = P/P_o$$

It serves as the easiest method to predict the stability of a food system. Water activity, solid mobility, and glass transition in the food system are fundamentally imperative to predict the stability of a food system during storage. A water activity of 0.90 to 1.00 favors the growth of microorganisms at room temperature. With the process of dehydration, this water activity level can be brought down to 0.60 for safe storage as at this level, the growth of microorganisms is inhibited [46]. The process of drying and dehydration principally works on the mechanism of simultaneous energy and mass exchange until the equilibrium moisture content (EMC) is reached. It involves a continuous supply of heat, either from a heat source of the surrounding environment to the moist food along with simultaneous loss of mass in the form of moisture from the moist product reducing the amount of available free water to give a dried product [53]. This process of dehydration can be broadly classified into three stages: (a) the transient early stage, (b) constant-rate, and (c) falling-rate period. In the first stage, the product is reaching the heating temperature followed by a constant phase. In a constant phase, the free water present in a product is easily removed. It is then followed by a falling rate period where removal of moisture is difficult as during this phase the water present is predominately bound or held within a solid matrix [25, 47].

The amount of moisture within a moist solid in equilibrium with air at a particular humidity and temperature is called equilibrium moisture content (EMC). A plot between EMC and relative humidity gives a sigmoid shaped curve called sorption isotherm. This sorption isotherm is important for designing a dehydration operation. When the partial pressure of water inside the food is greater than that of the surrounding air, then it loses its moisture to attain equilibrium by a process called "desorption." In contrast, when the vapor pressure of water in food is lower than that of the surrounding air, then it gains moisture to acquire an equilibrium state through a process called "absorption." The process of dehydration brings about many structural changes in the product such as shrinkage, porosity, cracks, case hardening, etc. Shrinkage is the most common phenomenon observed during dehydration in which the viscoelastic matrix contracts to space previously taken by water which is removed [1].

4.4 FOOD DRYING AND DEHYDRATION TECHNOLOGIES

Food dehydration is an intricate process. Many different drying systems are commercially used by the food industry. The selection of a correct type of dryer is an important step governing the efficiency of the process. Amongst various conventional methods of food dehydration solar drying, osmotic dehydration and heated-air drying are the most common ones. From these, heated-air drying and sun drying are the frequently used method for drying of agricultural produce on the farm level. Though, longer drying times, higher temperatures, and loss of nutritional and sensory properties aroused

Fundamentals of Advanced Drying Methods 75

the need for the development of other improved and more efficient drying technologies. Given this, drying technologies such as freeze-drying, vacuum drying, spray drying has emerged as alternative technologies to retain product quality and nutritional value but have been often characterized by high energy requirement. Higher operating cost due to continuous maintenance of vacuum in freeze and vacuum drying hinders the proficient use of this technique. Insight of the several limitations associated with conventional drying technologies such as non-uniform product quality, low efficiency, longer dehydration times, high energy requirement as well as the significant alteration in product's sensorial and textural quality have aroused the need for expansion of dehydration techniques with the advent of advance techniques [42]. Due to this, there have been extensive advancements made in various drying technologies to enhance the quality of product, process control, and process efficiency. These advanced techniques utilize distinct scientific phenomena to improve the performance of existing dehydration technologies [40]. These advanced technologies prevent the chances of cross-contamination and quality loss. They generally require lesser dehydration times (3–6 min) and comparatively low dehydration temperatures (30–70°C) that permit enhanced maintenance of color, vitamins, and other heat-sensitive nutrients [11].

Dehydration technologies can be divided into four generations: first, second, third, or fourth generation. Hot-air drying technologies such as kiln, tray, cabinet, and tunnel dryers are amongst the first-generation technologies generally employed for grains and agricultural commodities. Amongst the second-generation technologies, spray, and drum dryers are used for slurries or paste. Freeze and vacuum drying belongs to the third generation technologies. IR drying, microwave-coupled drying, heat pump drying and refractance window (RW™) drying belongs to the fourth generation of dehydration technologies based on the kind of product they can handle and the preservation of the quality attributes intended [51].

4.5 ADVANCED FOOD DEHYDRATION TECHNIQUES

4.5.1 REFRACTANCE WINDOW (RW) DRYING

Refractance window (RW) technique of drying is an advanced contact, indirect, film-drying method of food dehydration. It works on the principle of refractive properties of IR radiations coming from the circulating hot water surface [58]. This technology has been commercially utilized by

MCD Technologies, Tacoma and Mt. Capra, Washington for dehydration of various products [51]. In this technique, a pulpy wet product is evenly applied over the exterior of an IR translucent mylar sheet (plastic sheet), made of polyester, and is held by the buoyancy force applied by the hot water stream at a temperature of 90 to 95°C to transmit thermal energy to the material to be dehydrated through conduction and IR radiation [54] (Figure 4.1). The transmissivity of these mylar films mainly depends upon the thickness of these films. Decreasing the thickness of films facilitates better transmission of IR radiations but decreases its mechanical strength and hence needs to be optimized for maximum efficiency. Generally, the optimum thickness of the film was kept as 0.25 mm [58]. These films can either be stationary or moving. In the case of a moving belt arrangement, a concurrent system with a belt velocity of 0.6 and 3 m/min (meter per minute) is chosen. All three heat transfer modes are involved. During the process, the water is rapidly evaporated to reach the stage of equilibrium. This technique carries out heat and mass transfer mechanisms through the creation of a "window" from which IR radiation can pass through the moist product placed on the film. These radiations transfer heat to the water molecules present in the moist product resulting in a temperature increase of product to about 74°C and initiated escape of moisture from the product. As soon as the product gives up its most moisture, this window shuts and refract back to the hot water stream. Moisture evaporation from the product continues through conduction until it reaches the critical moisture content level. Latter stages of drying are also supported by evaporative cooling which keeps the temperature of the product less than the temperature of heated water and prevents the product from overheating. After the completion of the dehydration process, the temperature of the product is lowered by passing it to a cold water conveyer to reach the temperature lower than glass transition to prevent sticking of products that can be easily collected with the help of scraper blades. The water used during the process can be reused to achieve a better heat efficiency of the process. The whole process is carried out under ambient conditions and prevents any chance of cross-contamination.

This technique offers inventive frontiers of shorter drying times, better thermal efficiency, retained product quality in terms of color, nutrients, etc., and consume half of the energy utilized in freeze-drying at a much lesser cost [6]. It also prevents the chances of oxidation and free radical formation with improved shelf-life [51]. This technique has been commercially utilized for drying of whey, micro-algae, and herbal formulations. It is effectively employed for dehydration of product slices, juices, and purees. It can be competently used for dehydration of products containing volatile compounds, pigments, and heat-sensitive components that govern the sensorial and nutritive value of the product, which can be retained using this technique. Most of the studies have shown the application of this technology for dehydration of slices, juices, and purees of fruits and vegetables till now. But there is limited information available on the application of this technology to structurally rigid foods such as nuts and tubers [51]. RW drying of food has been effectively utilized for dehydration of several fruits such as carrots, squash, mangoes, and strawberries with much-retained quality attribute [45]. The color of RW dried pomegranate juice after rehydration was found to be comparable to freeze-dried and better in comparison to spray-dried juice with much retained sensorial properties [7]. Product consistency and its uniform application are some of the most important parameters governing effective dehydration outcome. The prospect of this technique relies on the exploration of its application for dehydration of products like green leafy vegetables, microorganisms, meat, and marine products. Further, the optimization of its process design and energy efficiency stands needful [6].

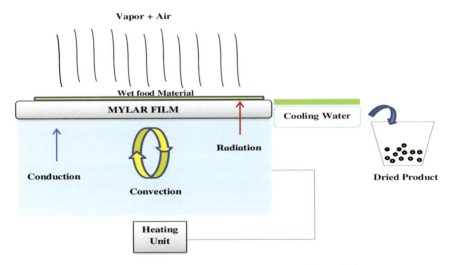

FIGURE 4.1 Schematic illustration of refractance window drying technique.
Source: Adapted from: Ref. [40].

4.5.2 MICROWAVE-ASSISTED DRYING

Microwaves are electromagnetic radiations with a wavelength range of 1 m–1 mm. Microwaves are non-ionizing radiations with frequencies 915 and 2,450 MHz commonly used for food applications. This technology comes under the fourth-generation technology. Microwave-assisted drying technology involves the generation of heat energy within the food product as a result of interaction between molecules and the electromagnetic field. The two major mechanisms involved in this di-electric heating are ionic polarization and dipolar rotation that is directly linked with the product's dielectric properties [40] (Figure 4.2). On exposure to microwaves, the water molecules' dipoles align themselves in the direction of the applied electric field. Once the field applied begins to oscillate, the dipoles also start oscillating leading to the conversion of electromagnetic energy to heat energy through the phenomenon of molecular friction and dielectric loss. Moreover, the excited ions also start moving to-and-fro under the influence of alternating fields. These ions start colliding with the adjacent molecules or atoms, generating heat. Thus, food with ions or salts takes lesser time to be heated as compared to those that contain only water [20]. Microwaves alone cannot accomplish the dehydration process and hence, it is generally coupled with forced air circulation and vacuum conditions to enhance the performance of the microwave process [55]. The advantages offered by microwave-assisted dehydration are mainly attributed to its tendency of volumetric heating and internal vapor generation. The volumetric heat generation within the product enables rapid heating with enhanced energy efficiency. The generation of heat inside the product enables internal vapor pressure buildup that releases moisture from the product. This causes a substantial lessening of drying time with leading enhanced product quality [20]. The technique utilizes the falling rate period to accomplish the drying process with enhanced control over the final moisture content of the product [40]. This technique shows improved dehydration kinetics, process control, product quality with reduced dehydration time. The one crucial step in this process is the continuous removal of water vapor. Combining convective drying with microwave drying facilitates the continuous removal of vapors through the system by passing hot air over the surface of the product. The temperature of the air can be modulated to lessen the drying time [55]. The temperature of drying air passing over the product can be modulated to reduce the drying time. Other techniques such as vacuum drying and freeze-drying can be coupled with microwave assisted-drying to overcome the constraints of uneven heating. Combining microwaves with vacuum drying offers improved color and texture to the dried product [42].

Microwave drying and microwave-assisted other hybrid drying techniques have been widely exploited for drying of various agricultural

Fundamentals of Advanced Drying Methods

commodities like apple, mango, garlic, banana, soybean, carrots, potatoes, etc. The associated constraints like difficulty in scaling up, lower penetration power, and non-uniform heating restrict its use for industrial implementation on a larger scale [71]. Drying of carrots using microwave-assisted drying technology reveals improved product quality in terms of color and rehydration characteristics [55, 67].

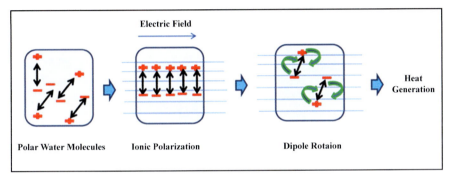

FIGURE 4.2 Heat generation mechanism in microwave heating.

4.5.3 SUPERHEATED STEAM DRYING

Superheated steam can be effectively utilized as a drying medium in an energy-efficient drying technique. It is essentially important for food products that offer hygienic processing through a closed system [41]. It utilizes steam at a temperature greater than the temperature of saturation under a stated pressure to remove excess moisture from a product. The evaporated vapor combines with the superheated steam and condensed as a heating medium for other applications [57] (Figure 4.3). This drying system works by extending the constant rate period and lower downs the critical moisture content in comparison to conventional drying [61]. Superheated steam is devoid of oxygen and hence prevents any chance of oxidative or combustion reactions. The degree of superheating extends with reduced operating pressures and predominates. It offers advantages such as higher drying efficiency, recycling of exhaust steam with no emission of hazardous gas, dirt into the surroundings [16]. Superheated steam can be employed for drying of commodities with heat-sensitive nutrients. The use of superheated steam at low pressures promotes faster drying at a lower temperature with more efficiency and hence recommended for heat sensitive fruits and vegetables. Superheated steam drying gives dried products with better texture, retained nutrients, and good quality. It also minimizes the chances of case hardening which is a major defect observed during drying operations [57]. Superheated steam drying technique has been effectively employed for drying of potatoes, rice, soybean, carrots, and cabbage under pilot operations [57].

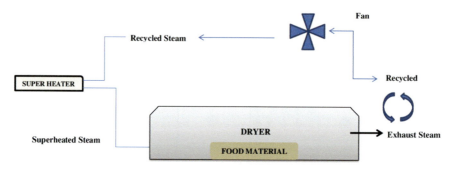

FIGURE 4.3 Schematic illustration of a superheated steam drying.

4.5.4 HEAT-PUMP COUPLED DRYING

Heat pump-coupled drying technique is a kind of hybrid drying system that requires only a little modification of combining heat pump to the existing drying process. In general, a heat pump unit operates on the thermodynamic principle of the vapor compression cycle of a refrigeration system. A heat pump generally works by drawing heat from a low-temperature medium to another high-temperature medium, thus cools the first and heats the second. An ideal heat pump dryer consists of a convective air dryer along with a refrigeration system (Figure 4.4). The systems evaporator dehumidifies the air, which is then heated using the heat pump condenser. The maximum temperature that can be attained using the condenser depends on the temperature of the refrigerant used in the system. It is generally operated at low drying temperatures of –20°C to 70°C based on the properties of the refrigerant, the capacity of the compressor, and airflow of the system [49]. The parameters such as drying temperature and air humidity can be controlled under this system and hence have been widely recommended for agricultural products and heat-sensitive commodities like spices. It offers numerous advantages like lower energy consumption, better energy recovery, good product quality, and retained nutrients [40]. A study indicated that the use of heat pump-assisted drying for drying of mango slices was found useful in lowering the energy consumption of the system as compared to conventional electrically heated

Fundamentals of Advanced Drying Methods

dryer [30]. Heat pump-dried apple, guava, and potatoes using cryogens like carbon dioxide (CO_2) and liquid nitrogen instead of air were found to be better in terms of color with improved porosity and rehydration ability in comparison to the product dried under vacuum [26]. Also, fluidized bed heat pump drying of green peas under ambient freeze-drying settings was found to have improved color, rehydration ability, and floatability [3, 55].

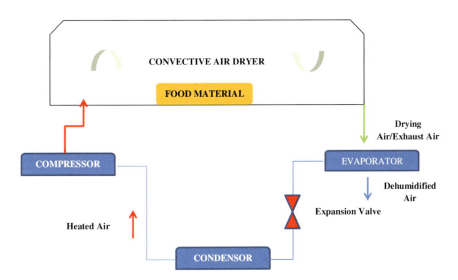

FIGURE 4.4 Schematic illustration of a heat-pump dryer.

4.5.5 SPRAY-FREEZE-DRYING (SFD)

Spray-freeze-drying (SFD) is an eccentric freeze-drying system that gives a powdered product with uniquely different properties while still possessing the benefits of the conventional freeze-dried process. This process is an amalgamation of both spray- and freeze-drying techniques. The process consists of three main stages: (i) spraying; (ii) freezing; and (iii) drying [29]. The homogenized liquid sample is fed into two spray nozzles using a peristaltic pump to convert the liquid stream to fine droplets (atomization) (Figure 4.5). It is then subjected to a freeze dryer where the tiny droplets come into contact with a cryogenic fluid that freezes the droplets. The frozen droplets were then sublimed under low temperature and pressure to give fine powdered product [23]. This technique has offered the advantages of better product quality and texture with much-retained volatiles and bioactive [29].

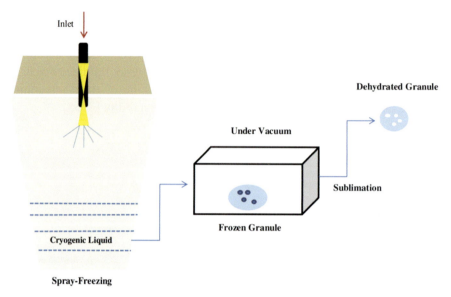

FIGURE 4.5 Spray-freeze drying technique.
Source: Adapted from Ref. [29].

4.5.6 ULTRASOUND (US) ASSISTED DRYING

Ultrasound-assisted drying technologies are most popular amongst the various novel drying technologies. It makes use of low-frequency sound waves (20–100 kHz) at higher intensities. The use of ultrasonic waves for the treatment of a product in a liquid medium encourages the development of micro-channels on the exterior as a result of the distortion of the solid structure of the product due to continuous contraction and release mechanism of ultrasonic waves [22]. This contraction and release mechanism assists in the easy escape of both free and bound moisture. The process of ultrasonic treatment aids in heat and mass transfer from a material [66]. The force applied during the ultrasonic treatment can be greater than the surface tension that confines the water within the microchannels and hence facilitates the easy removal of moisture from the product [14]. The ultrasonic waves can be used as assisting technology for various dehydration techniques like convective [24]; osmotic [21]; vacuum [9]; IR drying [17]; and freeze-drying [56] to increase the efficiency of these technologies with reduced drying time and cost. US can be used to accelerate the rate of heat and mass transfer mechanism in case of vacuum drying as vacuum results in quick moisture removal from the surface but restricts the

Fundamentals of Advanced Drying Methods

mass transfer. Using US in combination enhances the evaporation rate under negative pressure [66]. Further, this process can prevent the chances of case hardening and food degradation due to the involvement of lower temperature using a high power US technique. The use of US as an assisting technology for dehydration results in lower dehydration time, lower resistance to mass transfer, retained product quality, and enhanced water diffusivity. Though, associated restraints like high mass load densities need to be worked on for industrial scale-up [40].

4.5.7 INFRARED (IR) DRYING

Infrared (IR) drying utilizes the IR radiation emitted from a heat source for raising the temperature that aids in easy removal of moisture from a product. It has been used as an effective technique for food processing and has also been utilized for drying applications. It offers several advantages of even heating, improved product quality, superior process control, better heat transfer potential [60], more cost-effectiveness, reduced drying times, and lesser energy consumption [17, 27]. Incidence of IR radiation on the wet food material induces certain changes in electronic, vibrational, and state at both atomic and molecular levels [22], without raising the temperature of surrounding air. This facilitates the diffusion rate of moisture through the wet material and modifies the radiation properties of the sample due to a decrease in moisture content which is diffused out from the material to air (Figure 4.6).

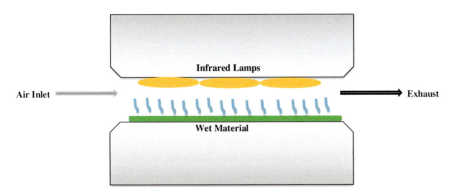

FIGURE 4.6 Schematic illustration of infrared drying technique.

Application of IR radiation had shown to exert notable effects on the porosity of banana [34], rehydration potential of onion [39], lower the impact on the color of pineapple and potatoes with comparatively higher drying rate [65]. This technique can be coupled with freeze drying to further enhance the quality of sweet potatoes [36] and strawberries [59]. The dehydration process of agricultural produce extensively occurs in the falling rate period and results in faster heat and moisture transfer as a function of applied IR power. The process of IR drying has been successfully employed for other food processing applications such as thawing, blanching, etc. This process of drying facilitates uniform or even heating of products with lesser residence and come-up time. Most of the studies were conducted for solid food materials and creates a huge scope for exploration in the drying of liquid food as well. Further, low rehydration efficiency offered by IR-dried food materials necessitates the further need for improvement in the process to overcome this constraint for proficient food applications [40]. Also, through the use of IR radiation for drying application hastens up the drying process, the radiation intensity if not efficiently utilized could damage the heat-sensitive components of the agricultural material [27].

4.5.8   *HIGH ELECTRIC FIELD (HEF) DRYING*

The technique of the high electric field (HEF) drying or electrohydrodynamic (EHD) drying involves the use of alternating or direct current at normal to high frequencies (around 60 Hz) to facilitate easy removal of moisture from a product. The difference in potential created directly affects the rate of the drying process. Induction of exothermic reactions inside the dielectric food system facilitates in quick moisture escape. The heat generated is an outcome of a secondary flow persuaded by the electric field called 'corona wind' [52]. The associated parameters like electrode size, field strength, and space between the electrodes are important factors governing the efficacy of the drying process. The HEF drying process is a convective technique that efficiently removes moisture while preserving most of the heat-sensitive nutrients and components like vitamin C. Being a non-thermal dehydration technique, this process minimizes the chances of browning reactions. A study revealed that Japanese radish dehydrated using HEF drying technique has shown to enhance rehydration ability with fewer product shrinkages [8]. This process of drying results in superior process efficiency, reduced drying time, and low power consumption in comparison to other conventional dehydration technologies [2]. However, no apparent limitations associated with the use of this technology have been documented, only satisfactory

Fundamentals of Advanced Drying Methods 85

research on diverse crops under variable process conditions can rationalize its proficient use on large-scale [40].

4.5.9 IMPINGING STREAM DRYING

Impinging stream drying is an alternative to flash drying with extensively high drying capacities. It utilizes the heat generated as a result of the collision between to opposite streams to evaporate moisture [35]. This technique is most widely used for lowering the moisture content of agricultural products with high moisture content. Also, it is widely employed for reducing the surface moisture of commodities like grains. The process of impinging stream drying takes lesser time and lowers the final moisture percent in food sample to a greater extent. Also, it offers higher drying rates with better process efficiency. The basic mechanism of impinging stream drying involves the collision of two oppositely charged solid-in-gas streams that are subjected to higher velocities towards each other. The particles are then accelerated to a velocity of an air stream. The collision of two-phase streams generates a zone called an impingement zone of high turbulence that contains a concentrated area of particles that facilitate faster heat and mass transfer. The difference in the densities of two-stream aids in the transfer of particles from one stream to another under the influence of inertia. The particles will then tend to slow down due to the action of friction force of the opposite stream until the final velocity of the particle is zero. Again, the particles accelerated by the opposite direction of the stream towards the impinging stream and repeated several times so that the particles lose their momentum and carried away from the impingement zone. This facilitates the transfer of heat and mass for an extended period and is found to be more effective than the other thermal drying techniques. This way this technique can be effectively utilized for drying of high moisture particles [64].

4.6 MONITORING OF EFFECT OF DRYING ON QUALITY ATTRIBUTES

The conventional dehydration system does not provide adequate control over the process parameters and leads to non-uniform product quality, high energy requirements, and over-drying of product. Thus, effective process control systems need to be coupled with the drying systems to continuously monitor the product quality attributes such as moisture content, product color as a result of operating parameters such as temperature. It is essential to study or monitor the effect of temperature on the quality aspect to enhance process efficiency [42]. The continuous and instantaneous monitoring of factors like temperature, air velocity, humidity of air and product characteristics like color, shrinkage, and water activity provides improved modulation of dryer outcome, product quality, and energy requirements. A dryer with all these desired qualities is considered as a smart drying system [63].

The use of advanced analytical techniques for process control and quality monitoring has been widely employed in drying operations. The use of various single-point spectroscopic techniques has been well documented to monitor the drying process. Use of visible and near-IR region, machine vision and hyper- or multi-spectral imaging offers a non-destructive assessment of parameters such as moisture content [18, 28], shrinkage [68], color [43] and heat-sensitive components such as vitamin C [69] and phenolic components [10] after adequate calibration. These techniques can be widely employed for monitoring the quality of agricultural produce such as fruits and vegetables during drying [53]. Near-IR sensors can be utilized for monitoring surface water activity as a function of air relative humidity [62]. Also, triboelectric sensors were used for the analysis of moisture content [50]. Acoustic sensors have been employed for monitoring of structural change in the food material during drying [31, 42].

4.7 FUTURE PROSPECTS

Advanced drying and dehydration technologies employed for agricultural produce treatment has emerged extensively over the years for the interest of industrial application to ensure better stability and extended shelf life of the fresh produce. These technologies have emerged as an alternative solution to various constraints associated with conventional drying technologies wholly or in parts. These technologies form an important part of numerous researches in the field of food engineering and development given increasing environmental concerns associated with conventional methodologies. The above-discussed technologies were found effective in dealing with different aspects of the drying process such as drying rates, process efficiency, energy conservation, cost economics, and product quality. Apart from these advantages, certain future inventions need to be put forward that can accelerate the adoption of these technologies for practical agri-food applications. These aspects are summarized below:

- Adoption of renewable energy sources for the development of the drying system;

Fundamentals of Advanced Drying Methods

- Automation interventions in the systems;
- Continuous systems development for the handling of high loads of fresh agricultural produce [40];
- Coupling of two or more of these technologies to give hybrid techniques;
- Development of environment-friendly system free of any environment polluting sources;
- Development of low-cost economic drying systems;
- Drying systems for the treatment of heat-sensitive products;
- Use of computational fluid dynamics (CFD) technique to demonstrate varying process factors that facilitate process efficiency.

4.8 SUMMARY

Advanced dehydration technologies have been developed for drying of agricultural produce, offering superior quality products coupled with higher energy efficiency and reduce dehydration times. Novel technologies like heat-pump dehydration system, RW drying, US-coupled drying system, microwave-coupled drying, impinging stream drying, superheated steam drying, HEF drying, SFD, and IR drying systems are amongst these technologies offering advantages of better process efficiency and product characteristics over the conventional drying technologies. However, it is essential to critically inquire and demonstrate the economic aspect as well as the practical aspect of these technologies for intensive application in the agri-food processing sector. Advance automation interventions should be coupled with these technologies to further increase the effectiveness of these technologies.

KEYWORDS

- **advanced drying techniques**
- **computational fluid dynamics**
- **dehydration**
- **efficiency**
- **electrohydrodynamic**
- **food quality**
- **heat and mass transfer**

REFERENCES

1. Aguilera, J. M., (2003). Drying and dried products under the microscope. *Food Science and Technology International, 9*(3), 137–143.
2. Alemrajabi, A. A., Rezaee, F., Mirhosseini, M., & Esehaghbeygi, A., (2012). Comparative evaluation of the effects of electrohydrodynamic, oven, and ambient air on carrot cylindrical slices during drying process. *Drying Technology, 30*(1), 88–96.
3. Alves-Filho, O., García-Pascual, P., Eikevik, T. M., & Strømmen, I., (2004). Dehydration of green peas under atmospheric freeze-drying conditions. In: *XIV Simposio Internacional de Secado* (pp. 1521–1528). São Paulo-Brazil.
4. Arabhosseini, A., Padhye, S., Huisman, W., Van, B. A., & Müller, J., (2011). Effect of drying on the color of tarragon (*Artemisia dracunculus L.*) leaves. *Food and Bioprocess Technology, 4*(7), 1281–1287.
5. Aversa, M., Curcio, S., Calabrò, V., & Iorio, G., (2012). Experimental evaluation of quality parameters during drying of carrot samples. *Food and Bioprocess Technology, 5*(1), 118–129.
6. Baeghbali, V., & Niakousari, M., (2018). A review on mechanism, quality preservation and energy efficiency in refractance window drying: A conductive hydro-drying technique. *Journal of Nutrition, Food Research and Technology, 1*(2), 50–54.
7. Baeghbali, V., Niakousari, M., & Farahnaky, A., (2016). Refractance window drying of pomegranate juice: Quality retention and energy efficiency. *LWT – Food Science and Technology, 66*, 34–40.
8. Bajgai, T. R., & Hashinaga, F., (2001). High electric field drying of Japanese radish. *Drying Technology, 19*(9), 2291–2302.
9. Başlar, M., & Ertugay, M., (2013). The effect of ultrasound and photosonication treatment on polyphenol oxidase (PPO) activity, total phenolic component and color of apple juice. *International Journal of Food Science and Technology, 48*(4), 886–892.
10. Bayili, R. G., Abdoul-Latif, F., Kone, O. H., Diao, M., Bassole, I. H. N., & Dicko, M. H., (2011). Phenolic compounds and antioxidant activities in some fruits and vegetables from Burkina Faso. *African Journal of Biotechnology, 10*(62), 13543–13547.
11. Bernaert, N., Van, D. B., Van, P. E., & De Ruyck, H., (2019). Innovative refractance window drying technology to keep nutrient value during processing. *Trends in Food Science & Technology, 84*, 22–24.
12. Chen, G., (2007). Drying operations: Agricultural and forestry products. In: Capehart, B. L., (ed.), *Encyclopedia of Energy Engineering and Technology* (Vol. 1, pp. 332–337). London–UK: Taylor & Francis.
13. Chung, D. S., & Chang, D. I., (1982). Principles of food dehydration. *Journal of Food Protection, 45*(5), 475–478.
14. De La Fuente-Blanco, S., Riera-Franco De, S. E., Acosta-Aparicio, V. M., Blanco-Blanco, A., & Gallego-Juárez, J. A., (2006). Food drying process by power ultrasound. *Ultrasonics, 44*, e523–e527.
15. Derossi, A., Severini, C., & Cassi, D., (2011). Mass transfer mechanisms during dehydration of vegetable food: Traditional and innovative approach; Chapter 15. In: El-Amin, M., (ed.), *Advanced Topics in Mass Transfer* (pp. 305–354). Rijeka–Croatia: InTech.
16. Devahastin, S., & Mujumdar, A. S., (2014). Superheated steam drying of foods and biomaterials; Chapter 3. In: Tsotsas, E., & Mujumdar, A. S., (eds.), *Modern Drying*

Fundamentals of Advanced Drying Methods

Technology: Process Intensification (pp. 57–84). Weinheim–Germany: Wiley-VCH Verlag GmbH & Co.

17. Dujmić, F., Brnčić, M., Karlović, S., Bosiljkov, T., Ježek, D., Tripalo, B., & Mofardin, I., (2013). Ultrasound-assisted infrared drying of pear slices: Textural issues. *Journal of Food Process Engineering, 36*(3), 397–406.

18. ElMasry, G., Wang, N., ElSayed, A., & Ngadi, M., (2007). Hyperspectral imaging for nondestructive determination of some quality attributes for strawberry. *Journal of Food Engineering, 81*(1), 98–107.

19. Falade, K. O., & Omojola, B. S., (2010). Effect of processing methods on physical, chemical, rheological, and sensory properties of okra (*Abelmoschus esculentus*). *Food and Bioprocess Technology, 3*, 387–394.

20. Feng, H., Yin, Y., & Tang, J., (2012). Microwave drying of food and agricultural materials: Basics and heat and mass transfer modeling. *Food Engineering Reviews, 4*(2), 89–106.

21. Fernandes, F. A. N., & Rodrigues, S., (2008). Application of ultrasound and ultrasound-assisted osmotic dehydration in drying of fruits. *Drying Technology, 26*(12), 1509–1516.

22. Fernandes, F. A. N., & Rodrigues, S., (2007). Ultrasound as pre-treatment for drying of fruits: Dehydration of banana. *Journal of Food Engineering, 82*(2), 261–267.

23. Filková, I., Huang, L. X., & Mujumdar, A. S., (2006). Industrial spray drying systems; Chapter 10. In: Majumdar, A. S., (ed.), *Handbook of Industrial Drying* (3rd edn., pp. 215–256). Boca Raton–FL: CRC Press.

24. Gamboa-Santos, J., Montilla, A., Carcel, J. A., Villamiel, M., & GarciaPerez, J. V., (2014). Air-borne ultrasound application in the convective drying of strawberry. *Journal of Food Engineering, 128*(3), 132–139.

25. Gunathilake, D. M. C. C., Senanayaka, D. P., Adiletta, G., & Senadeera, W., (2018). Drying of agricultural crops; Chapter 14. In: Chen, G., (ed.), *Advances in Agricultural Machinery and Technologies* (1st edn., pp. 331–365). Boca Raton–FL: CRC Press.

26. Hawlader, M. N. A., Perera, C. O., & Tian, M., (2006). Properties of modified atmosphere heat pump dried foods. *Journal of Food Engineering, 74*(3), 392–401.

27. Hnin, K. K., Zhang, M., Mujumdar, A. S., & Zhu, Y., (2018). Emerging food drying technologies with energy-saving characteristics: A review. *Drying Technology, 37*(12), 1465–1480.

28. Huang, M., Wang, Q., Zhang, M., & Zhu, Q., (2014). Prediction of color and moisture content for vegetable soybean during drying using hyperspectral imaging technology. *Journal of Food Engineering, 128*, 24–30.

29. Ishwarya, S. P., Anandharamakrishnan, C., & Stapley, A. G., (2015). Spray-freeze-drying: A novel process for the drying of foods and bioproducts. *Trends in Food Science & Technology, 41*(2), 161–181.

30. Kohayakawa, M. N., Silveria-Junior, V., & Telis-Romero, J., (2004). Drying of mango slices using heat pump dryer. In: *Proceedings of the 14th International Drying Symposium* (Vol. B, pp. 884–891). Saõ Paulo, Brazil.

31. Kowalski, S. J., & Mielniczuk, B., (2006). Drying-induced stresses in macaroni dough. *Drying Technology, 24*(9), 1093–1099.

32. Krishnamurthy, K., Khurana, H. K., Soojin, J., Irudayaraj, J., & Demirci, A., (2008). Infrared heating in food processing: An overview. *Comprehensive Reviews in Food Science and Food Safety, 7*(1), 2–13.

33. Lamidi, R. O., Jiang, L., Pathare, P. B., Wang, Y., & Roskilly, A. P., (2019). Recent advances in sustainable drying of agricultural produce: A review. *Applied Energy, 233*, 367–385.

34. Lee, D. J., Jangam, S., & Mujumdar, A. S., (2013). Some recent advances in drying technologies to produce particulate solids. *KONA Powder and Particle Journal, 30,* 69–83.

35. Léonard, A., Blacher, S., Nimmol, C., & Devahastin, S., (2008). Effect of far-infrared radiation assisted drying on microstructure of banana slices: An illustrative use of x-ray microtomography in microstructural evaluation of a food product. *Journal of Food Engineering, 85*(1), 154–162.

36. Lin, Y. P., Tsen, J. H., & King, V., (2005). Effects of far-infrared radiation on the freeze-drying of sweet potato. *Journal of Food Engineering, 68*(2), 249–255.

37. López, J., Uribe, E., Vega-Gálvez, A., Miranda, M., Vergara, J., Gonzalez, E., & Di Scala, K., (2010). Effect of air temperature on drying kinetics, vitamin C, antioxidant activity, total phenolic content, non-enzymatic browning and firmness of blueberries variety O´Neil. *Food and Bioprocess Technology, 3,* 772–777.

38. Maisnam, D., Rasane, P., Dey, A., Kaur, S., & Sarma, C., (2017). Recent advances in conventional drying of foods. *Journal of Food Technology and Preservation, 1*(1), 25–34.

39. Mongpraneet, S., Abe, T., & Tsurusaki, T., (2002). Far infrared-vacuum and -convection drying of Welsh onion. *Transactions of the ASAE, 45*(5), 1529–1535.

40. Moses, J. A., Norton, T., Alagusundaram, K., & Tiwari, B. K., (2014). Novel drying techniques for the food industry. *Food Engineering Reviews, 6*(3), 43–55.

41. Mujumdar, A. S., & Jangam, S. V., (2011). Some innovative drying technologies for dehydration of foods. In: *Proceedings of ICEF* (pp. 555–556). Athens, Greece.

42. Mujumdar, A. S., & Law, C. L., (2010). Drying technology: Trends and applications in postharvest processing. *Food and Bioprocess Technology, 3,* 843–852.

43. Nahimana, H., & Zhang, M., (2011). Shrinkage and color change during microwave vacuum drying of carrot. *Drying Technology, 29*(7), 836–847.

44. Nieto, A., Castro, M. A., & Alzamora, S. M., (2001). Kinetics of moisture transfer during air drying of blanched and/or osmotically dehydrated mango. *Journal of Food Engineering, 50*(3), 175–185.

45. Nindo, C. I., & Tang, J., (2007). Refractance window dehydration technology: A novel contact drying method. *Drying Technology, 25*(1), 37–48.

46. Oliveira, S. M., Brandao, T. R., & Silva, C. L., (2016). Influence of drying processes and pretreatments on nutritional and bioactive characteristics of dried vegetables: A review. *Food Engineering Reviews, 8*(2), 134–163.

47. Onwude, D. I., Hashim, N., & Chen, G., (2016). Recent advances of novel thermal combined hot air drying of agricultural crops. *Trends in Food Science & Technology, 57*(Part A), 132–145.

48. Orsat, V., Changrue, V., & Raghavan, G. S., (2006). Microwave drying of fruits and vegetables. *Stewart Postharvest Review, 2,* 1–7.

49. Patel, K. K., & Kar, A., (2012). Heat pump assisted drying of agricultural produce: An overview. *Journal of Food Science and Technology, 49*(2), 142–160.

50. Portoghese, A., Berrutia, F., & Briens, C., (2007). Continuous on-line measurement of solid moisture content during fluidized bed drying using triboelectric probes. *Powder Technology, 181*(2), 169–177.

51. Raghavi, L. M., Moses, J. A., & Anandharamakrishnan, C., (2018). Refractance window drying of foods: A review. *Journal of Food Engineering, 222,* 267–275.

52. Ramachandran, M. R., & Lai, F. C., (2010). Effects of porosity on the performance of EHD-enhanced drying. *Drying Technology, 28*(12), 1477–1483.

Fundamentals of Advanced Drying Methods 91

53. Raponi, F., Moscetti, R., Monarca, D., Colantoni, A., & Massantini, R., (2017). Monitoring and optimization of the process of drying fruits and vegetables using computer vision: A review. *Sustainability, 9*(11), 2009.

54. Rostami, H., Dehnad, D., Jafari, S. M., & Tavakoli, H. R., (2018). Evaluation of physical, rheological, microbial, and organoleptic properties of meat powder produced by refractance window drying. *Drying Technology, 36*(9), 1076–1085.

55. Sagar, V. R., & Kumar, P. S., (2010). Recent advances in drying and dehydration of fruits and vegetables: A review. *Journal of Food Science and Technology, 47*(1), 15–26.

56. Santacatalina, J. V., Fissore, D., Carcel, J. A., Mulet, A., & Garcia-Perez, J. V., (2015). Model-based investigation into atmospheric freeze-drying assisted by power ultrasound. *Journal of Food Engineering, 151*(2), 7–15.

57. Sehrawat, R., Nema, P. K., & Kaur, B. P., (2016). Effect of superheated steam drying on properties of foodstuffs and kinetic modeling. *Innovative Food Science & Emerging Technologies, 34*, 285–301.

58. Shende, D., & Datta, A. K., (2019). Refractance window drying of fruits and vegetables: A review. *Journal of the Science of Food and Agriculture, 99*(4), 1449–1456.

59. Shih, C., Pan, Z., McHugh, T., Wood, D., & Hirschberg, E., (2008). Sequential infrared radiation and freeze-drying method for producing crispy strawberries. *Transactions of the ASAE, 51*(1), 205–216.

60. Skjöldebrand, C., (2001). Infrared heating; Chapter 11. In: Richardson, P., (ed.), *Thermal Technologies in Food Processing* (1st edn., pp. 208–228). Cambridge–England: Woodhead Publishing Limited.

61. Speckhahn, A., Srzednicki, G., & Desai, D. K., (2010). Drying of beef in superheated steam. *Drying Technology, 28*(9), 1072–1082.

62. Stawczyk, J., Muñoz, I., Collell, C., & Comaposada, J., (2009). Control system for sausage drying based on on-line NIR a_w determination. *Drying Technology, 27*(12), 1338–1343.

63. Su, Y., Zhang, M., & Mujumdar, A. S., (2015). Recent developments in smart drying technology. *Drying Technology, 33*(3), 260–276.

64. Supakumnerd, K., & Sathapornprasath, K., (2013). A review of the impinging stream dryer. *SWU Engineering Journal, 8*(1), 16–23.

65. Tan, M., Chua, K. J., Mujumdar, A. S., & Chou, S. K., (2001). Effect of osmotic pre-treatment and infrared radiation on drying rate and color changes during drying of potato and pineapple. *Drying Technology, 19*(9), 2193–2207.

66. Tekin, Z. H., Başlar, M., Karasu, S., & Kilicli, M., (2017). Dehydration of green beans using ultrasound-assisted vacuum drying as a novel technique: Drying kinetics and quality parameters. *Journal of Food Processing and Preservation, 41*(6), e13227.

67. Wang, J., & Xi, Y. S., (2005). Drying characteristics and drying quality of carrot using a two-stage microwave process. *Journal of Food Engineering, 68*(4), 505–511.

68. Yadollahinia, A., & Jahangiri, M., (2009). Shrinkage of potato slice during drying. *Journal of Food Engineering, 94*(1), 52–58.

69. Yang, H., & Irudayaraj, J., (2002). Rapid determination of vitamin C by NIR, MIR, and FT-Raman techniques. *The Journal of Pharmacy and Pharmacology, 54*(9), 1247–1255.

70. Zhang, M., Chen, H. Z., Mujumdar, A. S., Tang, J. M., Miao, S., & Wang, Y. C., (2017). Recent developments in high-quality drying of vegetables, fruits, and aquatic products. *Critical Reviews in Food Science and Nutrition, 57*(6), 1239–1255.

71. Zhou, X., & Wang, S., (2019). Recent developments in radio frequency drying of food and agricultural products: A review. *Drying Technology, 37*(3), 271–286.

CHAPTER 5

ENCAPSULATION METHODS IN THE FOOD INDUSTRY: MECHANISMS AND APPLICATIONS

REKHA RANI, PAYAL KARMAKAR, NEHA SINGH, and RONIT MANDAL

ABSTRACT

With increasing health consciousness among consumers, more inclination is observed towards foods having specific health benefits. Various bioactive compounds are hence used for fortification of different food products. Bioactive compounds are present in very small quantities in their source, be it plant, animal, or several species of microorganisms. However, the very low bioavailability, bioaccessibility, and membrane permeability of the bioactive compounds due to the unfavorable environment of the gut poses a challenge to their full utilization by the human body; the other challenge being the low stability of these compounds in the heterogeneous food microstructure when exposed to various unfavorable conditions during food processing and storage. To overcome this challenge, bioactive compounds can be encapsulated which is an effective way of protecting and controlling their release. The use of the compounds in encapsulated form instead of free form enhances the benefits obtained from the consumption of the foods fortified with the same. Use of edible matrices for encapsulation can also help in masking the off-odors and flavors of the core material. The encapsulated system is seen as a suspension of colloidal particles with the active material embedded in the coat. This technique of stabilization of functional components finds application in food, pharmaceutical as well as cosmetic industries. This chapter deals with various technologies used for encapsulation and their promising applications in food industries.

Advances in Food Process Engineering: Novel Processing, Preservation, and Decontamination of Foods.
Megh R. Goyal, N. Veena, & Ritesh B. Watharkar (Eds.)
© 2023 Apple Academic Press, Inc. Co-published with CRC Press (Taylor & Francis)

5.1 INTRODUCTION

Encapsulation is a technology where a sensitive material is embedded within an outer material forming small particles of the very small molecular order and size ranges, which provides protection to the former [132]. The process involves two components: the first one is an active ingredient which is termed as the core material, payload phase, internal phase, or fill. This is protected inside the second component, which is known as coat material, shell, membrane, external phase, wall material, encapsulant or matrix. This coating protects the core ingredient from destabilization when subjected to various unfavorable external conditions [96]. Also, such material ensures controlled delivery of core materials at the required targets. These encapsulated compounds are categorized into numerous types depending upon the type of core material, coat material or the different techniques used for encapsulation (discussed later in this chapter) [56]. This technology of encapsulation finds application in various sectors including the pharmaceutical and drug industry, cosmetic industry and most importantly in food industries [63].

This chapter focuses on the technology of encapsulation, its governing principles, benefits, and techniques used for encapsulation, their characterization, recent developments, and applications in the food industry, as well as challenges faced in this field.

5.2 BASIC PRINCIPLES FOR ENCAPSULATION OF BIOACTIVE COMPOUNDS

People of the 21st century are more aware about their health and welfare. With the constant development in the field of science and technology, and upgradation in the lifestyle of common people, there has been a recent visible trend of consumption of natural bioactive compounds and products rather than artificial substances. Emphasizing on the food industry, it has been observed that there has been an increase in the demand among the customers of functional food products which impart definite health benefits along with extended shelf-life and better safety and quality to the products [114]. The source of these bioactive compounds ranges a wide variety including mainly vegetables, fruits, pulses, or other plant sources [105]. However, a majority of these compounds pose stability challenges owing to having unsaturated bonds in their molecular structures [140] which make them highly susceptible to oxidation and more so when being subjected to heat treatment or when exposed to light, thereby rendering instability to these compounds. Fluctuations in pH and moisture during processing treatments also contribute towards their unstable nature [51]. Food industry is thus looking for solutions to tackle the problems posed by these unstable compounds.

On the other hand, probiotic bacteria, which are a species of bacteria that render health benefits upon consumption, pose a survivability challenge due to the acidic conditions (pH ~ 2.0) of the gastric environment. They need to be viable and proliferate in the human gut to impart their benefits.

These problems can be addressed by an efficient technology of encapsulation which helps in coating of the bioactive compounds or bacteria and increasing their bioavailability. The preservation of functional properties of the encapsulated ingredients during their storage and processing is the main purpose of this technology [87]. Various textural and rheological changes in their properties can also be brought about with the motto of promoting easy handling. For instance, polyunsaturated fatty acids-enriched oil was encapsulated in starch and proteins and made into a free-flowing powder [95]. This ensured the protection of oil from oxidation and made it suitably usable as a food ingredient. This contributes an important role in aiding the controlled delivery of the bioactive compounds at specific targets [13]. Thus, the targeted delivery of the bioactive compounds is achieved so that they are more bioaccesible. Encapsulation of bioactive compounds with edible coating can mask off-flavors [88]. Shelf-life of encapsulated probiotic bacteria is more than their non-encapsulated counterparts. Solubility is also enhanced to a great extent by the encapsulation. These will be discussed further in subsequent section on the benefits of the encapsulation processes.

5.3 ENCAPSULATION OF DIFFERENT BIOACTIVE COMPOUNDS AND CELLS

A large group of publications have emerged in the recent years which deal with the encapsulation of several natural bioactive compounds like essential fatty acids, antioxidant compounds like polyphenols, essential oils, carotenoids, as well as probiotics, etc. It is also required that the products for encapsulation comprising the wall materials and core materials be fabricated from food-grade compounds. This section briefly deals with the selection of core and wall materials. More discussion on these aspects, the readers are suggested to read Sections 5.8 and 5.10.

5.3.1 CORE MATERIAL

Core materials which are also called internal phase comprises of the active ingredient that can either be a pure compound or a group of compounds that are extracted from fruits, vegetables, flowers, leaves, and other aerial parts, stems, and seeds. The major compounds that have been used in encapsulation are antioxidants (phenolic compounds), carotenoids, flavors, vitamins, enzymes, and microbial cells (like probiotics).

One of the major group of phytochemicals, that proved beneficial and promising candidates in the functional foods are phenolic compounds. They represent one of the largest groups of plant secondary metabolites. These biomolecules have the potential to play wide roles like antioxidant activity, anti-inflammatory, and anti-allergic effects [83] and reduce the risk of cancer and heart-related problems. These compounds are found in fruits (like berries such as grapes, blueberries, strawberry, and raspberry), vegetables (onion, potato, lettuce, spinach, etc.), beverages (coffee, tea, fruit juices, and wine), and cereals. However, these polyphenolic compounds are typically poorly soluble in water. Also, their stability decreases with high temperature treatment. Encapsulation offers a good opportunity for their protection.

Carotenoids represent a large group of natural yellow or orange coloring pigments. They are ubiquitous in nature and found in carrots, tomatoes, spinach, broccoli, watermelon, orange, etc. Like phenolic compounds, they also show benefits like anticancer, prevention of cardiovascular diseases, hypertension, and diabetes. They are also good for eyes health. The carotenoids are classified into carotenes and their oxygenated derivates called xanthophylls. Some examples include β-carotene, lutein, zeaxanthin, lycopene. As their structure has conjugated double bonds, they are prone to oxidative changes and lose their effect.

Vitamins are organic molecules such as fat and water-soluble vitamins exert benefits for proper functioning of the body. They are important for physiological functions like tissue formation, cell functions, body growth and hormonal functions. However, they tend to be unstable due to heat, oxygen or even light. They have also been encapsulated and fortified into food products which offer a great delivery system of these molecules.

Essential oils are a mixture of complex volatile molecules obtained from plants which are known by their characteristic aroma. These volatile molecules are produced by plants like mint, cloves, basil, eucalyptus, lemongrass. They show antimicrobial effects against several microorganisms and are employed in the pharmaceutical and food industry. However, they are lost during handling of products [20].

Encapsulation Methods in the Food Industry

Probiotic bacteria, as per FAO/WHO (2006) are living microbial cells "which when administered in adequate amounts confer a health benefit on the host." To extract and utilize their health benefits need to reach their designated targets before they can do so. Their viability is greatly influenced by the environmental conditions they are in. Heat and acidic pH (in the gastric juices) can inhibit them to a great extent. For their efficient delivery, encapsulating in a suitable matrix is helpful. Probiotic microorganisms like *Lactobacillus* and *Bifidobacterium* species have been mostly encapsulated in research [80].

5.3.2 WALL MATERIAL

Wall materials or coating materials are act as the containment of the bioactive molecules within. Choice of appropriate wall material is necessary based on the intended application. Wall materials that are mostly employed are obtained from proteins (caseins, caseinates, whey protein concentrates (WPC), gelatin, etc.), polysaccharides (alginates, starch, maltodextrin, etc.), lipids (mono- and di-glycerides of fatty acids). Based on application, these materials are combined together to enhance the functionality desired. These groups differ in their chemical structure and properties, which govern their encapsulation efficacies. The main criteria of their proper selection are essential as they exert different effects on the containment and preservation of the bioactive compounds by hindering these from deteriorative changes. These materials should also be able to release the compounds at the specific sites effectively. Criteria of selection of proper wall materials and their classifications have been discussed in detail later in the chapter.

5.4 DIFFERENT FORMS OF ENCAPSULATES

Mainly two types of configurations exist on which of encapsulates are classified: the matrix type and the reservoir type [67, 142].

5.4.1 RESERVOIR TYPE ENCAPSULATES

Reservoir type system also called the capsule type or single-core type consist of a shell of wall material around the active ingredient in the form of a capsule or a core (Figure 5.1(a)). Some of the characteristics of these type of systems are given in Table 5.1. These type of encapsulates may be vulnerable to breaking of shell and oozing of the content when shearing action, pressure, heating during processing are applied [142].

TABLE 5.1 Characteristics of Reservoir Type Encapsulates

Processing Technology and Steps	Loading (%)	Particle Size (μm)
Fluidized Bed Coating:	5–50	5–5,000
Fluidization of the active ingredient and coating by spraying		
Emulsification:	1–90	0.2–5,000
Preparation of emulsions with oil phase and ionic emulsifier; mixed with aqueous solution		
Coacervation:	40–90	10–800
Preparation of emulsions under turbulent mixing condition; allowing of crosslinking		
Co-Extrusion:	70–90	150–8,000
Use of a concentric nozzle through which internal phase is passed through an inner nozzle and external passed by the outer one		

5.4.2 MATRIX TYPE ENCAPSULATES

Matrix type also called poly- or multi-core encapsulates, which include numerous reservoir chambers in a particle. In this system, the active ingredient is present as small dispersed droplets (Figure 5.1(b)).

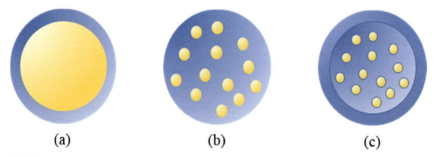

FIGURE 5.1 Configurations of encapsulates: (a) reservoir; (b) matrix; and (c) coated matrix types.

For some cases, the active ingredient phase is adsorbed on the matrix surface. The matrix act like a solvent in these systems. Usually, the loading is low, which increases the material costs. The encapsulation efficiency depends

Encapsulation Methods in the Food Industry

on the adequate dispersion and mixing of the active ingredient with the coating material. Characteristics of matrix type systems is given in Table 5.2.

TABLE 5.2 Characteristics of Matrix Type Encapsulates

Processing Technology and Steps	Loading (%)	Particle Size (µm)
Spray Drying:	5–50	10–400
Mixing of the internal phase and external phase, atomize, and dehydrate.		
Spray Cooling:	10–20	2–200
Mixing of the internal phase and external phase, atomize the solution and cool.		
Melt Injection:	5–20	200–2,000
Dissolving of internal phase in the external, extrusion through filter followed by cooling.		
Melt Extrusion:	5–40	300–5,000
Dissolving internal phase in external, extrusion through twin-screw extruder followed by cooling		
Microspheres by Extrusion or Dropping:	20–50	200–5,000
Dissolving the internal phase in external phases and dropping into gelling solution		
Rapid Expansion of Supercritical Fluid (SCF):	20–50	10–400
Mixture of both phases in a SCF, then rapid expansion to allow precipitation of shell onto internal phase		
Freeze-Drying:	Numerous	20–50,000
Mixing of the internal phase and external phase, freeze-drying, and grinding		

5.4.3 COATED MATRIX TYPE ENCAPSULATES

Coated matrix type of encapsulates are also multi-core type system that include an outer coating on the matrix type encapsulates (Figure 5.1(c)). They can be conceived as the mixture of matrix type and reservoir type.

5.5 SIZE OF PARTICLES FOR ENCAPSULATION

The size of encapsulates varies with different applications. Terms like nanoencapsulation and microencapsulation technology have emerged in the recent decades which is based on the size of the fabricated encapsulates. The term microcapsules, microparticles or microspheres are used to denote the microencapsulation technology include particle size range between 1 µm and 800 µm. In nanoencapsulation technology the range of particle size exists between 1 nm and 1,000 nm and the particles are called nanoparticles/nanocapsules [67].

5.6 BENEFITS OF ENCAPSULATION IN FOOD PRODUCTS

5.6.1 PROTECTION OF INGREDIENTS

Bioactive compounds like minerals, vitamins, polyphenol possess numerous health benefits and by applying heat or in the presence of light and moisture, they may get degraded. By applying encapsulation technique, stability of active components present in food or feed ingredients can be protected by slowing down the degradation process and prevent undesirable changes in flavor, texture, color, and sensory attributes during processing and storage [89].

5.6.2 CONTROLLED DELIVERY

In food product two types of release takes place, viz.: 1) delayed; 2) sustained. Delayed-release process involves release after a certain condition. For example, in ready-to-eat food, flavor can be released by applying heat treatment, like if sodium bicarbonate is encapsulated, the release will take place during the baking process. In case of probiotic bacteria microencapsulation, probiotic microorganism is secured from gastric acidity and the discharge takes place only in the small intestine.

Sustained-release mechanism of bioactives takes place at a steady rate at the target area. For instance, in chewing gum, the rate of release of flavor and sweeteners are maintained throughout the chewing process [89].

5.6.3 MASKING TASTE AND FLAVOR

Polyphenols, vitamins, and minerals possess various health benefits. However, their tastes are unpleasant which can be masked off by encapsulating ingredients.

Encapsulation Methods in the Food Industry

5.6.4 EASY HANDLING

Encapsulation also helps the core material to be easily handled during the manufacturing of food products. This could be done by providing the uniform position to the core, enhancing blending, and mixing, and changing to solid form from liquid phase for products like essential oils and plant extracts [89, 134].

5.7 TECHNIQUES FOR ENCAPSULATION IN FOOD PRODUCTS

In the recent years, there has been a surge in the interests in the target release of bioactives in the body by novel encapsulation technologies which use food-grade encapsulants at commercial scale. Various technologies are being explored and developed for the encapsulation of bioactive compounds. The encapsulated or core material is generally in liquid phase but can also be in solid or a gaseous phase. Many technologies based on drying have emerged for encapsulating compounds which are in liquid form. Some other technologies have also emerged in the recent years. This section discusses some of the technologies that are used for manufacture of encapsulates.

5.7.1 SPRAY DRYING

Conventionally, the commonly used technique for encapsulating agro-food constituents has been through spray drying which transforms liquids into stabilized, free flowing and facilely applied powders. Nearly 90% of microencapsulates via spray drying usually comprises of dissipation, emulsifying, or dissemination of the active composites, annihilation, and spraying [40]. In spray drying, only aqueous dispersions are applicable; thus, the matrix used for encapsulation must be readily soluble in water.

The first phase of the spray drying process includes diffusion of the active agent in an aqueous suspension of the carrier thereby, development, and evaporation of tiny droplets having a diameter of a few microns. Subsequently, the droplets are dehydrated in the heated chamber (160–220°C) of the spray drier and enclosed by capsules of diameters ranging from 10 to 400 µm [15, 142]. The drying process lasts only for a time scale of a few milliseconds to a few seconds. It is due to this reason even thermosensitive components such as proteins, enzymes, flavors, living cells, etc., are retained during spray drying without any substantial losses. The materials used for matrix are hydrophilic carbohydrate molecules whereas for core, the materials used are hydrophilic and hydrophobic active molecules [26]. The foremost reasons of the wide usage of spray drying in the agro-food industries is its high production rate as well as operating costs is also low. Although, the technique has certain drawbacks such as irregular size and shape of the obtained fragments, complexity of the equipment, as well as agglomeration of the particles. Thus, powders with a minimum surface oil content and a maximum bioactive compounds retention are the two main requirements for the efficient encapsulation through spray drying [11].

Spray drying is normally employed for the manufacturing of stabilized dry food additives and flavors [28]. A coating material for obtaining good spray drying results must possess the following characteristics: form stable emulsion and be good emulsifier; display appropriate dissolving characteristics; have ability to aggregate; and form viscous emulsion at high concentrations [40]. This technique has been used for the encapsulation of a large number of bioactive food components viz., proteins (dairy by-products, gelatin), vitamins, minerals, flavors, food pigments, unsaturated fats, endoenzymes, exoenzymes, and gums (guar gum, gum acacia) and modified starch are frequently used as wall materials [88].

In a spray dryer, the key criterion for determining the encapsulation efficacy is the temperature of air entering the heated chamber as well as the wall/core mass ratio. Numerous researches have exhibited that as the quantity of core material increases, the efficiency of encapsulation becomes lesser [107, 115]. Further, there is effective evaporation of the fugitive components such as ethanol and water from the particles surface with the rise in temperature of inlet gas and its content is also reduced. Besides, rate of evaporation as well as the amount of particle formation are also disproportionated with the rise in inlet air temperature resulting in the reduction in the efficacy of encapsulation [115]. Additionally, there is an increase in the amount of dried substance, i.e., the active agent and wall matrix as the feeds of spray dryer, which enhances the output of the produce [10, 38].

Another associated outcome of high temperatures during the whole process is the alteration or loss of volatile active composites, subsequently causing the reduction in the productivity of the process [107]. The higher the temperature of air entering the heating chamber, the higher will be the temperature of outlet air. However, as mentioned, during spray drying, the active ingredients are exposed to the inlet temperature for only short duration. Hence, owing to the storage of the resulted microcapsules in the collecting reservoir/container

Encapsulation Methods in the Food Industry 103

for a longer duration, the temperature of outlet gas and the overall drying period may be regarded as more significant throughout the process.

5.7.2 SPRAY COOLING/CHILLING

Spray chilling and spray cooling technologies are used to manufacture encapsulates coated by lipids. Application of spray chilling or spray cooling relies greatly on the melting properties of lipids. In spray chilling, the encapsulation is characteristically done with fractionated or hydrogenated fat having melting ranges of 32–42C. While, in spray cooling, the coating material is generally comprised of hydrogenated fat having a melting range of 45–122°C, though additional materials can also be employed. Thus, both the procedures vary only in the melting ranges of the coating material used and are frequently employed to encapsulate solid constituents, for example, minerals, vitamins, or acidifier.

With spray chilling/cooling technology it is possible to attain high output and is like spray drying but unlike spray drying, there is usually no water evaporation. These processes can be carried out in continuous mode or batch mode. The principle of spray chilling/cooling is different from the spray drying. The agent that exists in dry form or as aqueous suspension is dissolved in lipids. After that, the internal and external phase are dried using cool or chilled air that leads the carrier to be solidified to surround the core material.

Spray cooling/chilling is a safe, rapid as well as consistent physical technique. Besides, it is regarded as ecologically safe procedure in relation to alternate procedures like spray drying. One more positive feature is the easiness in the expansion of the production owing to the continuous operation of the technique. Generally, it is employed for dried commodities to preserve flavors, minerals, proteins, and enzymes [26]. This technique is also used for cost-effective microencapsulation of probiotic bacteria, as this process is neither utilizes organic solvents nor operated at high temperature that may be lethal to probiotic cells. Because the coating membrane comprises of lipids, the discharge of the bacteria takes place straight in the intestine due to the presence of lipases in the intestinal walls [92].

5.7.3 FLUIDIZED BED COATING

Fluidized bed coating is another important technique used in the manufacture of microencapsulates which boost the productiveness of the food commodity

[85, 86]. The technique generally involves adding coating to pulverized particles in a continuous mode or a batch mode. The powdered components are floated in a stream of fluidizing air and atomized wall material is spurted on them (Figure 5.2).

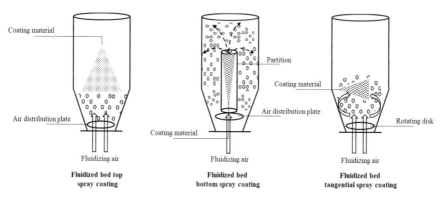

FIGURE 5.2 Fluidized bed coating techniques for the microencapsulation of cells.

The particles subjected to such coating should be spherical in shape (lowest surface area) with a narrow particle size distribution and good flowability [142]. The core is always solid and the carrier material used is an aqueous solution of cellulose or modified starch, gelatin, or other proteins. This technique is carried out to give another layer of coating to previously spray dried products, particularly to encapsulates with a subtle core, like oils [26]. A main advantage of encapsulation via fluidized bed is that it can be performed at comparatively lower temperatures in comparison with spray drying which is attributed to an optimal heat and mass transfer as well as equal temperature distribution.

Water evaporation is affected by various factors like moisture content, air current, spraying rate, humidity of the air inlet as well as temperature [89, 142]. The main drawback of this process is difficulty in controlling the agglomeration or clumping of particles which happens when the particle surface temperature is more than the glass transition temperature (T_g) or the melting point of the wall material or coating substance. Some studies exhibited that atomization pressure, inlet temperature, and concentrations of coating solution are the foremost factors that need to be monitored to reduce agglomerations [100].

Fluidized bed coating has been applied for the encapsulation of vitamin B, ascorbate, ferrous salts, and a few other minerals. Lactic acid and sorbic

Encapsulation Methods in the Food Industry

acid encapsulate, as well as of potassium sorbate and calcium propionate manufactured by fluidized bed coating have found extensive usage in bakery products. It can also be used for the encapsulation of probiotic bacteria, thus greatly enhancing their survival while further processing. Furthermore, application of fluidized bed coating is done for intended release in the gut by generating secondary coating of molecules [26, 89]. Likewise, encapsulates of acidulants, food pigments, and flavors were also employed in the meat industry.

5.7.4 *MICROGEL ENCAPSULATION EXTRUSION PROCESS*

Extrusion is an extensively exploited technique employed in the food industry and was patented in 1957 by Swisher. The basic principle behind all the extrusion techniques encompasses extrusion of an encapsulating liquid substance and biologically active core substance by means through a die and thereafter droplets formation at the expulsion point of the nozzle. Extrusion methods involve dispersing of a liquid suspension of polymer (generally 0.6–3.0% by weight sodium alginate) as well as an active constituent into droplets dripping onto "a gelling bath" comprising of 0.05–1.5 M calcium chloride solution [142]. The dripping tool can merely be a pipette, spraying nozzle, syringe, coaxial air flow, jet cutter, vibrating nozzle, electrostatic field, or atomizing disk (Figure 5.3).

For large-scale/industrial applications, jet cutter has been established to surpass all other extrusion technologies. Whereas, electrostatic extrusion has emerged as a very good technique for the manufacture of very small particles, i.e., less than 50 μm by the application of electrostatic forces at the endpoint of a capillary/needle to mix liquid suspensions and produce electrically charged tiny drops. Usually, all extrusion methods have the advantage that they do not involve the use of solvents as well as harsh circumstances. Though, reduced pace of production is the only limiting constraint for an extensive implementation of extrusion techniques in the food sector.

Among the ingredients, polysaccharides such as κ-carrageenan, alginate, chitosan, and gellan gum are mainly used to produce microgels due to their status as "generally recognized as safe (GRAS)" as well as their technological properties. Besides, even though several polymeric substances involve of organic solvents for their dissolving and consequent electrospinning, certain macromolecules or biopolymers can be electrospun from an aqueous suspension through customization of the processing parameters and/ or changing the aqueous medium characteristics by means of appropriate additives such as whey protein. On the whole, the alteration of shape and size of microgels generated by atomization techniques can be accomplished via alteration in the constitution, consistency, and surface tension of the polymer solutions, as well as other functional factors for a certain method. Therefore, the dimension of microgel particles for coaxial airflow process depends upon the feed flow speed as well as compressed air flow [98].

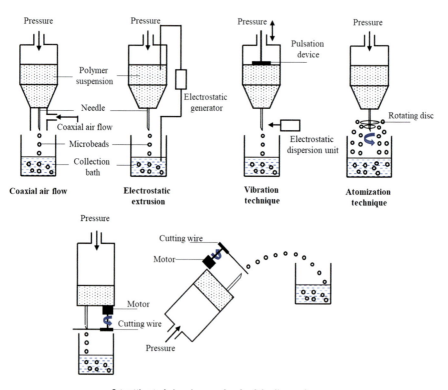

FIGURE 5.3 Extrusion techniques.

Source: Adapted from Ref. [89].

5.7.5 EMULSIFICATION PROCESS

Another recurrently used encapsulation technology is emulsification in which an oil-in-water (O/W) emulsion is produced and dehydrated by various drying procedures namely spray drying or freeze-drying to obtain a powder. There are two kinds of emulsions viz., water-in-oil (W/O) emulsions and

O/W emulsions. An extension of those two emulsions is a water-in-oil-in-water double-emulsions. The size of emulsified particles relies on the design of the reactor as well as mixer geometries and can be regulated on fluctuating the mixing speed and water-to-oil ratio [57]. Hence, it is feasible to regulate the development of micelles, diverse configuration of glycerides (hexadic, cuboidal, lamellar) by means of an easy and facile method that can cover one or several biologically active compounds. This technique is extensively adapted for the controlled release of flavors and aromas.

In the formation of emulsion, the droplets are created by dispersion of an oil phase (generally vegetable oil) within a biopolymeric suspension, along with the incorporation of a gelling medium which favors gelation. However, certain alteration in the process exists. Firstly, both polymer as well as gelation agent exist in the aqueous phase of the emulsion and secondly, for alginates, to permit a discharge of calcium ions, a liposoluble acid (acetic acid) is added owing to the decrease in pH as they start networking with alginate.

Gelling ingredients aside from alginate might also be utilized, for instance κ-carrageenan (gelation in the course of cooling together with potassium ions), gelatin (cross-linking in conjunction with anionic polysaccharides, i.e., gellan gel at neutral pH), pectin (physical or chemical cross-linked) and chitosan (cross-linking by inserting anions). Furthermore, emulsions are mostly employed for the enmeshment of lipophilic actives that could be amalgamated with foods with high lipid content as well as beverages like non-alcoholic drinks, energy drinks, fortified waters, sauces, dips, etc. [130].

5.7.6 EMULSION TECHNOLOGIES

On a daily basis, many foodstuffs that are consumed are basically the emulsions. For instance, milk, ice cream, and mayonnaise are oil-in-water (O/W) emulsion where particles of oil remain suspended in an aqueous phase, whereas butter or margarine is water-in-oil (W/O) emulsion in which water droplets are dispersed in the butterfat. As earlier mentioned, multiple emulsions comprise of tiny droplets of the first phase are surrounded by bigger droplets of a secondary phase, that alone is distributed inside a continuous phase.

The simplest type of multiple emulsion is double emulsion and can be perceived in food such as creams, mayonnaise, and salad dressings [120]. The characteristics of O/W emulsions can also be altered by developing another coating surrounding the droplets of oil via multilayer emulsions or "layer-by-layer deposition method." Solid lipid particle emulsions (SLPEs) encompass lipid particles in solid-state (generally covered with an emulsifier) and scattered in a liquid continuous phase (Figure 5.4).

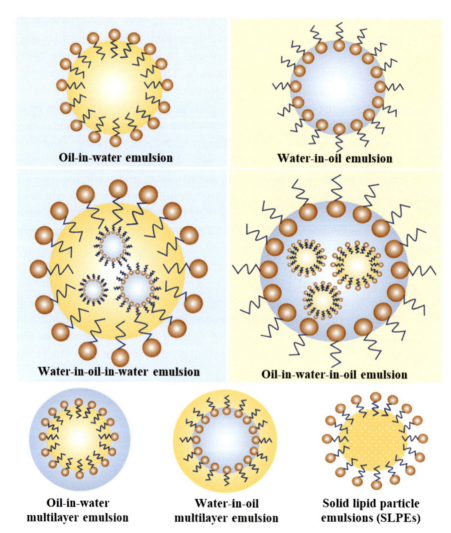

FIGURE 5.4 Schematic representations of different emulsion-based delivery systems. *Source:* Adapted from Ref. [33].

Emulsions-based delivery carriers fundamentally comprise of an oil, surface-active agent, and water. Based on particle size, they are classified into emulsions or macroemulsions, microemulsions, and mini-emulsions.

Encapsulation Methods in the Food Industry

The parameters that influence the size of emulsion droplets are selection of oil as well as surfactant, surfactant/oil ratio, and energy involved. Macro-emulsions or emulsions are kinetically stable and have particle size more than 200 nm, though, it is thermodynamically not stable. Whereas, micro-emulsions are transparent low viscosity, thermodynamically stable isotropic dispersions with particle sizes ranging from 5–100 nm which are prepared easily. Although, a higher concentration of surfactant is required generally in combination with a co-surfactant. Nanoemulsions or miniemulsions have droplet sizes ranging between macroemulsions and microemulsions, i.e., 20–200 nm [103]. Since, nanoemulsions have a tend to break down during storage and are thermodynamically stable but are more resilient to oxidation as compared to microemulsions [81].

Usually, the properties of emulsion (size of particle, stability, and interfacial tension) rely upon the type of structural complexes as well as oil phase/surfactant/water phase ratio. Hence, operationalization of a screening process for ideal emulsion constituents through the creation of a pseudo ternary phase diagram of kinetic stability is a vital step in defining the optimal formulation.

5.7.7 COMPLEX COACERVATION

In 1950, the first microencapsulation technique developed was coacervation [121]. The simple coacervation bind only a single type of polymer, whereas, for complex coacervation it is obligatory to use two or more polymeric substances. In most cases, coacervates are protein/polysaccharide composites. Nevertheless, the protein-protein combinations have also been subjected to investigation. Polymers like gum Arabic and Gelatin and are generally known and widely used combination in complex coacervation. Complex coacervation process might be separated into three stages viz., development of three immiscible phases, sedimentation of coating, and lastly, solidification of the coating. The innovative process comprises of incorporation of a cross-linker such as tripolyphosphate [9], transglutaminase [78], or genipin [97, 135] in the midst of the third stage of coacervation.

An important limiting factor of complex coacervation process is the challenges faced in encapsulating hydrophilic compounds. Taking this into account, the procedure is more suitable for hydrophobic compounds as core materials. Thus, for the encapsulation of hydrophilic substance, few variations to the process are necessary, such as inclusion of double emulsion stage at the commencement of the coacervation procedure [18, 108, 109].

The aforementioned encapsulation process is productive but costly and has limited encapsulating materials that are appropriate for food application.

The coacervate microcapsules are generally circular, multinucleate, and have defined walls. Though, an excess of polymer can induce surplus charge thereby hampering the development of the complex [109]. Since, the concentration of polymer in the solution is more (i.e., highly viscous), the displacement of macromolecules will be less owing to the degree of opposition as the molecules of solvent become bigger. Usually, the drying of coacervated capsules is done in the last stage of the operation process either via mechanical dehydrator, freeze-drying, or spray drying process. The mechanism employed for drying process too has an influence on the properties of the end product [117]. The microcapsules formed through coacervation are resilient to heat, possess brilliant controlled release features depending upon temperature, mechanical stress, pH shifts as well as sustained release [7, 31, 91, 110].

Complex coacervation is mainly carried out for the essential oil encapsulation. Vanilla in gum Arabic/chitosan combination [135], mustard seed and jasmine oil in gelatin/gum acacia [78, 97], sweet orange oil in gum acacia/soy protein isolate [55], Jamaica pepper or allspice (*Pimenta dioica* L. *Merr.*) essential oil in κ-carrageenan/chitosan polymer blend [30] are few studies on encapsulation of essential oils by researchers. Findings have shown that whey protein isolate (WPI) coacervates together with carrageenan appear to be appropriate encapsulation material for encapsulation of probiotics [45].

5.7.8 LIPOSOME ENTRAPMENT

Liposomes are sphere-shaped phospholipid particles that are commonly employed for intended drug release. Their recent demand in functional food has aroused the interest of many researchers. The capability of liposomes to capture hydrosoluble, liposoluble as well as amphipathic compounds make them appropriate carriers. The feasibility of manufacturing liposomes on industrial scale apart from their enormously distinctive properties offers an enormous benefit for their applications in food sector [37]. Liposomes can be produced with the aid of natural constituents that are non-hazardous, biocompatible, eco-friendly aside from allowing easy and rapid application of liposomes in agri-food industries thereby reducing the monitoring impediments [21]. There are various conventional procedures for the preparation of liposomes, among which are the thin-film hydration method, solvent injection method, reversed-phase evaporation, and heating-based method [2, 125].

Encapsulation Methods in the Food Industry   111

The restricted use of liposomes in the food sector is because of time taking manufacturing procedures and is highly expensive. Furthermore, most of the conventional processes utilize noxious substances and organic solutions, which must be evaded in the food industry. Thus, certain alternate approaches are being evolved that helps in overcome drawbacks of conventional methods.

Over the past few years, the liposome entrapment technique has been studied to develop encapsulating system for proteins [123], enzymes [49, 119], antioxidants [48, 60], vitamins [42], minerals [35], flavors [89, 138], and antimicrobials [22, 141]. Rashidinejad et al. [104] confirmed the retention of the polyphenols in the low-fat hard cheese in which liposomes encapsulated tea catechin as well as epigallocatechin gallate were precisely incorporated. Liposomes are also utilized in cheese ripening for the encapsulation of enzymes [50]. Additionally, liposomes are employed for the encapsulation of antimicrobial peptides or host defense peptides, various bacteriocins, delivery of hormones, vaccines, enzymes as well as vitamins into the body.

5.7.9 ENCAPSULATION BY CYCLODEXTRINS

Cyclodextrins are cyclic oligosaccharides, which have an oleophilic inner pouch of about 5–8, wherein particle of an active ingredient of appropriate dimension can be reversibly embroiled in an aqueous solution. The chief characteristics of encapsulation via cyclodextrin is that it upsurges the water solvability of numerous moderately soluble substances [61, 118, 139]. Besides, they have the ability to form inclusion complexes by means of discrete molecular interactions. Apart from it they also form intricate "mega-molecular" compositions which tend to enhance the solubility of hydrophobic substances in order to ensure that the solubility exceeds than that of free molecules [72].

Cyclodextrins compounds have been employed for conserving the active agents like essential oils, vitamins, food pigments, fragrances, and flavors against light, degradation by the action of heat as well as oxidation process [17, 118]. The astringent and bitter constituents of foods stuff such as soybean and beverages like catechins in green tea, chlorogenic acid and polyphenols in cocoa/hot chocolate, naringin in citrus fruit juice, or caffeine in coffee can be aggregated with cyclodextrin to reduce or fully get rid of their taste since the intricate molecules could not respond with the gustatory cells in the mouth and subsequently no bitter taste is observed. Moreover, cyclodextrin has been employed for the reduction of enzymatic browning in apple juice through combining it with the enzyme polyphenol oxidase (PPO), responsible for browning in apple juice and thus, ameliorating the antioxidant activity of the juice attributed to the protective effect of maltosyl-β-cyclodextrin that opposes the oxidation of ascorbic acid [75].

In addition, declination of anthocyanins content in grape juice can be lowered by incorporating cyclodextrin encapsulated polyphenol (such as chlorogenic acid) to the juice owing to the co-pigmentation effect [113]. Another application of cyclodextrin can be seen in the packaging of food products. López-de-Dicastillo et al. [74] reported that using of cyclodextrin helps in better retention of aldehydes as well as the active packaging films (ethylene-vinyl alcohol (EVOH) copolymer) comprising of β-cyclodextrins and effected a substantial decrease of 1-hexanal in fried peanuts, attaining 50% reduction over brief period of 7–35 days.

5.7.10   SUPERCRITICAL FLUID (SCF) BASED ENCAPSULATION

From the late 19th century, supercritical fluid (SCF) technology has been in use as an implement to know the natural mineralization. Although, commercial exploitation of SCF technology has commenced in the 1970s due to its capability to replace toxic industrial solvents regarding the concern for the environmental issues, and lastly, this process might be economical to liquid extraction and distillation methods [3]. SCFs (for example, CO_2, N_2) have some specific properties intervening liquids and gases that can be effortlessly altered by varying pressure as well as temperature. In encapsulation, the most commonly employed SCF is CO_2 as a matter of fact that it is the least expensive, naturally occurring solvent on earth with supercritical region at controlled temperatures as well as pressures (critical pressure, $P_c = 7.39$ MPa and critical temperature, $T_c = 31.1°C$). Also, CO_2 is an inert agent appropriate for fabrication of compounds that are sensitive to heat.

During the encapsulation process, the SCF can be served as a solvent, co-solvent, antisolvent, solute, or propellant gas and is used for precipitation process that leads to the formation of particles and can be removed by depressurization at the end of processing. The apparatus (nozzles and spray chambers) used for the formation of particles are nearly similar to other encapsulation processes. The formation of particles during the encapsulation process via SCF mainly relies on the composite morphological behavior of an admixture of polymers, active agent as well as SCF involved in the process. The particle development by precipitation with SCF avert the widespread utilization of organic solvents, high operating temperatures, eliminate

Encapsulation Methods in the Food Industry   113

the water consumption as well as mechanical force that can damage sensitive compounds. Thus, SCF are exploited in various operations associated with therapeutic biomedical compounds as well as for the encapsulation of delicate food components like essential oils, vitamins, natural flavors, and enzymes. The wall material or matrix used for this process are acylglycerols, ester of fatty acid and a combination of those, i.e., waxes.

5.7.11 ENCAPSULATION IN YEAST

Presently, the foremost difficulty for effective implementation of immobilized yeast cell technique on commercial scale is physiological of monitoring yeast and graduate modification of flavor development throughout the fermentation process [89]. The first research related to the use of yeast cells as a coating membrane for the encapsulation of liposoluble compounds was introduced in the 1970s. Even though other yeast strains such as *Yarrowia lipolytica* have been proposed for encapsulation, *Saccharomyces cerevisiae* (occurs as a by-product of fermentation process) appear to be most appropriate bio-capsules for the encapsulation of functional foods [137] and can be manufactured conveniently in the industry. Besides, yeast is considered for human consumption and highly nutritious. This eventually highlights the economic efficiency of yeast in the encapsulation technique. Apart from yeast cells, the encapsulation technology entails water as well as food bioactive component.

The yeast encapsulates are heat resistant delivery systems that can tolerate the heat even at temperature of 250°C. Yeast encapsulation provides active dispersion of encapsulated bioactive and its defense against oxidation and evaporation [142]. Hence, this procedure might be useful for the encapsulation of core materials that are utilized as additives in the dehydrated foodstuffs. The withholding of bioactives is considerably influenced by the moisture content in the surroundings as high moisture elevates the delivery of encapsulated bioactives from yeast cells [142]. Water activity is considered as a key element for the delivery of core material from yeast cells.

Application of yeast encapsulation in the food sector is mainly used for fragrance (essential oils) and antioxidant compounds. For example, encapsulation of limonene within yeast cells exhibit successful outcome, i.e., it protects limonene during cooking. Dardelle et al. [23] reported a higher sensory score in instant noodles and battered products having yeast-based delivery systems with encapsulated limonene when compared with encapsulated spray dried product. Yeast encapsulation is employed

114   Advances in Food Process Engineering

for a hydro-soluble antioxidant and chlorogenic acid. It has also found its application in encapsulation of curcumin (a bioactive compound that also imparts bright yellow color and produced by *Curcuma longa* plants; used as food color) via yeast cells [94]. Additionally, encapsulation of enzymes within yeast cells is achievable. Endoenzyme of lactic acid bacteria (LAB) (cell free extract), liable to carry out the formation of amino acid as well as flavors in Cheddar cheese, had been effectively encapsulated via yeast and incorporated at the time of Cheddar cheese manufacturing [137].

5.8 MATERIALS FOR ENCAPSULATION

5.8.1 PROTOCOL FOR CHOOSING ENCAPSULATION MATERIALS

Substances employed for the fabrication of the defensive shell of capsule must be fit for human consumption, environmentally friendly as well as competent of creating a boundary in the midst of core material and the external phase. Various substances may be employed to encapsulate or coat solid substances, fluids, or gases of varied forms as well as characteristics. Nevertheless, guidelines for food additives are obdurate as the materials used in the food industry should be authorized as "generally recognized as safe (GRAS)." Essentially, the entire food manufacturing operation should be planned with the purpose of attaining the safety provisions of statutory bodies like the United States Food and Drug Administration (FDA) or the European Food Safety Authority (EFSA).

Furthermore, there are some important factors that should be considered before choosing an encapsulation material, viz., purpose of encapsulate that it should impart to the resultant product, probable constraints for the matrix, concentration of wall materials, stability prerequisites, nature of discharge and price constraints. The maximum of ingredients employed in the agro-food industries for the encapsulation process are biomolecules.

Moreover, components utilized for the encapsulation needs to be natural product, offer utmost defense to the internal phase against surroundings, uphold core material inside the encapsulate throughout the operation or retention period subjected to different circumstances, must not collaborate with the coating material or matrix, own acceptable textural characteristics at high concentrations whenever required and should possess good processability throughout the encapsulation process. The physicochemical features of the materials should be taken into consideration before making the ultimate decision as well as anticipate its performance under various

Encapsulation Methods in the Food Industry 115

circumstances existing in food production [142]. Regardless of the varied variety of encapsulated commodity produced and efficaciously commercialized in the pharmaceutic as well as beauty companies; microencapsulation has established a distinctly lesser demand in the agro-food sector.

5.8.2 MATERIALS USED

5.8.2.1 POLYSACCHARIDES

Polysaccharides are increasingly commonly encapsulated material utilized in food industries. Besides, they exhibit excellent performance for its application in the encapsulation process by virtue of being a fundamental component of various food process, come in various molecular weights and sizes, are cost-effective, and have preferable physicochemical characteristics like dissolvability, melting, phase modification, etc. Conventionally employed carbohydrates are starch and their by-products, for instance, amylopectin, amylose, dextrins, polydextrose, maltodextrin, in addition to cellulose as well as their derived products.

Amylose and amylopectin are intricate macromolecules having the ability to interact intramolecularly as well as intermolecularly to create organized structures. Encapsulation via modified food starch is executed by numerous techniques such as fluidized bed granulation, spray drying, fluidized bed spray drying, and extrusion. Maltodextrins are considered as a replacement in encapsulation as they are less expensive than gum Arabic.

Besides, the inimitable spiral configuration of amylose renders starches a remarkable competent agent for encapsulating constituents such as fats, flavors, etc. Additionally, a prebiotic component, fructans can increase the viability of probiotic bacteria whereas trehalose aids as auxiliary nutritive substance for yeast. Ethylated cellulose acts as great encapsulating membrane. Moreover, with the increase in the thickness of the membrane, the water penetrability of core material (e.g., water-soluble vitamins) is curtailed [8].

Plant exudates and extracts viz., gum acacia, pectins, and soluble soybean polysaccharides, galactomannans, gum karaya, gum tragacanth, mesquite gum along with extracts from ocean water like carrageenan and alginate (processed from brown seaweeds) have been employed as well. Gum acacia is a frequently used coating material owing to its emulsification properties in a wide pH range, viscosity, and dissolvability, training film around the droplets and binding properties. Though, high cost is its chief drawback.

Carrageenan is broadly used as stabilizing and thickening agents, thus, making it potential for use in probiotic food owing to its gelation property with the alteration in temperature from 40 to 45°C. Besides, alginate can also be employed for microencapsulation as well as cell immobilization by means of ionotropic gelation, which encompasses reducing the concentrated alginate solution into calcium chloride solution that causes the polymer to gel externally into a microcapsule. Moreover, hydrocolloids, and alginates are employed to capture flavor and essential oils at room temperatures. de Roos [25] reported that utilization of alginate beads enhances the preservation of aroma in crackers in the process of baking.

Polysaccharides extracted from microbes and animals such as chitosan, dextran, gellan as well as xanthan gum are also been utilized. Chitosan is naturally derived cationic copolymer that are biodegradable and biocompatible, have ability to form hydrogels as well as edible film and possesses the property of heat-shrinking. Hence, chitosan has been widely considered for feasibility of using it for the fabrication of thermosensitive carriers in the food as well as pharmaceutical industry [142]. Carboxymethylcellulose (CMC), an anionic water-soluble polymer, is another component that is used in the industrial sector because of its capability to act as a thickener, stabilizer, suspending agent and ability to form resistant films against oils, organic solvents, and greases.

5.8.2.2 PROTEINS

Other than polysaccharides, another suitable component for the encapsulation process are proteins and can be used to produce microcapsules via coacervation techniques or double emulsions by creating a network using glutaraldehyde or heat gelation which act as an excellent barrier against oxygen and aroma. Milk proteins (casein, WPC, and WPI), gelatin as well as gluten are the few frequently employed protein coatings for microencapsulation. Besides, protein has the capability to produce stable emulsions along with fugitive flavoring compounds, although, there are some of the drawbacks in their application viz., sparingly soluble in cold water, heat instability, probability to react chemically with carbonyls, sensitivity to organic solvent and high costs. Amalgamation of WPI/gum acacia, chitosan/whey protein/gum Arabic, pectin/β-lactoglobulin (β-Lg), milk protein products/xanthan and β-Lg/κ-carrageenan are most often used for encapsulation process [4].

5.8.2.3 LIPIDS

Lipid materials viz., fatty acids, fatty alcohols, esters (candelilla wax, carnauba wax, beeswax), glycerides as well as phospholipids are used for food applications. Glycerides are insoluble in water. In addition, the heterogeneity of the combination of di- and monoglycerides regulates the extent of the melting range; and therefore, make it suitable for encapsulation owing to its emulsifying characteristics.

Generally, waxes are water-insoluble and have a melting point varying between 60°C and 80°C are regarded as a satisfactory material for the encapsulation of aromas. Some of the chief benefits of these materials comprises of easier handling, stability, and protection. Animal waxes like bees wax as well as plant waxes such as candelilla and palm wax or carnauba wax, are the food additives approved by the European Union [86]. Beeswax, secreted by young honeybees, are compatible with most of the other fatty acids, waxes, glycerides as well as hydrocarbons. Some other waxes are also used, like palm wax (queen of waxes) that is collected from the palm *Copernicia prunifera* leaves, native to Brazil. Candelilla wax, obtained from the *Candelilla* shrub leaves produced in the northern Mexico is compatible with all fatty acids, glycerides, animal, and vegetable waxes, a wide array of synthetic or natural resins, and hydrocarbons in specific amounts.

5.8.2.4 POLYMERS

Food-grade polymers are also used in food encapsulation, for instance, polypropylene, polyvinyl acetate, polyvinylpyrrolidone (PVP), paraffin, polystyrene, and polybutadiene [8, 142]. PVP powder (synthetic natural polymer) is readily soluble in water as well as organic solvents. Further, it is an excellent coating polymer owing to its efficient film-forming ability and excellent stability to the variation in temperature. Paraffin is a colorless, odorless, bland, waxy material and accessible in food-grade quality although are indigestible. It is solid at ambient temperature and begins to melt at temperature ranging between 48°C and 95°C.

5.8.2.5 INORGANIC MATERIALS

Several inorganic food-grade constituents are employed as coating material in food encapsulation, individually or in conjunction with other components such as tripolyphosphate, silicon oxides, or aluminum oxides (Al_2O_3) [4].

5.9 CHARACTERIZATION OF ENCAPSULATED PARTICLES

Characterization of encapsulated particles is a mandatory step for ascertaining the possible benefits and the expected toxicology of the encapsulated particles [111]. The particle size, morphology, structure, Zeta potential, dispersion, aggregation, and sorption are some of the basic properties used to characterize the encapsulates. Properties like encapsulation efficiency and stability are also important. Techniques like scanning electron microscopy (SEM), atomic force microscopy (AFM) provide high resolution pictures of encapsulated product's surface. High resolution transmission electron microscopy (TEM) is generally employed to study the morphology of encapsulates. Capillary electrophoresis, Hydrodynamic chromatography, Field flow fractionation and Size exclusion chromatography (SEC) help in separation of nanoparticles be it on the basis of size and charge distribution or their hydrodynamic radius [44, 82]. For preparative size fractionation of samples, tools such as ultrafiltration, nanofiltration (NF) and ultracentrifugation are used. Other techniques used to characterize encapsulated particles size are static and dynamic light scattering (DLS), neutron scattering, small-angle neutron scattering, X-ray diffraction (XRD), and nuclear magnetic resonance.

DLS is a method employed to measure emulsions, nanoparticles, colloids particle sizes and determines their state of aggregation in suspensions [127]. The principle that exists behind this is that when a sample is illuminated by a laser beam, the fluctuation of light scattered at a given angle is detected as a function of time. The changes of intensity of scattered light occur due to Brownian motion of dispersed particles, which is a function of the size of those particles. Thus, the particle size is obtained from the diffusion coefficient and size relation based on Stokes-Einstein equation (Eqn. (1)) [142]. Particle size of encapsulates affects their texture, mouthfeel, and viscosity.

$$D = \frac{k_B}{6\pi\mu R} \tag{1}$$

where; D is the diffusion coefficient; (m^2/s); k_B is the Boltzmann constant; μ is viscosity (Pa-s); R is the particle hydrodynamic radius (m).

Using DLS, another property related to particle size is the PDI, which indicates the uniformity in size of encapsulation. Zeta potential, which denoted the electrical potential that exists at the boundary surrounding the

Encapsulation Methods in the Food Industry 119

strongly bonded ions associated with the particle surfaces, is also measured using DLS method. However, other methods like electroacoustic technique are also used to measure Zeta potential.

TEM is used to study the particles morphology. TEM produces high resolution images of the nano- or microparticles. The main principle behind the technique is that when electrons (100–1,000 keV) are transmitted through the sample, the internal and external features of particles are visualized based on the difference between the electrons and the atomic constituents. TEM gives valuable information about the particle size, shape, and internal structure of the particles. Confocal laser scanning microscopy has also been used to study the particle morphology.

XRD is mainly helpful in determining the identity of crystalline solids based on their atomic structure [77].

Stability of particles against sedimentation, coalescence, Ostwald, and aggregation is often tested by storage stability test, whereby samples are kept under normal storage conditions or temperature or shearing stress for some period and tested for particle size, zeta potential, etc.

Encapsulation efficiency, which denotes the amount of active ingredient encapsulated to the initial amount, is also a parameter characterizing the encapsulates. This determines their profitability. Some of the methods used for measuring the encapsulation efficiency include centrifugation, filtration, SEC for separation and high performance liquid chromatography (HPLC) for measuring the amount of separated active ingredients.

5.10 TYPES OF ENCAPSULATED FOOD INGREDIENTS

5.10.1 ENCAPSULATION OF AROMA

Flavors are one of the most important characteristics of food. Flavors are expensive and after some time their intensity get decrease so food manufacturers need solution to preserve aroma. By employing encapsulation, aroma is protected by wall material and protected by the effect of chemicals reactions, heat treatment, evaporation, or migration into food. Flavor encapsulation can be carried out by spray drying and spray cooling techniques. The material used for encapsulation of flavors are maltodextrin, gum Arabic, corn syrup solids, milk, or soy proteins, mono, and disaccharides, hydrolyzed gelatin. Cinnamon was encapsulated using maltrodextrin and propylene glycol and biscuits flavored with cinnamon-maltrodextrin showed higher sensory scores [36].

5.10.2 COLORING AGENT ENCAPSULATION

Natural and synthetic food colors are generally used by the food industry. The major problem with colors is migration inside the food sample or into the surroundings of the product like cakes, desserts with jelly, meat product like surimi. In these products, migration of colors into non-colored layers. To overcome from this problem, coloring materials can be encapsulated with carbohydrate-based wall material [46].

5.10.3 ENCAPSULATION OF CAROTENOIDS

Carotenoids (such as lycopene, astaxanthin, β-carotene, lutein, zeaxanthin and astaxanthin) are heat sensitive, prone to light and oxidation due to having a conjugated structure. These compounds can be encapsulated by freeze-, spray-, and drum drying techniques. Carotenoids can be encapsulated using liposomes and cyclodextrins. Maltodextrins also proved as good protecting material and bioavailability and solubility of carotenoids improved by using this [46, 142]. Gum, gum Arabic, gellan, and mesquite are also good wall material for encapsulation of carotenoids.

For encapsulation of synthetic pigments, chitosan matrix cross-linked with glutaraldehyde was used by preparing multiple emulsion and solvent evaporation technique [46].

Encapsulation of β-carotene using different methods such as nanoemulsions [73, 102], microemulsion [32], liposome [70], solid lipid nanoparticles [19], and complex assemblies with macromolecules [93] are reported [43].

Curcumin, the natural colorant from turmeric has a polyphenolic nature. Several nanoemulsions preparation methods are used to encapsulate curcumin like phase inversion composition and phase inversion temperature are low-energy methods; microfluidization, high pressure homogenization, and ultrasonication are high-energy methods and membrane emulsification as intermediate energy method [53].

5.10.4 VITAMINS ENCAPSULATION

In food matrix, stability is the major problem with fat and water-soluble vitamins and can be improved by encapsulating them. For water-soluble vitamins, spray cooling and fluidized bed coating technique is generally used whereas for fat-soluble vitamins, emulsions are spray dried. Coating of

Encapsulation Methods in the Food Industry

vitamins are desirable in food products like bakery products, biscuits, cereal bars, dry soups and in dairy products [134].

5.10.5 PHENOLIC COMPOUNDS ENCAPSULATION

Phenolic compounds containing products are unpleasant while eating due to their bitter taste and astringency. The unpleasant taste of these compounds limits their use in food products at higher concentration. By encapsulating polyphenolic compounds can improve the *in vivo* and *in vitro* bioavailability. By using spray drying technique, polyphenol containing plant extracts were encapsulated such as amaranthus betacyanin extracts, β-carotene, and olive leaf extract [27].

5.10.6 ESSENTIAL OILS ENCAPSULATION

Essential oils are readily water-soluble and when added to water, they impart their aroma and taste. They possess a wide range of antimicrobial activity besides a broad range of food-borne pathogens, and these contain phenols, terpenes, alcohols, aldehyde, esters, and ketones. *Pimpinella saxifraga* essential oil (PSEO) was encapsulated by using (1–3%) sodium alginate coating and the effect of refrigerated temperature on oxidative stability of cheese added with encapsulated PSEO and bacterial viability was carried out [66]. The incorporation of 3% sodium alginate coating with PSEO, enhanced preservation of cheese by decreasing the weight loss, stabilizing the color and pH, and increasing bacterial viability and oxidative stability, and free from any unpleasant flavor while consumption. Oregano essential oil was encapsulated using nanoemulsion technique and added in low-fat cut cheese and packed in mandarin fiber edible coating material. A reduction in *Staphylococcus* count in coated cheese cubes having 2.0% or 2.5% w/w of essential oil was obtained, within 15 days of refrigerated storage at 4±1°C [5].

5.10.7 ENCAPSULATION OF PROBIOTIC ORGANISMS

Nowadays, probiotic microorganisms are incorporated in various functional food formulations and especially in fermented dairy products. Probiotic encapsulation is related to capsule formation by entrapping the probiotics surrounded by polymeric matrix with emulsion or extrusion methods. This technique enhanced the delivery of these sensitive probiotic microorganism and target delivery in gastrointestinal tract including easy addition in food, increased sustainability, and stability at the time of storage [142]. Selection of encapsulated material should be like that it should not change the taste and texture of food.

The secondary materials used for probiotics encapsulation are gelatin, κ-carrageenan, alginate, gellan, chitosan, agarose, xanthan, locust bean gum and starch [99, 142]. The bead size should be < 100 μm to decrease the unwanted sensory effect of microcapsules in foods [128]. For encapsulation technique, two points should be considered; first, the size should be in the range of 1 to 5 μm diameter and the second viable form. Fluidized bed, freeze-dried, spray dried and spray coating techniques are also used for encapsulation of probiotic organisms. During the spray drying process, probiotic strains may not survive. Supporting materials like starch, protein, skim milk powder, whey protein, maltodextrin, gum Arabic, gelatin, gum acacia, sugar, non-fat dry milk solids, and prebiotics are used.

5.11 FOOD PRODUCTS ACTING AS CARRIER OF ENCAPSULATED BIOACTIVE COMPOUNDS AND PROBIOTICS

As the research has progressed, the encapsulated bioactive compounds and probiotics have been introduced into several food matrices, so as to enhance or complement their nutritional properties. Milk and dairy products, fruits, and vegetable products, cereals, and legumes have been incorporated with encapsulated bioactives. Table 5.3 summarizes some of the applications of encapsulated ingredients in various food matrices.

5.12 ENCAPSULATION: CHALLENGES AHEAD AND SAFETY ISSUES

While encapsulating a specific compound, it must be kept in mind that its bioavailability is not hindered due to the encapsulating material as it depends on various parameters such as size, composition, concentration, and characteristics. Moreover, these parameters keep on altering while passing through mouth, stomach, and small intestine due to variations in temperature, pH, molecular environment, and ionic strength, which, in turn, affects the behavior and extent of bioavailability of the active compound [131]. For example, the dilution of the encapsulation system starts in the mouth,

TABLE 5.3 Applications of Encapsulation in Various Food Matrices

Food Matrix	Encapsulated Ingredient	Encapsulation Conditions and Characteristics	References
Milk and Dairy Products			
Milk	Chitooligosaccharides	Process – Spray cooling Coating material – Polyglycerol monostearate Coating: Core ratio – 10:1 Encapsulation efficiency – 88.08% Release rate – 7.6%	[16]
	Isoflavones	Process – Spray cooling Coating material – Medium-chain triglyceride Coating: Core ratio – 15:1 Encapsulation efficiency – 70.2% Release rate – 8%	[52]
	Mistletoe extract (lectin)	Process – Spray cooling Coating material – Medium-chain triglyceride Coating: Core ratio – 15:1 Encapsulation efficiency – 66.1% Release rate – 17.2% (pH 2.00) Particle size – 19.5 μm	[64]
	L-Ascorbic acid	Process – Spray cooling Coating material – Polyglycerol monostearate Coating: Core ratio – 5:1 Encapsulation efficiency – 94.2%	[69]

TABLE 5.3 *(Continued)*

Food Matrix	Encapsulated Ingredient	Encapsulation Conditions and Characteristics	References
	Peanut sprout extract (resveratrol)	Process – Spray drying	[71]
		Coating material – Medium-chain triglyceride	
		Encapsulation efficiency – 98.74%	
		Particle size – 3–10 µm	
	Nano ginseng	Coating material – Liposome	[1]
		Coating: Core ratio – 6:1	
Yogurt	Omega-3 oil	Process – Complex coacervation	[124]
		Coating material – Gelatin/acacia gum	
	Iron	Process – Spray cooling	[112]
		Coating material – Alginate	
	Probiotic bacteria (*Lb. acidophilus*, *B. lactis*)	Process – Emulsification	[58]
		Coating material – Calcium-induced alginate-starch	
Cheese	Encapsulated proteinases	Process – Emulsification	[62]
		Coating material – Liposome	
		Encapsulation efficiency – >30%	
	Iron	Process – Spray cooling	[68]
		Coating material – Polyglycerol monostearate	
		Coating: Core ratio – 5:1	
		Encapsulation efficiency – 72%	

TABLE 5.3 *(Continued)*

Food Matrix	Encapsulated Ingredient	Encapsulation Conditions and Characteristics	References
	L-Ascorbic acid	Process – Spray cooling	[68]
		Coating material – Polyglycerol monostearate	
		Coating: Core ratio – 5:1	
		Encapsulation efficiency – 94%	
Milk powder	Iron	Coating material – Phospholipid SFE-171	[79]
		Meat Products	
Salami batter	Probiotic (*Lactobacillus plantarum* and *Pediococcus pentosaeccus)*	Coating material – Alginate microcapsules	[59]
Dry fermented sausages	Probiotic (*L. reuteri*)	Process – Immobilization	[116]
		Coating material – Wheat grains	
		Fruit and Vegetable Juice Products	
Apple juice	Carvacrol	Process – Homogenization	[14]
		Coating material – Tween 20	
		Encapsulation efficiency – 72%	
		Particle size – 2.26 μm	
	Orange oil	Process – Ultrasonication	[122]
		Coating material – Tween 80	
		Coating: Core – 1:1	
		Particle size – 20–30 μm	

TABLE 5.3 *(Continued)*

Food Matrix	Encapsulated Ingredient	Encapsulation Conditions and Characteristics	References
	Steppogenin	Process – Microemulsion Coating material – Tween 80 and polyethylene glycol Particle size – 36.56 nm	[126]
	Trans-cinnamaldehyde	Process – Ionic gelation Coating material – Alginate chitosan nanoparticles Coating: Core – 37.5:1 Encapsulation efficiency – 73.24% Particle size – 166.26 μm	[76]
	Pectinase	Process – Entrapment Coating material – Poly(lactic-co-glycolic acid)	[12]
Orange juice	L-Ascorbic acid	Process – Freeze-thaw Coating material – Liposomes Encapsulation efficiency – 5.4% Particle size – 1.3 μm	[133]
	Cinnamic acid	Process – Freeze-drying Coating material – Cyclodextrin inclusion complex	[129]
	Lycopene	Process – High-pressure homogenization Coating material – Tween 20 Particle size – <1 μm	[106]

Encapsulation Methods in the Food Industry 127

TABLE 5.3 *(Continued)*

Food Matrix	Encapsulated Ingredient	Encapsulation Conditions and Characteristics	References
	Eugenol	Process – Ultrasonication Coating material – Tween 80 Particle size – 13 nm	[41]
Grape juice	Pimaricin	Process – Ionic gelation Coating material – Poly(N-isopropylacrylamide)/acrylic acid nanohydrogel	[39]
Watermelon juice	Trans-cinnamaldehyde	Process – High energy homogenization Coating material – Alginate chitosan nanoparticles Coating: Core – 5:4 Encapsulation efficiency – <70% Particle size – <200 μm	[54]
Tomato juice	Nisin	Process – Ionic gelation Coating material – Alginate chitosan nanoparticles Particle size – 130–178 μm	[6]
Carrot juice	Isoeugenol	Process – Spray drying Coating material – β-Lactoglobulin Particle size – 0.49 μm	[65]
Cereal Products			
Breakfast cereals	Omega-3 fatty acids	Process – Fluidized bed Coating material– Fish gelatin/Corn starch suspension Particle size – 254 nm	[29]

TABLE 5.3 (Continued)

Food Matrix	Encapsulated Ingredient	Encapsulation Conditions and Characteristics	References
Cereal bar	Omega-3 fatty acids	Process – Spray drying Coating material – Maltodextrin/chickpea protein isolate Encapsulation efficiency – <88.72% Particle size – 24 μm	[90]

Encapsulation Methods in the Food Industry

which contains mucin, digestive enzymes, and salts and is again diluted in the stomach because of highly acidic fluids contains enzymes like proteases and lipases. Further dilution occurs in the small intestine containing buffers, digestive enzymes, and bile salts [84, 101].

Proteins are most affected by this phenomenon as they are hydrolyzed by pepsin and trypsin in the stomach and small intestine, respectively. Lipids are also hydrolyzed by pancreatic lipases and solubilized by bile salts [136]. The encapsulate should be capable of protecting the inner ingredient from chemical degradation which is also responsible for color degradation [24]. In case of applications where optical clarity is needed, encapsulation systems should be designed such that the particles have a diameter less than 50 nm whereas, in cases where it is not required, particles having a diameter greater than 50 nm can be used [130].

The cost factor can be a potential challenge to this technology. The use of high shear mixers and homogenizers required for emulsion formation involves high initial cost as well as high repair and maintenance cost. On the other hand, low-cost techniques such as spontaneous emulsification pose a threat to environmental safety as they use high levels of synthetic surfactant. Another safety issue can be the use of nanostructures in encapsulation, mainly in the case of food industries as the small dimensions of these structures have increased reactivity with the gastrointestinal tract [47]. Less explored areas of research include knowledge about biotransformation of nanoparticles after oral administration as well as their excretion [34]. However, this field needs more research as not much is known about the potential toxicity imposed by these kinds of compounds.

5.13 SUMMARY

Encapsulation is a system that enables to create of tailormade-protected compounds which will cater to specific health benefits owing to their specific functionalities. However, various parameters need to be kept in mind while judiciously choosing material to be used as coat for respective core materials. Even though there are many possibilities of encapsulating an active ingredient, the appropriateness actually relies on various factors such as the kind of active ingredient of interest, required particle size distribution and particle size, the functionality, and additionally, the types of food matrices where such ingredients are to be organized, controlled when combined with a variety of constituents, its compatibility with foods; rheological and textural influences, revenue on investment are basically the utmost important issues. Nanotechnology is a new technique used for encapsulation which shows promising results. It is said that the better the drying technique, the more efficient is the ultimate encapsulation. Although further research is warranted on the behavior of encapsulates while their transit in the human gastrointestinal tract and probable toxicity of nanoparticles, but it can be clearly foreseen that the encapsulation technology will continue to establish itself further as a very efficient mode of delivery of bioactive compounds with improved bioavailability and as a curative measure for various targeted illnesses and will also be able to face the challenges of several deficiency diseases mankind is prone to. It will be safe to say that in the coming days, encapsulated bioactive compounds will show a noteworthy role in enhancing the benefits of consuming functional and fortified foods. As the innovative researches for the stabilization of bioactive compounds as well as the development of new proposals come up, it will be feasible to upgrade health benefits and improve the nutritional value of food components.

KEYWORDS

- **bioactive compounds**
- **controlled release**
- **generally recognized as safe**
- **microencapsulation**
- **nanoencapsulation**
- **wall materials**

REFERENCES

1. Ahn, S. I., Chang, Y. H., & Kwak, H. S., (2010). Optimization of microencapsulation of *Inonotus obliquus* extract powder by response surface methodology and its application into milk. *Food Science of Animal Resources, 30*(4), 661–668.
2. Akbarzadeh, A., Rezaei-Sadabady, R., Davaran, S., Joo, S. W., Zarghami, N., Hanifehpour, Y., Samiei, M., et al., (2013). Liposome: Classification, preparation, and applications. *Nanoscale Research Letters, 8*, Article No. 102.
3. Al-fagih, I. M., Alanazi, F. K., Hutcheon, G. A., & Saleem, I. Y., (2011). Recent advances using supercritical fluid techniques for pulmonary administration of macromolecules via dry powder formulations. *Drug Delivery Letters, 1*(2), 128–134.

Encapsulation Methods in the Food Industry

4. Amaral, P. H. R., Andrade, P. L., & De Conto, L. C., (2019). Microencapsulation and its uses in food science and technology: A review; Chapter 6. In: Salaün, F., (ed.), *Microencapsulation-Processes, Technologies, and Industrial Applications* (pp. 1–18). London–United Kingdom: Intech Open Ltd.

5. Artiga-Artigas, M., Acevedo-Fani, A., & Martín-Belloso, O., (2017). Improving the shelf-life of low-fat cut cheese using nanoemulsion-based edible coatings containing oregano essential oil and mandarin fiber. *Food Control, 76,* 1–12.

6. Bernela, M., Kaur, P., Chopra, M., & Thakur, R., (2014). Synthesis, characterization of nisin loaded alginate-chitosan-pluronic composite nanoparticles and evaluation against microbes. *LWT: Food Science and Technology, 59,* 1093–1099.

7. Briones, A. V., & Sato, T., (2010). Encapsulation of glucose oxidase (GOD) in polyelectrolyte complexes of chitosan–carrageenan. *Reactive and Functional Polymers, 70*(1), 19–27.

8. Brownlie, K., (2007). Marketing perspective in encapsulation technologies in food applications. In: Lakkis, J. M., (ed.), *Encapsulation, and Controlled Release Technologies in Food Systems* (pp. 213–235). Iowa, United States: Blackwell.

9. Butstraen, C., & Salaün, F., (2014). Preparation of microcapsules by complex coacervation of gum arabic and chitosan. *Carbohydrate Polymers, 99,* 608–616.

10. Çam, M., Içyer, N. C., & Erdogan, F., (2014). Pomegranate peel phenolics: Microencapsulation, storage stability and potential ingredient for functional food development. *LWT-Food Science and Technology, 55,* 117–123.

11. Carneiro, H. C. F., Tonon, R. V., Grosso, C. R. F., & Hubinger, M. D., (2013). Encapsulation efficiency and oxidative stability of flaxseed oil microencapsulated by spray drying using different combinations of wall materials. *Journal of Food Engineering, 115,* 443–451.

12. Cerreti, M., Markošová, K., Esti, M., Rosenberg, M., & Rebroš, M., (2017). Immobilization of pectinases into PVA gel for fruit juice application. *International Journal of Food Science & Technology, 52*(2), 531–539.

13. Champagne, C. P., & Fustier, P., (2007). Microencapsulation for the improved delivery of bioactive compounds into foods. *Current Opinion in Biotechnology, 18*(2), 184–190.

14. Char, C., Cisternas, L., Pérez, F., & Guerrero, S., (2016). Effect of emulsification on the antimicrobial activity of carvacrol. *CyTA–Journal of Food, 14*(2), 186–192.

15. Chen, Q., McGillivray, D., Wen, J., Zhong, F., & Quek, S. Y., (2013). Coencapsulation of fish oil with phytosterol esters and limonene by milk proteins. *Journal of Food Engineering, 117,* 505–512.

16. Choi, H. J., Ahn, J., Kim, N. C., & Kwak, H. S., (2006). The effects of microencapsulated chitooligosaccharide on physical and sensory properties of the milk. *Asian-Australasian Journal of Animal Sciences, 19*(9), 1347–1353.

17. Ciobanu, A., Mallard, I., Landy, D., Brabie, G., Nistor, D., & Fourmentin, S., (2012). Inclusion interactions of cyclodextrins and crosslinked cyclodextrin polymers with linalool and camphor in *Lavandula angustifolia* essential oil. *Carbohydrate Polymers, 87,* 1963–1970.

18. Comunian, T. A., Thomazini, M. A., Alves, A. J. G., De Matos, Jr. F. E., De Carvalho, B. J. C., & Favaro-Trindade, C. S., (2013). Microencapsulation of ascorbic acid by complex coacervation: Protection and controlled release. *Food Research International, 52,* 373–379.

19. Cornacchia, L., & Roos, Y. H., (2011). Stability of β-carotene in protein-stabilized oil-in-water delivery systems. *Journal of Agricultural and Food Chemistry, 59*(13), 7013–7020.

20. Corrêa-Filho, L. C., Moldão-Martins, M., & Alves, V. D., (2019). Advances in the application of microcapsules as carriers of functional compounds for food products. *Applied Sciences, 9*, 571.
21. Da Silva, M. P., Daroit, D. J., & Brandelli, A., (2010). Food applications of liposome encapsulated antimicrobial peptides. *Trends in Food Science & Technology, 21*(6), 284–292.
22. Da Silva, M. P., Sant'Anna, V., Barbosa, M. S., Brandelli, A., & Franco, B. D. G. M., (2012). Effect of liposome-encapsulated nisin and bacteriocin-like substance p34 on *listeria monocytogenes* growth in minas frescal Cheese. *International Journal of Food Microbiology, 156*(3), 272–277.
23. Dardelle, G., Normand, V., Steenhoudt, M., Bouquerand, P. E., Chevalie, M., & Baumgartner, P., (2007). Flavor-encapsulation and flavor-release performances of a commercial yeast-based delivery system. *Food Hydrocolloids, 21*, 953–960.
24. Davidov-Pardo, G., Gumus, C. E., & McClements, D. J., (2016). Lutein-enriched emulsion based delivery systems: Influence of pH and temperature on physical and chemical stability. *Food Chemistry, 196*, 821–827.
25. De Roos, K. B., (2003). Effect of texture and microstructure on flavor retention and release. *International Dairy Journal, 13*(8), 593–605.
26. de Vos, P., Faas, M. M., Spasojevic, M., & Sikkema, J., (2010). Review: Encapsulation for preservation of functionality and targeted delivery of bioactive food components. *International Dairy Journal, 20*(4), 292–302.
27. Deladino, L., Anbinder, P. S., Navarro, A. S., & Martino, M. N., (2008). Encapsulation of natural antioxidants extracted from *Ilex paraguariensis*. *Carbohydrate Polymers, 71*, 126–134.
28. Desai, K. G. H., & Park, H. J., (2005). Recent developments in microencapsulation of food ingredients. *Drying Technology, 23*(7), 1361–1394.
29. Diguet, S., Feltes, K., Kleemann, N., Leuenberger, B., & Ulm, J. U. S., (2008). *Patent Application No. 10/592, 860.*
30. Dima, C., Cotârlet, M., Alexe, P., & Dima, S., (2014). Microencapsulation of essential oil of pimento [*Pimenta dioica* (L) Merr.] by chitosan/κ-carrageenan complex coacervation method. *Innovative Food Science & Emerging Technologies, 22*, 203–211.
31. Dong, Z., Maa, Y., Hayat, K., Jia, C., Xia, S., & Zhang, X., (2011). Morphology and release profile of microcapsules encapsulating peppermint oil by complex coacervation. *Journal of Food Engineering, 104*(3), 455–460.
32. Donhowe, E. G., Flores, F. P., Kerr, W. L., Wicker, L., & Kong, F., (2014). Characterization and *in vitro* bioavailability of β-carotene: Effects of microencapsulation method and food matrix. *LWT—Food Science and Technology, 57*, 42–48.
33. Đorđević, V., Balanč, B., Belščak-Cvitanović, A., Lević, S., Trifković, K., Kalušević, A., Kostić, I., et al., (2015). Trends in encapsulation technologies for delivery of food bioactive compounds. *Food Engineering Reviews, 7*, 452–490.
34. European Food Safety Authority, (2009). Scientific opinion on 'the potential risks arising from nanoscience and nanotechnologies on food and feed safety.' Scientific opinion of the scientific committee. *The EFSA Journal, 958*, 1–39.
35. Evens, R., Schamphelaere, K. A. C., Balcaen, L., Wang, Z., Roy, K., Resano, M., Flórez, M., et al., (2012). The use of liposomes to differentiate between the effects of nickel accumulation and altered food quality in *Daphnia magna* exposed to dietary nickel. *Aquatic Toxicology, 109*, 80–89.

Encapsulation Methods in the Food Industry

36. Fadel, H. H. M., Hassan, I. M., Ibraheim, M. T., Abd, E. L., Mageed, M. A., & Saad, R., (2019). Effect of using cinnamon oil encapsulated in maltodextrin as exogenous flavoring on flavor quality and stability of biscuits. *Journal of Food Science and Technology, 56*(10), 4565–4574.

37. Fathi, M., Mozafari, M. R., & Mohebbi, M., (2012). Nanoencapsulation of food ingredients using lipid based delivery systems. *Trends in Food Science & Technology, 23*, 13–27.

38. Fioramonti, S. A., Martinez, M. J., Pilosof, A. M. R., Rubiolo, A. C., & Santiago, L. G., (2014). Multilayer emulsions as a strategy for linseed oil microencapsulation: Effect of pH and alginate concentration. *Food Hydrocolloids, 43*, 8–17.

39. Fuciños, C., Fuciños, P., Fajardo, P., Amado, I. R., Pastrana, L. M., & Rúa, M. L., (2015). Evaluation of antimicrobial effectiveness of pimaricin-loaded thermosensitive nanohydrogels in grape juice. *Food and Bioprocess Technology, 8*, 1583–1592.

40. Gharsallaoui, A., Roudaut, G., Chambin, O., Voilley, A., & Saurel, R., (2007). Applications of spray-drying in microencapsulation of food ingredients: An overview. *Food Research International, 40*(9), 1107–1121.

41. Ghosh, V., Mukherjee, A., & Chandrasekaran, N., (2014). Eugenol-loaded antimicrobial nanoemulsion preserves fruit juice against microbial spoilage. *Colloids and Surfaces B: Biointerfaces, 114*, 392–397.

42. Gonnet, M., Lethuaut, L., & Boury, F., (2010). New trends in encapsulation of liposoluble vitamins. *Journal of Controlled Release, 146*(3), 276–290.

43. Gul, K., Tak, A., Singh, A. K., Singh, P., Yousuf, B., & Wani, A. A., (2015). Chemistry, encapsulation, and health benefits of β-carotene: A review. *Cogent Food and Agriculture, 1*(1), 1018696.

44. Hassellov, M., Kammer, F. V., & Beckett, R., (2007). Characterization of aquatic colloids and macromolecules by field-flow fractionation; Chapter 5. In: Wilkinson, K. J., & Lead, J. R., (eds.), *Environmental Colloids and Particles: Behavior, Structure, and Characterization* (Vol. 10, pp. 223–276). Chichester–England: John Wiley & Sons Ltd.

45. Hernández-Rodríguez, L., Lobato-Calleros, C., Pimentel-González, D. J., & Vernon-Carter, E. J., (2014). *Lactobacillus plantarum* protection by entrapment in whey protein isolate: κ-carrageenan complex coacervates. *Food Hydrocolloids, 36*, 181–188.

46. Higuera-Ciapara, I., Felix-Valenzuela, L., Goycoolea, F. M., & Argüielles-Monal, W., (2002). Microencapsulation of astaxanthin in a chitosan matrix. *Carbohydrate Polymers, 56*(1), 41–45.

47. Hornyak, G. L., Tibbals, H. F., Dutta, J., & Moore, J. J., (2008). *Introduction to Nanoscience and Nanotechnology* (1st edn., p. 1694). Boca Raton: CRC Press.

48. Isailović, B., Kostić, I., Zvonar, A., Đorđvić, V., Gašperlin, M., Nedović, V., & Bugarski, B., (2013). Resveratrol loaded liposomes produced by different techniques. *Innovative Food Science and Emerging Technologies, 19*, 181–189.

49. Jahadi, M., Khosravi-Darani, K., Ehsani, M. R., Mozafari, M. R., Saboury, A. A., Seydahmadia, F., & Vafabakhsh, Z., (2012). Evaluating the effects of process variables on protease-loaded nano-liposome production by Plackett–Burman design for utilizing in cheese ripening acceleration. *Asian Journal of Chemistry, 24*(9), 3891–3894.

50. Jahadi, M., Khosravi-Darani, K., Ehsani, M. R., Mozafari, M. R., Saboury, A. A., & Pourhosseini, P. S., (2013). The encapsulation of flavorzyme in nanoliposome by heating method. *Journal of Food Science and Technology, 52*(4), 2063–2072.

51. Janiszewska-Turak, E., (2017). Carotenoids microencapsulation by spray drying method and supercritical micronization. *Food Research International, 99*(Part 2), 891–901.

52. Jeon, B. J., Kim, N. C., Han, E. M., & Kwak, H. S., (2005). Application of microencapsulated isoflavone into milk. *Archives of Pharmacal Research, 28*(7), 859–865.
53. Jiang, T., Liao, W., & Charcosset, C., (2020). Recent advances in encapsulation of curcumin in nanoemulsions: A review of encapsulation technologies, bioaccessibility, and applications. *Food Research International, 132*, 109035.
54. Jo, Y., Chun, J., Kwon, Y., Min, S., Hong, G., & Choi, M. J., (2015). Physical and antimicrobial properties of trans-cinnamaldehyde nanoemulsions in water melon juice. *LWT–Food Science and Technology, 60*, 444–451.
55. Jun-xia, X., Hai-yan, Z., & Jian, Y., (2011). Microencapsulation of sweet orange oil by complex coacervation with soybean protein isolate/gum Arabic. *Food Chemistry, 125*(4), 1267–1272.
56. Jyothi, S., Seethadevi, A., Prabha, K. S., Muthuprasanna, P., & Pavitra, P., (2012). Microencapsulation: A review. *International Journal of Pharma and Bio Sciences, 3*(1), 509–531.
57. Kailasapathy, K., (2009). Encapsulation technologies for functional foods and nutraceutical product development. *CAB Reviews: Perspectives in Agriculture, Veterinary Science, Nutrition, and Natural Resources, 4*(6), 1–19.
58. Kailasapathy, K., (2006). Survival of free and encapsulated probiotic bacteria and their effect on the sensory properties of yoghurt. *LWT-Food Science and Technology, 39*(10), 1221–1227.
59. Kearney, L., Upton, M., & McLoughlin, A., (1990). Enhancing the viability of *Lactobacillus plantarum* inoculum by immobilizing the cells in calcium-alginate beads. *Applied Environmental Microbiology, 56*(10), 3112–3116.
60. Kerdudo, A., Dingas, A., Fernandez, X., & Faure, C., (2014). Encapsulation of rutin and naringenin in multilamellar vesicles for optimum antioxidant activity. *Food Chemistry, 159*, 12–19.
61. Kfoury, M., Auezova, L., Greige-Gerges, H., Ruellan, S., & Fourmentin, S., (2014). Cyclodextrin, an efficient tool for trans-anethole encapsulation: Chromatographic, spectroscopic, thermal, and structural studies. *Food Chemistry, 164*, 454–461.
62. Kheadr, E. E., Vuillemard, J. C., & El Deeb, S. A., (2000). Accelerated cheddar cheese ripening with encapsulated proteinases. *International Journal of Food Science & Technology, 35*(5), 483–495.
63. Kidd, P., (2011). Astaxanthin, cell membrane nutrient with diverse clinical benefits and anti-aging potential. *Alternative Medicine Review, 16*(4), 355–364.
64. Kim, N. C., Kim, J. B., & Kwak, H. S., (2008). Microencapsulation of Korean mistletoe (*Viscum album* var. Coloratum) extract and its application into milk. *Asian-Australasian Journal of Animal Sciences, 21*(2), 299–306.
65. Krogsgård, N. C., Kjems, J., Mygind, T., Snabe, T., Schwarz, K., Serfert, Y., & Meyer, R. L., (2016). Enhancing the anti-bacterial efficacy of isoeugenol by emulsion encapsulation. *International Journal of Food Microbiology, 229*, 7–14.
66. Ksouda, G., Sellimi, S., Merlier, F., Falcimaigne-Cordin, A., Thomasset, B., Nasri, M., & Hajji, M., (2019). Composition, anti-bacterial, and antioxidant activities of *Pimpinella saxifraga* essential oil and application to cheese preservation as coating additive. *Food Chemistry, 288*, 47–56.
67. Kwak, H. S., (2014). *Nano-and Microencapsulation for Foods* (p. 432). West Sussex, UK: John Wiley & Sons, Ltd.

Encapsulation Methods in the Food Industry

68. Kwak, H. S., Ju, Y. S., Ahn, H. J., Ahn, J., & Lee, S., (2003). Microencapsulated iron fortification and flavor development in cheddar cheese. *Asian-Australasian Journal of Animal Science, 16*(8), 1205–1211.

69. Lee, J. B., Ahn, J., Lee, J., & Kwak, H. S., (2004). L-ascorbic acid microencapsulated with polyacylglycerol monostearate for milk fortification. *Bioscience, Biotechnology, and Biochemistry, 68*(3), 495–500.

70. Lee, S. C., Yuk, H. G., Lee, D. H., Lee, K. E., Hwang, Y., & Ludescher, R. D., (2002). Stabilization of retinol through incorporation into liposomes. *Journal of Biochemistry and Molecular Biology, 35*(4), 358–363.

71. Lee, Y. K., Ganesan, P., & Kwak, H. S., (2013). Properties of milk supplemented with peanut sprout extract microcapsules during storage. *Asian-Australasian Journal of Animal Sciences, 26*(8), 1197–1204.

72. Liang, H., Yuan, Q., Vriesekoop, F., & Lv, F., (2012). Effects of cyclodextrins on the antimicrobial activity of plant-derived essential oil compounds. *Food Chemistry, 135*(3), 1020–1027.

73. Liang, R., Huang, Q., Ma, J., Shoemaker, C. F., & Zhong, F., (2013). Effect of relative humidity on the store stability of spray-dried beta carotene nanoemulsions. *Food Hydrocolloids, 33*(2), 225–233.

74. López-de-Dicastillo, C., Catalá, R., Gavara, R., & Hernández-Muñoz, P., (2011). Food applications of active packaging EVOH films containing cyclodextrins for the preferential scavenging of undesirable compounds. *Journal of Food Engineering, 104*(3), 380–386.

75. López-Nicolás, J. M., Núñez-Delicado, E., Sánchez-Ferrer, Á., & García-Carmona, F., (2007). Kinetic model of apple juice enzymatic browning in the presence of cyclodextrins: The use of maltosyl-β-cyclodextrin as secondary antioxidant. *Food Chemistry, 101*(3), 1164–1171.

76. Loquercio, A., Castell-Perez, E., Gomes, C., & Moreira, R., (2015). Preparation of chitosanalginate nanoparticles for trans-cinnamaldehyde entrapment. *Journal of Food Science, 80*(10), N2305–2315.

77. Luykx, D. M. A. M., Peters, R. J. B., Van, R. S. M., & Bouwmeester, H., (2008). A review of analytical methods for the identification and characterization of nano delivery systems in food. *Journal of Agricultural and Food Chemistry, 56*(18), 8231–8247.

78. Lv, Y., Yang, F., Li, X., Zhang, X., & Abbas, S., (2014). Formation of heat-resistant nanocapsules of jasmine essential oil via gelatin/ gum Arabic based complex coacervation. *Food Hydrocolloids, 35*, 305–314.

79. Lysionek, A. E., Zubillaga, M. B., Salgueiro, M. J., Piñeiro, A., Caro, R. A., Weill, R., & Boccio, J. R., (2002). Bioavailability of microencapsulated ferrous sulfate in powdered milk produced from fortified fluid milk: A prophylactic study in rats. *Nutrition, 18*(3), 279–281.

80. Martín, M. J., Lara-Villoslada, F., Ruiz, M. A., & Morales, M. E., (2015). Microencapsulation of bacteria: A review of different technologies and their impact on the probiotic effects. *Innovative Food Science & Emerging Technology, 27*, 15–25.

81. Maswal, M., & Dar, A. A., (2014). Formulation challenges in encapsulation and delivery of citral for improved food quality. *Food Hydrocolloids, 37*, 182–195.

82. Mavrocordatos, D., Pronk, W., & Boller, M., (2004). Analysis of environmental particles by atomic force microscopy, scanning, and transmission electron microscopy. *Water Science and Technology, 50*(12), 9–18.

83. Mazza, G., (1998). *Functional Foods: Biochemical and Processing Aspects* (1ˢᵗ edn., Vol. 1, p. 480). Boca Raton: CRC Press.

84. McClements, D. J., Li, F., & Xiao, H., (2015). The nutraceutical bioavailability classification scheme: Classifying nutraceuticals according to factors limiting their oral bioavailability. *Annual Review of Food Science and Technology, 6*, 299–327.

85. Meiners, J. A., (2012). Fluid bed microencapsulation and other coating methods for food ingredient and nutraceutical bioactive compounds; Chapter 7. In: Garti, N., & McClements, D. J., (eds.), *Encapsulation Technologies and Delivery Systems for Food Ingredients and Nutraceuticals* (pp. 151–176). Cambridge–UK: Woodhead Publishing Limited.

86. Milanovic, J., Manojlovic, V., Levic, S., Rajic, N., Nedovic, V., & Bugarski, B., (2010). Microencapsulation of flavors in carnauba wax. *Sensors, 10*(1), 901–912.

87. Moschakis, T., & Biliaderis, C. G., (2017). Biopolymer-based coacervates: Structures, functionality, and applications in food products. *Current Opinion in Colloid & Interface Science, 28*, 96–109.

88. Nedovic, V., Kalusevic, A., Manojlovic, V., Levic, S., & Bugarski, B., (2011). An overview of encapsulation technologies for food applications. *Procedia Food Science, 1*, 1806–1815.

89. Nedovic, V., Kalusevic, A., Manojlovic, V., Petrovic, T., & Bugarski, B., (2013). Encapsulation systems in the food industry. In: Yanniotis, S., Taoukis, S., Stoforos, N., & Karathanos, V. T., (eds.), *Advances in Food Process Engineering Research and Applications* (pp. 229–253). New York: Springer Science+Business Media.

90. Nickerson, M. T., Low, N., & Karaca, A. C., (2014). *Microencapsulation Using Legume Proteins. European Patent Office.* Patent no. WO/2014/064591.

91. Nori, M. P., Favaro-Trindade, C. S., De Alencar, S. M., Thomazini, M., De Camargo, B. J. C., & Contreras, C. C. J., (2011). Microencapsulation of propolis extract by complex coacervation. *LWT-Food Science and Technology, 44*(2), 429–435.

92. Okuro, P. K., Thomazini, M., Balieiro, J. C. C., Liberal, R. D. C. O., & Fávaro-Trindade, C. S., (2013). Co-encapsulation of *Lactobacillus acidophilus* with inulin or polydextrose in solid lipid microparticles provides protection and improves stability. *Food Research International, 53*(1), 96–103.

93. Pan, X., Yao, P., & Jiang, M., (2007). Simultaneous nanoparticle formation and encapsulation driven by hydrophobic interaction of casein-graft-dextran and β-carotene. *Journal of Colloid and Interface Science, 315*(2), 456–463.

94. Paramera, E. I., Konteles, S. J., & Karathanos, V. T., (2010). Microencapsulation of curcumin in cells of *Saccharomyces cerevisiae. Food Chemistry, 125*(3), 892–902.

95. Pattnaik, M., & Mishra, H. N., (2020). Effect of microwave treatment on preparation of stable PUFA enriched vegetable oil powder and its influence on quality parameters. *Journal of Food Processing and Preservation, 44*(4), e14374.

96. Paulo, F., & Santos, L., (2017). Design of experiments for microencapsulation applications: A review. *Material Science and Engineering: C, 77*, 1327–1340.

97. Peng, C., Zhao, S. Q., Zhang, J., Huang, G. Y., Chen, L. Y., & Zhao, F. Y., (2014). Chemical composition, antimicrobial property and microencapsulation of mustard (*Sinapis alba*) seed essential oil by complex coacervation. *Food Chemistry, 165*, 560–568.

98. Perrechil, F. A., Sato, A. C. K., & Cunha, R. L., (2011). κ-carrageenan–sodium caseinate microgel production by atomization: Critical analysis of the experimental procedure. *Journal of Food Engineering, 104*(1), 123–133.

Encapsulation Methods in the Food Industry

99. Petrovic, T., Nedovic, V., Dimitrijevic-Brankovic, S., Bugarski, B., & Lacroix, C., (2007). Protection of probiotic microorganisms by microencapsulation. *Chemical Industry and Chemical Engineering Quarterly, 13*(3), 169–174.

100. Prata, A. S., Maudhuit, A., Boillereaux, L., & Poncelet, D., (2012). Development of a control system to anticipate agglomeration in fluidized bed coating. *Powder Technology, 224*, 168–174.

101. Primozic, M., Duchek, A., Nickerson, M., & Ghosh, S., (2018). Formation, stability, and in vitro digestibility of nanoemulsions stabilized by high-pressure homogenized lentil proteins isolate. *Food Hydrocolloids, 77*, 126–141.

102. Qian, C., Decker, E. A., Xiao, H., & McClements, D. J., (2012). Physical and chemical stability of β-carotene-enriched nanoemulsions: Influence of pH, ionic strength, temperature, and emulsifier type. *Food Chemistry, 132*(3), 1221–1229.

103. Rao, J., & McClements, D. J., (2011). Food-grade microemulsions, nanoemulsions, and emulsions: Fabrication from sucrose monopalmitate and lemon oil. *Food Hydrocolloids, 25*(6), 1413–1423.

104. Rashidinejad, A., Birch, E. J., Sun-Waterhouse, D., & Everett, D. W., (2014). Delivery of green tea catechin and epigallocatechin gallate in liposomes incorporated into low-fat hard cheese. *Food Chemistry, 156*, 176–183.

105. Reineccius, G., (2019). Use of proteins for the delivery of flavors and other bioactive compounds. *Food Hydrocoloilds, 86*, 62–69.

106. Ribeiro, H. S., Ax, K., & Schubert, H., (2003). Stability of lycopene emulsions in food systems. *Journal of Food Science, 68*(9), 2730–2734.

107. Rocha, G. A., Fávaro-Trindade, C. S., & Ferreira, G. C. R., (2012). Microencapsulation of lycopene by spray drying: Characterization, stability, and application of microcapsules. *Food and Bioproducts Processing, 90*, 37–42.

108. Rocha-Selmi, G. A., Bozza, F. T., Thomazini, M., Bolini, H. M. A., & Fávaro-Trindade, C. S., (2013). Microencapsulation of aspartame by double emulsion followed by complex coacervation to provide protection and prolong sweetness. *Food Chemistry, 139*(1–4), 72–78.

109. Rocha-Selmi, G. A., Fávaro-Trindade, C. S., & Ferreira, G. C. R., (2013). Morphology, stability, and application of lycopene microcapsules produced by complex coacervation. *Journal of Chemistry, 2013*, 982603.

110. Rocha-Selmi, G. A., Theodoro, A. C., Thomazini, M., Bolini, H. M. A., & Fávaro-Trindade, C. S., (2013). Double emulsion stage prior to complex coacervation process for microencapsulation of sweetener sucralose. *Journal of Food Engineering, 119*(1), 28–32.

111. Royal Society, (2004). *Nanoscience and Nanotechnologies: Opportunities and Uncertainties* (p. 127). Plymouth, UK: Royal Society & the Royal Academy of Engineering.

112. Sadiq, I. H., & Doosh, K. S., (2019). Study of the physochemical, rheological, and sensory properties of yoghurt fortified with microencapsulation iron. *The Iraqi Journal of Agricultural Science, 50*(4), 1345–1355.

113. Shao, P., Zhang, J., Fang, Z., & Sun, P., (2014). Complexing of chlorogenic acid with β-cyclodextrins: Inclusion effects, antioxidative properties and potential application in grape juice. *Food Hydrocolloids, 41*, 132–139.

114. Shishir, M. R. I., Xie, L., Sun, C., Zheng, X., & Chen, W., (2018). Advances in micro and nano-encapsulation of bioactive compounds using biopolymer and lipid-based transporters. *Trends in Food Science & Technology, 78*, 34–60.

115. Shu, B., Yu, W., Zhao, Y., & Liu, X., (2006). Study on microencapsulation of lycopene by spray-drying. *Journal of Food Engineering, 76*(4), 664–669.

116. Sidira, M., Karapetsas, A., Galanis, A., Kanellaki, M., & Kourkoutas, Y., (2014). Effective Survival of immobilized *Lactobacillus casei* during ripening and heat treatment of probiotic dry-fermented sausages and investigation of the microbial dynamics. *Meat Science, 96*(2), 948–955.

117. Silva, D. F., Fávaro-Trindade, C. S., Rocha, G. A., & Thomazini, M., (2012). Microencapsulation of lycopene by gelatin–pectin complex coacervation. *Journal of Food Processing and Preservation, 36*(2), 185–190.

118. Silva, F., Figueiras, A., Gallardo, E., Nerín, C., & Domingues, F. C., (2014). Strategies to improve the solubility and stability of stilbene antioxidants: A comparative study between cyclodextrins and bile acids. *Food Chemistry, 145*, 115–125.

119. Smith, A. M., Jaime-Fonseca, M. R., Grover, L. M., & Bakalis, S., (2010). Alginate-loaded liposomes can protect encapsulated alkaline phosphatase functionality when exposed to gastric pH. *Journal of Agricultural and Food Chemistry, 58*(8), 4719–4724.

120. Souilem, S., Kobayashi, I., Neves, M. A., Sayadi, S., Ichikawa, S., & Nakajima, M., (2014). Preparation of monodisperse food-grade oleuropein-loaded W/O/W emulsions using microchannel emulsification and evaluation of their storage stability. *Food and Bioprocess Technology, 7*, 2014–2027.

121. Srivastava, Y., Semwal, A. D., & Sharma, K., (2013). Application of various chemical and mechanical microencapsulation techniques in food sector: A review. *International Journal of Food and Fermentation Technology, 3*(1), 1–13.

122. Sugumar, S., Singh, S., Mukherjee, A., & Chandrasekaran, N., (2016). Nanoemulsion of orange oil with non ionic surfactant produced emulsion using ultrasonication technique: Evaluating against food spoilage yeast. *Applied Nanoscience, 6*, 113–120.

123. Sun-Waterhouse, D., & Wadhwa, S. S., (2013). Industry-relevant approaches for minimizing the bitterness of bioactive compounds in functional foods: A review. *Food and Bioprocess Technology, 6*, 607–627.

124. Tamjidi, F., Nasirpour, A., & Shahedi, M., (2012). Physicochemical and sensory properties of microencapsulated fish oil. *Food Science and Technology International, 18*(4), 381–390.

125. Tan, C., Xia, S., Xue, J., Xie, J., Feng, B., & Zhang, X., (2013). Liposomes as vehicles for lutein: Preparation, stability, liposomal membrane dynamics, and structure. *Journal of Agricultural and Food Chemistry, 61*(34), 8175–8184.

126. Tao, J., Zhu, Q., Qin, F., Wang, M., Chen, J., & Zheng, Z. P., (2017). Preparation of steppogenin and ascorbic acid, vitamin E, butylated hydroxytoluene oil-in-water microemulsions: Characterization, stability, and antibrowning effects for fresh apple juice. *Food Chemistry, 224*, 11–18.

127. Tiede, K., Boxall, A., Tear, S. P., Lewis, J., David, H., & Hassellov, M., (2008). Detection and characterization of engineered nanoparticles in food and the environment. *Food Additives and Contaminants Part A, 25*(7), 795–821.

128. Truelstrup-Hansen, L., Allan-Wojotas, P. M., Jin, Y. L., & Paulson, A. T., (2002). Survival of Ca-alginate microencapsulated *Bifidobacterium* spp. in milk and simulated gastrointestinal conditions. *Food Microbiology, 19*(1), 35–45.

129. Truong, V. Y. T., Boyer, R. R., McKinney, J. M., O'Keefe, S. F., & Williams, R. C., (2010). Effect of α-cyclodextrin-cinnamic acid inclusion complexes on populations of *Escherichia coli* O157:H7 and *Salmonella enterica* in fruit juices. *Journal of Food Protection, 73*(1), 92–96.

Encapsulation Methods in the Food Industry

130. Velikov, K. P., & Pelan, E., (2008). Colloidal delivery systems for micronutrients and nutraceuticals. *Soft Matter, 4*(10), 1964–1980.

131. Verkempinck, S. H. E., Salvia-Trujillo, L., Moens, L. G., Charleer, L., Van, L. A. M., Hendrickx, M. E., & Grauwet, T., (2018). Emulsion stability during gastrointestinal conditions effects lipid digestion kinetics. *Food Chemistry, 246*, 179–191.

132. Vincekovíc, M., Viskić, M., Jurić, S., Giacometti, J., Bursác, K. D., Putnik, P., Donsì, F., et al., (2017). Innovative technologies for encapsulation of Mediterranean plants extracts. *Trends in Food Science & Technology, 69*, 1–12.

133. Wechtersbach, L., Ulrih, N. P., & Cigic, B., (2012). Liposomal stabilization of ascorbic acid in model systems and in food matrices. *LWT–Food Science and Technology, 45*(1), 43–49.

134. Wilson, N., & Shah, N. P., (2007). Microencapsulation of vitamins. *ASEAN Food Journal, 14*(1), 1–14.

135. Yang, Z., Peng, Z., Li, J., Li, S., Kong, L., Li, P., & Wang, Q., (2014). Development and evaluation of novel flavor microcapsules containing vanilla oil using complex coacervation approach. *Food Chemistry, 145*, 272–277.

136. Yao, M. F., Xiao, H., & McClements, D. J., (2014). Delivery of lipophilic bioactives: Assembly, disassembly, and reassembly of lipid nanoparticles. *Annual Review of Food Science and Technology, 5*, 53–81.

137. Yarlagadda, A. B., Wilkinson, M. G., Ryan, S. P., Doolan, I. A., O'Sullivan, M. G., & Kilcawley, K. N., (2014). Utilization of a cell-free extract of lactic acid bacteria entrapped in yeast to enhance flavor development in cheddar cheese. *International Journal of Dairy Technology, 67*(1), 21–30.

138. Yoshida, P. A., Yokota, D., Foglio, M. A., Rodrigues, R. A. F., & Pinho, S. C., (2010). Liposomes incorporating essential oil of Brazilian cherry (*Eugenia uniflora* L.): Characterization of aqueous dispersions and lyophilized formulations. *Journal of Microencapsulation, 27*(5), 416–425.

139. Yuan, C., Lu, Z., & Jin, Z., (2014). Characterization of an inclusion complex of ethyl benzoate with hydroxypropyl-β-cyclodextrin. *Food Chemistry, 152*, 140–145.

140. Zheng, L., Ding, Z., Zhang, M., & Sun, J., (2011). Microencapsulation of bayberry polyphenols by ethyl cellulose: Preparation and characterization. *Journal of Food Engineering, 104*(1), 89–95.

141. Zou, Y., Lee, H. Y., Seo, Y. C., & Ahn, J., (2012). Enhanced antimicrobial activity of nisin-loaded liposomal nanoparticles against foodborne pathogens. *Journal of Food Science, 77*(3), M165–M170.

142. Zuidam, N. J., & Shimoni, E., (2010). Overview of microencapsulates for use in food products or processes and methods to make them; Chapter 2. In: Zuidam, N. J., & Nedović, V. A., (eds.), *Encapsulation Technologies for Active Food Ingredients and Food Processing* (pp. 3–29). Verlag–New York: Springer.

PART II

MEMBRANE TECHNOLOGY IN FOOD PROCESSING

CHAPTER 6

MEMBRANE PROCESSING IN THE FOOD INDUSTRY

S. ABDULLAH, RAMA C. PRADHAN, and SABYASACHI MISHRA

ABSTRACT

Membrane processing (MP) technologies nowadays play a very crucial role in food processing operations. They are introduced to replace thermal processing technologies since membrane technologies have several advantages over thermal technologies. Currently, MP technologies are extensively applied in various food industry applications such as cold pasteurization, purification, concentration, desalination, de-acidification, clarification, bioactive compound separation, demineralization, etc. The potential use of a membrane technique is determined primarily by various properties such as driving force, pressure, flow characteristics, and size of the membrane pores. This chapter contains an overview of basic MP technologies, their advantages and disadvantages, membrane modules, types of membranes and membrane technologies as well as various applications of MP technologies.

6.1 INTRODUCTION

Major conventional food processing technologies involve thermal treatment. These thermal technologies have several disadvantages, such as toxicity, less recovery efficiency and high energy consumption. It may also affect negatively on the volatile nutritious compounds of the food, such as vitamins, anthocyanin, biologically active proteins, and carotenoids. Furthermore, it may impact on the sensory properties of the foods by degrading its color

---

Advances in Food Process Engineering: Novel Processing, Preservation, and Decontamination of Foods.
Megh R. Goyal, N. Veena, & Ritesh B. Watharkar (Eds.)
© 2023 Apple Academic Press, Inc. Co-published with CRC Press (Taylor & Francis)

and flavor and also by imparting "cooked" taste into it [6]. To overcome these problems, membrane processing (MP) was introduced as an alternate technology to the traditional thermal processing for food processing. Currently, MP emerged as one among the significant branches of food processing, with almost 20–30% of the total worldwide production of membranes today are utilized for food processing applications, and has an annual growth of approximately 7.5% [26].

The word 'membrane' is a derivative of the Latin word 'membrana' meaning 'thin skin.' The membrane is defined as "a selective physical barrier that retains unwanted materials on the surface and allows certain compounds to pass through, depending on their physical and chemical properties, when a driving force is applied across the membrane" [14]. Membranes are generally semi-permeable, thin sheets made using polymeric materials. It is 'semi-permeable' because it will not allow all, but only selected species to pass through it. Membranes help to separate single or multiple compounds (fine particles, colloids, or dissolved solutes) from a gas or liquid solution. The MP is a non-thermal technique without any involvement of chemical agents and alterations of pH, temperature, and phase [12]. Moreover, this technology can work on the continuous process and is more energy-efficient and economical (reduces labor and production costs) as compared to other similar conventional technologies such as precipitation, centrifugation, gravity sedimentation, adsorption, solvent extraction, evaporation, flocculation, and coagulation. Another critical advantage of MP is that it is exceptionally selective against a compound of target with lower energy loss, minimum number of operations, smooth execution, and feasible scaling up [10, 40, 48].

In recent times, MP technologies are considered for multiple food processing applications such as the development of non-alcoholic beverages, recovery, separation, and concentration of bioactive compounds (for example, polysaccharides, antioxidants, carotenoids, anthocyanin, and phenolic compounds), recovery, and extraction of value-added components and aromas from processed and non-processed commodities, and concentration and clarification of foods [10]. Although MP is relatively a novel separation technology, unfamiliar to the ordinary people, various types of MP technologies have already been commercialized at different parts of the world. Also, currently, an immense number of different varieties of membranes are manufactured across the globe with versatile applications in the food processing sector.

In this chapter, membrane process technology (MPT) has been highlighted and discussed in detail for the processing and preservation of food items.

Membrane Processing in the Food Industry   145

6.2 CLASSIFICATION OF MEMBRANES

The primary characteristic of membranes includes solvent permeability, pore diameter, porosity, and thickness. The usual classification of membranes are based on their symmetry, morphology, the material of construction and technological evolution [23].

6.2.1 BASED ON SYMMETRY

Here, the membrane is classified as symmetric and asymmetric. The size of symmetric membrane pores will typically be well defined and uniform, whereas asymmetric membranes have a more substantial, non-uniform pore size [15].

6.2.2 BASED ON MORPHOLOGY

The size, quantity, and pore distribution on the membrane surface is referred to as membrane morphology [31]. Based on morphology, the membranes are classified as heterogeneous (porous) and homogeneous (non-porous). The heterogeneous membranes have large loosely packed pores (predominantly micropores), where the movement of particles through the membrane occurs based on size. It is more like standard filtration, where the separation happens by restricting minimum one, and allowing other compounds to pass through the membrane. On the other hand, the homogeneous membranes consist of densely packed pores of molecular size (10 to 200 mm). The principle of component separation in a homogeneous membrane is the 'solution-diffusion' mechanism. Here, the separation happens by disintegration and diffusion of the compounds along the membrane. These types of membranes are mostly used to separate compounds at the molecular level [18, 39].

6.2.3 BASED ON MATERIAL OF CONSTRUCTION

In this type of classification, membranes are classified as organic and inorganic. Organic membranes are those synthetic membranes made using biological polymers such as polyvinylidene difluoride, polysulfone (PS), polyamide (PA) and cellulose acetate (CA). Whereas, inanimate ingredients such as ceramics, glass, carbon, and metals are used for the manufacturing

146   Advances in Food Process Engineering

of inorganic membranes. Organic membranes are more prevalent in the commercial sector since it is cheaper, easy to fabricate and exists in a wide variety than inorganic membranes. Consequently, inorganic membranes currently exist only for MF and UF applications. However, inorganic membranes have high longer life and mechanical, chemical, and thermal resistance than organic membranes [5, 8].

6.2.4   BASED ON TECHNOLOGICAL EVOLUTION

Membranes are further classified into first, second, third, and fourth-generation membranes based on technological evolution. Membranes belonging to the first generation are elaborated from CA. They are temperature (up to 50°C) and pH (3–8) sensitive membranes which are vulnerable to disinfectants and microorganisms. The membranes derived from synthetic polymers, such as polyolefin or PS and their derivatives, are termed as second-generation membranes. These are presently the most popular membranes since it shows resistance to high temperature, strong bases, strong acids, and hydrolysis. However, they are susceptible to mechanical compacting [6, 31, 33].

The membranes made by alumina oxide or zirconium-based ceramic materials, accumulated over the surface constructed using graphite or similar substances, will come under third-generation membranes. They are costly and chemically inert membranes, which are resistant to high temperature (over 400°C), pressure, pH, and mechanical stresses. Similarly, the membranes derived from carbon fiber come under fourth-generation membranes. These are the most recently developed membranes which help to perform various hybrid processes [21, 23].

6.3   MEMBRANE MODULES

Membrane modules or membrane geometries are the devices in which a synthetic membrane is packed at a commercial level. Essential features of various membrane modules are tabulated in Table 6.1.

6.3.1   HOLLOW-FIBER MODULE (HFM)

The hollow-fiber module (HFM) involves compactly packed bundles of thin tubular membranes. Each bundle is generally made using numerous

Membrane Processing in the Food Industry

single hollow fibers (usually in the range of 50–3,000), thickly sealed on both ends inside a shell tube. This arrangement makes the HFMs resistant to higher pressures. The typical HFM diameter is in the range of 0.2–3 mm and has a length in the range of 18–120 cm. However, the RO fibers are exceptional with a diameter as low as 0.04 mm but have a higher resistance to pressure than other fibers. These modules can operate in both inside-outside and outside-inside modes. That is, in inside-outside, the feed flows along the inner side of the membrane fiber, and the final product comes out through its outer side. Whereas, in outside-inside, the movement of feed is from the outer area of the membrane fiber, and the permeate collects at its inner side.

TABLE 6.1 Different Membrane Modules and its Properties

Parameter	Hollow-Fiber	Spiral-Wound	Tubular	Plate and Frame
Channel spacing (cm)	0.02–0.25	–	1–2.5	0.03–0.25
Comfort to clean	Poor	Moderate	Excellent	Good
Comparative cost	Low	Low	High	High
Density of packing (m^2/m^3)	500–9,000	200–800	30–200	30–500
Fouling resistance	Poor	Moderate	Very good	Good
Hold up volume	Low	Low	High	Low
Major application	UF, GD, RO, DI	MF, UF, GD, RO, DI	UF, RO	MF, UF, PV, RO, DI
Mechanical damage resistance	Good	Good	Poor	Good
Particulate plugging	High	Very high	Low	Moderate

Abbreviations: UF: ultrafiltration; GD: gas diffusion; PV: pervaporation; MF: microfiltration; DI: dialysis; RO: reverse osmosis.

The HFMs consume only the minimum energy and have a thick packing density. However, it shows a fouling tendency, hence makes it difficult to treat high viscous feeds or feeds with high solid content. Also, these modules are very hard to maintain and clean [31].

6.3.2 SPIRAL WOUND MODULES (SWM)

In spiral wound modules (SWM), two flat sheets of the membrane and a mesh-like spacer (typically 0.03–0.1 cm in thick) are wrapped together and surrounded on a porous central tube. One edge of the membrane sheets has an opening near the central tube for collecting the permeate, while the remaining three edges are closed together. The liquid feed flows in an axial direction along the membrane surface, whereas the permeate passes through the center in a radial direction and collects outside the central tube.

The SWM have an average fouling tendency, compact design, more energy efficiency, acceptable packing density and higher surface area-volume ratio. The SWM are capable of treating feeds with large volumes, but, necessary pre-treatment should be done for high viscous feeds and feeds with large particulate matters [3, 6].

6.3.3 TUBULAR MODULES

Tubular modules have a shell-and-tube configuration, where, the tube-shaped membrane is fit inside two cold and hot fluid cylindrical chambers. It typically works in a cross-flow or tangential mode. In this module, the feed pumps along the membrane's center at high velocity, which develops a layer of concentration polarization (CP) along the surface of the membrane promoting to more comfortable cleaning and stable and high flux.

The tubes of the membranes generally have a length in between 0.6 m and 6 m and a large inner diameter between 5 mm and 25 mm, typically fabricated using perforated plastic or fabric support. These modules have a good effective area, low fouling tendency and are best suited to treat high viscous feeds and feed with high solid content [3, 31, 48].

6.3.4 PLATE AND FRAME MODULES (PFM)

The configuration of plate and frame modules (PFM) is similar to that of SWM. In this module, the spacers and the membranes are arranged together in layers (planar configuration), between flat sheets (or plates). Hence it is also known as flat sheet modules. They are more comfortable to maintain and clean but a bit costlier, therefore, has broader application in laboratory scales. The plate and frame membranes can be used to treat high viscous feeds and feed with high solid content. It generally has a length of about 10–60 cm and a channel gap in the range of 0.5–10 mm [3, 31].

Membrane Processing in the Food Industry 149

6.4 CLASSIFICATION OF MPT

The various technologies associated with membranes include membrane filtration, pervaporation (PV), dialysis, electrodialysis (ED), gas separation, forward osmosis (FO), and membrane distillation (MD) [31]. The principle difference between all these technologies is the mass transfer mechanism (driving force) and the kind of membrane involved in the process (Table 6.2).

6.4.1 MEMBRANE FILTRATION

In order of decrease in membrane pore size, membrane filtration technology is subdivided into four different processes namely MF, UF, NF, and RO. The driving force behind all these processes is the hydraulic pressure difference. The fundamental differences between all these filtration technologies are its membrane pore size, the difference in pressure, type of components passes through the membrane, kind of components to be separated and molecular weight cut off (MWCO). Figure 6.1 portrays the rejection of compounds by different membranes in an MPT. The membrane's particular characteristic and the hydraulic pressure gradient leads to the penetration of specific components through the membrane, and retainment of specific other components from passing the membrane [31].

The MF is a technique involving a larger membrane pore size (0.1–10 μm), lower operating pressure (110–300 kPa) and a MWCO higher than 100 kDa. This filtration technique can be used to separate or concentrate larger particles and molecules such as proteins, blood cells, and some colloids. It can also be used as a cold sterilization technique by separating specific yeast and bacterial species from the solution [10]. MF is typically made using polymeric materials such as PAs, chlorinated polyolefins, fluorinated polyolefins, polyesters, and cellulose esters and inanimate constituents such as silica, stainless steel, carbon/carbon composites, zircona/carbon composites, and alumina [23]. The primary commercial application of MF includes clarification of juice, beer, wine, and fermentation broth, microbial load reduction, separation of oil/water emulsion and wastewater treatment. MF also uses as a pre-treatment process before UF/NF/RO for fouling mitigation [15].

The UF is a technique involving a smaller membrane pore size of 1–100 nm, operating pressure of 150–500 kPa and a MWCO in the range of 0.3–1,000 kDa. This filtration technique can be used to eliminate or concentrate particles such as micelles, lipids, colloids, polymers, biomolecules, proteins,

TABLE 6.2 MPT: Classification and Properties

Driving Force	Membrane Process	Type of Membrane	Pore Size	Pressure Range	Molecular Weight Cut-Off	Particles Separated
Difference in hydraulic pressure	Microfiltration (MF)	Microporous	0.1–10 μm	110–300 kPa	>100 kDa	Suspended solids, bacteria, fungi
	Ultrafiltration (UF)	Microporous	1–100 nm	150–500 kPa	0.3–1,000 kDa	All macromolecules, bacteria, fungi, viruses
	Nanofiltration (NF)	Nanoporous	0.5–2 nm	500–1,500 kPa	200–2,000 Da	Sugars, divalent ions
	Reverse osmosis (RO)	Dense, non-porous	<1 nm	2,000–6,500 kPa	<100 Da	Monovalent ions
Difference in concentration and pressure	Gas diffusion (GD)	Polymer, homogeneous	–	–	–	Gas mixtures separation
Difference in concentration	Dialysis (DI)	Microporous	0.01–0.1 μm	–	–	Separating micro molecules and salts from macromolecules
Difference in electrical potential	Electrodialysis (ED)	Cation- and anion-exchange membrane	–	–	–	Separation of salts
Difference in vapor pressure and temperature	Pervaporation (PV)	Polymer, homogeneous	–	–	–	Azeotropic mixtures separation
Difference in vapor pressure	Membrane distillation (MD)	Porous, hydrophobic	–	–	–	Separation of salts

and emulsions. However, along with water, small constituents like salt, sugar, and vitamins will pass along the UF membranes. Like MF, this is also a cold sterilization technique since certain viruses and bacteria can be eliminated using this process [26, 28]. UF membranes are usually made using hydrophilic materials such as PA, polyethersulfone (PES), polycarbonate, cellulose esters, polyether amide, polyether ether ketone and PS [23]. UF has significant application for concentrating milk protein and fruit juices, processing of blood plasma and plant extract, production of antibiotics, eliminating endotoxins, removing oil during wastewater processing. Besides, it can be used as a pre-treatment process before performing the NF/RO process [15].

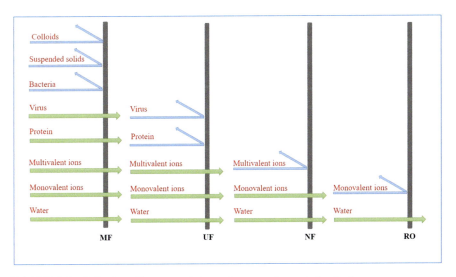

FIGURE 6.1 Rejection of compounds by various membrane technologies.

The NF is a comparatively new technique, intermediate to UF and RO [31]. It involves a 0.5–2 nm pore size, operating pressure of 500–1,500 kPa and a MWCO in the range of 200–2,000 Da. NF removes components with a small molecular weight, for example, amino acids, sugars, and divalent ions, along with all microorganisms. However, it allows the passing of certain monovalent ions (like sodium chloride), along with water, through the membrane [15]. NF can successfully eliminate heavy metals, for example, lead, copper, arsenic, cadmium, nickel, and chromium from the liquid feed. In food industries, NF is used for demineralization of sugars, the concentration of dyes, milk, and juices, and recovery of bioactive compounds from liquid foods [6, 15]. It has higher efficiency, reliability, ease of operation and lower energy consumption than RO [28].

The RO technique, also known as hyperfiltration, consists of a dense membrane with a pore size <1 nm, an operating pressure of 2,000–6,500 kPa and a MWCO of 100 Da [14]. It is used for desalination, purification, concentration, and separation. RO is recognized as the most effective method for the elimination of chemical residues, dissolved salts, and inorganic components from water. It can also combine with other filtration techniques such as UF and MF [6]. RO is commercially used for the desalination of seawater, treatment of wastewater and for eliminating heavy metals and salts. The food industry application of RO includes drinking water purification, whey, and milk preconcentration, alcoholic beverage dealcoholization and juice concentration [23]. However, higher energy consumption, fouling tendency and high pressure are some of the disadvantages of RO process [28].

These technologies can either be performed in cross-flow or dead-end configurations. In the former arrangement, the feed is supplied tangential, and in later it is provided perpendicular across the surface of the membrane. The probability of cake-layer formation on the surface of the membrane is high in dead-end than in cross-flow filtration. Therefore, dead-end filtration can be only performed in a batch process, whereas cross-flow filtration can be implemented as a continuous process. Hence the most popular and the standard configuration among food industries is the cross-flow process [10].

6.4.2 GAS DIFFUSION (GD)

The gas diffusion (GD) process, otherwise known as gas separation process, is best suited for separating or extracting components from a liquid solution or a gaseous phase. Hydraulic pressure difference is its driving force of separating compounds. The separation happens by the relative transport of the various constituents present in the feed. The component which has the quick diffusion characteristic will separate at the membrane's low-pressure region. Component's driving force and membrane discrimination are the main reasons for the degree of separation [31]. The major commercial applications of the GD process include dehydration of compressed air, separation of hydrogen from refinery gases, extraction of carbon dioxide (CO_2) from natural gases and separation of nitrogen and oxygen from the air [21].

Membrane Processing in the Food Industry

6.4.3 DIALYSIS (DI)

DI is the technique of passing the solvents and selected solutes through the membranes. The extent of component separation is dependent on the variations in the component's diffusion rate, which occurs due to the size difference of the molecules. Its driving force for the separation of components is the difference in concentration. The membranes used for the DI process are typically porous made using PS, PES or cellulosic UF and NF membrane materials [18]. The most popular example for DI is the hemodialysis, where compounds have higher molecular weight, such as blood cells and proteins got retained, and compounds having lower molecular weight, such as urea pass through the membrane [21]. In the food industry, it is used to remove alcohol from beer and microsolutes and salts from liquid samples [31].

6.4.4 ELECTRODIALYSIS (ED)

Maigrot and Sabates conducted the first reported research on ED in 1890. ED is a MPT which uses the difference in electrical potential as a driving force for separating the particles [22]. The ion-exchange membrane is the principal element of an ED technique [26]. The properties of this principle element are determined using their concentration, charge, and the chemical nature of the ions fixed in the polymer matrix. The different classification of these membranes are anion-exchange membrane (AEM), cation exchange membrane (CEM) and bi-polar membranes (BPM). The AEM is negatively charged that restricts cations and allows anions to permeate through the membrane. Similarly, the CEM is positively charged that restricts anions and permits cations to pass through. However, the BPM has both the charges, and they limit ions of positive and negative charges from moving through it [17]. The ED has different advantages such as higher chemical and thermal stability, higher mechanical strength, low electrical resistance, and high selectivity. Major food industry applications of the ED process includes reducing acidity from fruit juices, demineralization of milk and cheese whey products and deashing of molasses and sugars [6].

6.4.5 PERVAPORATION (PV)

Pervaporation (PV) is an extremely selective MPT with the driving force for separation is vapor pressure difference. The term 'pervaporation' since the permeate components initially gets evaporated and then permeates through the membrane [18]. It is reported to be more efficient in separating heat-sensitive compounds, isomers, and azeotropic and close-boiling mixtures [22]. In this technique, the GD process and the evaporation process combine to form a single unit for selectively separating azeotropic mixtures of multiple components. That is, when the feed mixture (in liquid phase) come close to the membrane, the volatile analytes (permeates) evaporates and diffuses through it and gets condenses and collected at the membrane's opposite side. The membranes used for PV process can be made using a non-porous inorganic (zeolite/ceramic) or a polymeric material [10]. In the food industry, PV technique can be used for removing alcohol from wine, recovering aroma from flowers, herbal extracts, beer, and fruit juices, and for restoring aroma compounds while fermentation [6].

6.4.6 FORWARD OSMOSIS (FO)

Forward osmosis (FO) is an emerging MPT, consists of a hydrophilic dense membrane with semi-permeable nature. Its driving force is osmotic pressure difference. The FO membranes are made using a thick active layer along with a loose supporting layer. It separates two liquid solutions (draw solution and feed) with an osmotic pressure gradient. The water transfers to the region of high concentration (draw solution region) from a region of low concentration (feed region) through a selectively permeable membrane until it reach a zero osmotic pressure gradient on both the region. The non-thermal nature of the FO process helps to retain the highly volatile components of the solution. Other major advantages of the FO process are easy scale-up, the capability to process high solid content products, lower tendency to fouling and low hydraulic pressure processing [6, 18, 31].

6.4.7 MEMBRANE DISTILLATION (MD)

Membrane distillation (MD) is a temperature-driven MPT where hydrophilic membranes with micropores are used to separate two aqueous solutions with varying temperature. An interface of liquid and vapor will be formed at either side of the membrane. It restricts the entry of liquid solutions but allows the movement of water vapor. Temperature gradient or the vapor pressure difference prevailing among the surfaces of the porous membrane

Membrane Processing in the Food Industry

are the driving force of the MD technique. It is called 'membrane distillation' since this process is similar to the combination of membrane separation and conventional distillation. The significant advantages of MD include low pressure and low-temperature operations, meager mechanical requirements, and cost. However, high energy-intensive, module design, flux decay due to temperature polarization and low permeate flux are some of the disadvantages which prevent the commercial use of this technology [3, 6, 31, 33].

6.5 FACTORS AFFECTING MPT

The performance of different MPT, primarily the pressure-driven MPT (such as MF, UF, and NF) is influenced by the following factors [10, 11]:

1. **Properties of the Membrane:** Different membrane features such as charge, size of pores, hydrophilicity/hydrophobicity, and surface topography can significantly impact the membrane fouling and solute-membrane interactions. Membrane made using polymeric hydrophobic materials, such as PS, polypiperazineamide, sulfonated polyether-sulfone and PA, are the most popular membranes for these processes.

2. **Operating Parameters:** Membrane selectivity, productivity, and fouling can be greatly impacted by various operational parameters such as concentration of the feed, cross-flow velocity (CFV), trans-membrane pressure (TMP) and temperature. For example, the raise in TMP can increase the rejection rate of certain components and can raise the membrane's permeate flux. Similarly, in a porous membrane, a rise in temperature can lead to viscosity reduction of the feed, which consequently decreases the flow resistance and increase turbulence. It also raises the diffusivity of the feed and solute-transfer rate through the membrane.

3. **Characteristics of the Feed Stream:** Feed's physicochemical characteristics is a crucial factor that can strongly impact membrane fouling, the main reason behind the drop of permeate flux. Fouling is the phenomenon of decline in flux that occurs because of the particle (solutes) depositions on the membrane surface, which can ultimately lead to membrane blockage. Hence, various properties of the feed, such as zeta potential, hydrophobic interactions, charge, conformation, etc., are vital to control membrane fouling.

6.6 MEMBRANE FOULING AND CLEANING METHODS

6.6.1 MEMBRANE FOULING

Fouling is one of the main concerns during MP since it can adversely affect membrane functioning [49]. In a membrane structure, there are different varieties of fouling, such as bio-fouling or biological fouling, inorganic fouling, organic fouling, and particulate or colloidal fouling. Bio-fouling is associated with the formation of biofilm on the membrane surface due to the activity of organisms such as fungi, bacteria, or algae. These biofilms produce biopolymers such as amino acid, protein, and polysaccharides, which causes fouling. These kinds of fouling are also known as microbial fouling. Organic fouling is formed when organic matters are accumulated at the pores. The adsorption or decomposition of magnesium carbonates or calcium salts, or other inorganic precipitates on the membrane pores is termed as inorganic fouling or scaling. Colloidal or particulate fouling is formed due to the decomposition of colloids or inert organic compounds on the membrane [33].

Fouling can be reduced or controlled by various methods such as pre-treating the feed with enzymes, centrifugation, ultrasound (US), and membrane heating and membrane cleaning [16, 29, 50].

6.6.2 MEMBRANE CLEANING

Membranes should be cleaned periodically to remove the blocks and pollutions caused by the decomposition of impurities and separation components on the membrane surface. An efficient and periodic cleaning not only reduces membrane fouling but also can extend the life of the membrane [48]. Panigrahi et al. [35]; and Cassano et al. [9] also used the same procedure with little modifications for membrane cleaning. Many researchers have reported the procedure of membrane cleaning using backward flushing coupled with chemical cleaning technique. The procedure followed by Ghosh et al. [20] is explained below:

- Initially, the membrane was rinsed using tap water at minimum TMP and a maximum flow rate for 30 min;
- Subsequently, 0.1% NaOH solution was heated to 50°C and flushed through the membrane for 30 min at the same TMP and flow rate (alkali wash);

Membrane Processing in the Food Industry   157

- Next, tap water was again rinsed to obtain a neutral pH;
- Then 0.1% nitric acid (HNO_3) solution was heated to 50°C and flushed through the membrane for 30 min at the same TMP and flow rate (acid wash);
- Finally, tap water was again rinsed until neutral pH.

The two most common methods of membrane cleaning methods are forward flushing and backward flush.

1. **Forward Flushing:** During forward flushing, permeate or feed water is flushed more rapidly into the membrane at high hydraulic pressure. This rapid flow results in high turbulence, and due to this, the impurities deposited in the membrane surface are discharged and released. However, the contaminants that blocked the membrane pores cannot be removed by forward flushing [23].

2. **Backward Flushing:** This is the most popular membrane cleaning method [42]. During backward flushing, the feed water or permeate is flushed in the direction opposite to the actual flow at high pressure. The flux during backward flushing will be twice as that of the flux during regular filtering. For restoring the flux and to improve the cleaning efficiency, a chemical cleaning technique (acid-alkali wash) is usually followed along with backward flushing. Citric, oxalic, and hydrochloric acids are the generally used acid solutions, with a pH in the range of 2–3, for the effective removal of the inorganic materials. Similarly, sodium hydroxide solution with a pH range of 10–12, is the commonly used alkali solution for chemical washing of membranes [23].

6.7 APPLICATIONS OF MPT

6.7.1 CLARIFICATION OF JUICES

Fruit juices extracted using conventional methods are generally turbid since these contain colloidal macromolecules (such as polyphenols, soluble starch, protein, and pectin) and other insoluble plant residues (such as insoluble starch, fat, hemicellulose, and cellulose). The presence of these dispersed particles could adversely affect the quality, sensory properties, and consumer acceptability of the juice. Clarification is the technique for eliminating those particles entirely/partially from the fruit juices by retaining the nutritional characteristics of the juice [38].

Conventional clarification processes such as flocculation, coagulation, enzyme addition, centrifugation, heating, sulfur dioxide addition, and alkalization cannot eliminate components such as high molecular weight proteins, non-sugar colloids, ash, gum, microorganisms, oxidative enzymes, and colorants. Moreover, these are slow, laborious techniques and has the risk of contamination [19, 35]. MPTs, being a non-thermal, easily scalable technology can be a successful alternative for these conventional clarification techniques since it gives a microbial free final product with high sensory properties in the absence of any foreign additives [16]. The application of MPT on clarification of various juices is summarized in Table 6.3.

Sagu et al. [43] conducted a comparative study on the efficiency of MF and centrifugation on clarifying banana juice and reported that clarified juice prepared using MF showed better quality with maximum clarity and minimum viscosity, acid insoluble ash and energy consumption. Biswas et al. [7] also reported that after working on clarified bottle gourd juice using MF and centrifugation, MF is the better clarification process for enhancing clarity and for reducing macromolecules and turbidity. Another similar study was conducted on jamun juice, using three hollow fiber MF membrane (with pore sizes 0.45, 0.1, and 0.2 μm) at different TMP (137.89, 103.45 and 68.9 kPa).

The membrane filtered juice was further compared with centrifuged juice and reported that MF with 0.45 μm pore size and 137.89 kPa TMP showed better performance than centrifugation and other membranes with minimum turbidity and maximum increase in clarity [20].

Pre-treatments, operating parameters, types of membrane treatments and membrane modules are the key factors that affect the efficient membrane functioning. MF and UF are the main membrane treatments used for juice clarification. Pre-treatments are carried out for reducing the presence of macromolecular compounds from the juice, which will prevent the accommodation of these macromolecules on the membrane surface. The most popular pre-treatment for juice clarification is enzymatic treatment. However, other pre-treatments such as centrifugation, US treatment, pulsed electric field treatment and addition of bentonite, gelatin, and filter aids are also performed widely [6].

Domingues et al. [16] carried out a research on the influence of different pre-treatments, such as chitosan coagulation, enzymatic liquefaction, and centrifugation, on the passion fruit juice clarification using MF. The MF was carried out using a hollow fiber membrane and reported that the most effective pre-treatment was chitosan coagulation since it resulted in the highest

TABLE 6.3 Examples for Clarification of Various Juices Using Membrane Technology

Filtered Product	Pre-Treatment	Membrane Treatment	Membrane Module	Operating Parameters	Salient Results	References
Apple juice	Gelatin and bentonite	UF	Hollow fiber	• Pore size – 10 kDa and 100 kDa	• Combination of UF along with gelatin and bentonite was found effective • 10 kDa membrane showed better performance in turbidity reduction and color improvement	[34]
Banana juice	Enzyme (pectinase) treatment	MF	Hollow fiber	• Pore size – 0.2 μm • TMP – 35 to 104 kPa • CFV – 10 to 20 L/h	• MF was obtained as the most suitable clarification method • Optimized condition: 76 kPa TMP and 20 L/h CFV	[43]
Jamun juice	Enzyme treatment, centrifugation, and membrane filtration	UF	Hollow fiber	• MF – 20 psi pressure and 0.45 μm pore size • UF – TMP (10, 15 and 20 psi) and pore size (10, 50 and 100 kDa)	• UF with pore size of 50 kDa and pressure of 20 psi gave the finest outcome • Observed maximum clarity and least color change even after two months • Juice treated with enzyme followed by centrifugation shows spoilage after 15 days	[19]
Jamun juice	Pectinase	MF	Hollow fiber	• Pore size – 0.45, 0.2, and 0.1 μm • TMP – 137.89, 103.45 and 68.9 kPa	• Optimized condition: pore size (0.45 μm) and TMP (137.89 kPa) • Flux decline after 130 min • Minimum turbidity of 8.56 NTU (nephelometric turbidity units) and highest clarity of 93.5 % Transmittance (at 660 nm) were obtained in microfiltered juice	[20]

TABLE 6.3 (Continued)

Filtered Product	Pre-Treatment	Membrane Treatment	Membrane Module	Operating Parameters	Salient Results	References
Passion fruit juice	Enzymatic liquefaction, chitosan coagulation and centrifugation	MF	Hollow fiber	• Filtration area – 0.056 m² • Membrane pore diameter – 0.40 µm	• Highest permeate flux and maximum turbidity and the color reduction were obtained in chitosan addition • Maximum viscosity reduction was obtained in enzyme-treated juice • Reduced color and turbidity of juice by MF	[16]
Passion fruit juice	Centrifugation and enzymatic treatment	MF	Tubular ceramic, hollow fiber	• TMP – 0.5 and 1 bar • Pore size – 0.3 µm	• The ceramic tubular membrane at 0.5 bar TMP and 0.3 µm was found most efficient	[13]
Sugarcane juice	Centrifugation	MF	Hollow fiber	• TMP (35, 69, 104, 138 kPa) • CFV (0.123, 0.246, 0.369 m/s) • Permeate flux (5.41–6.23 L/m².h)	• Optimized condition: 104 kPa TMP and 0.369 m/s CFV with flux 6.04 L/m².h • Turbidity and total solids reduced by 98% and 26%, respectively • Clarity improved 3 times	[35]

Membrane Processing in the Food Industry 161

permeate flux and maximum turbidity and color reduction. However, enzymatic treatment and centrifugation also reduced the viscosity and turbidity of the juice.

Moreover, a turbid free passion fruit juice was obtained after the MF process. A similar research was carried out by Ghosh et al. [19] on jamun juice using hollow fiber UF membrane with enzyme treatment, centrifugation, and MF as pre-treatments. This study showed that the enzyme-treated and centrifuged sample without UF got spoiled after 15 days. The UF juice filtered at a 20 psi TMP and a pore size of 50 kDa gave the best result with minimum turbidity of 0.03 NTU and maximum TSS of 16.06°Brix. The UF juice also showed a shelf-life of over 60 days. All these results demonstrate the superiority of MPT as a better technology for the clarification of juices.

6.7.2 CONCENTRATION OF FRUIT JUICE

Concentration is an essential unit operation in juice processing aimed to ensure a lesser packaging, storage, and transportation cost and longer shelf-life (Table 6.4). Concentrated juice has less water activity, which makes it resistant and more stable to chemical and microbial spoilage than the normal juice [34]. Traditionally, juice concentration is carried out using evaporation techniques which involve high energy consumption and high temperature resulting in quality depletion due to the formation of off-flavor, loss of volatile beneficial components and degradation of color. However, concentration using MPT is a non-thermal, low-cost technology which doesn't use any chemical agents and phase change [9].

The concentration process is highly influenced by the operating parameters, pre-treatments, membrane treatment and membrane modules used. The application of MPT on the concentration of various juices is summarized in Table 6.4. Currently, RO, NF, MD, FO, and OD are the most popular MPT employed for concentrating fruit juices. To make the concentration process easier without fouling, clarification of juice as a pre-treatment before concentration is essential. Clarification is usually performed using gelatin, bentonite, MF or UF. Quist-Jensen et al. [41] assessed the capacity of direct contact MD for concentrating ultrafiltered orange juice. During this study, the performance of the MD process with regards to total polyphenol content, antioxidant activity, TSS, and total suspended solids were estimated. The results observed that combining UF and MD could be used as a successful technology for obtaining concentrated juice having rich antioxidant, nutritional, and organoleptic properties.

TABLE 6.4 Examples of Concentration of Various Juices Using Membrane Technology

Filtered Product	Pre-Treatment	Membrane Treatment	Operating Parameters	Salient Results	References
Apple juice	UF	OD MD	• Recycle flow rate – 30 L/h	• Evaporation resulted in the 5-hydroxymethylfurfural formation • OD and MD preserved original product quality • Combination of OD and MD was found most efficient	[34]
Apple juice	Evaporation	MD	• Feed temperature – 30 to 50°C • Draw solution – $K_4P_2O_7$ • Flow rate of draw solution–2–4 mL/s • Temperature of the draw solution – 14°C • Concentration of draw solution – 545, 592, and 630 g/L	• Vitamin C and total phenols deteriorated at the higher temperature • Optimum parameters obtained – feed temperature (30°C) and draw solution flow rate (3 mL/s), concentration (630 g/L), temperature (14°C) • Achieved juice concentration – 35°Brix	[25]
Blackcurrant juice	Depectinization, centrifugation	RO	• Tubular membrane • TMP – 60 bar • Time – 120 min	• TSS of 28.6°Brix was obtained after concentration • 60 bar TMP and 25°C temperature was found to be the best operating condition	[37]
Blood orange juice	Pectinase treatment and UF	OD	• Stripping solution – 11.2 mol/L $CaCl_2$ • Feed flow rate – 478 mL/min • Stripping solution flow rate – 573 mL/min	• TSS content of 61.4°Brix was achieved • No significant differences in phenolic compounds after clarification • Other antioxidant compounds were well preserved	[14]

TABLE 6.4 *(Continued)*

Filtered Product	Pre-Treatment	Membrane Treatment	Operating Parameters	Salient Results	References
Jaboticaba juice	Vacuum filtration	FO	• Draw solution – NaCl • Feed flow rate – 50 mL/min • Draw solution flow rate – 200 mL/min • Draw solution concentration – 6–2 mol/L • Temperature – 25°C	• The concentration of NaCl significantly influenced salt and water transmembrane fluxes • Solution's flow rate only influenced the water permeate flux • Sensory and nutritious characteristics were well conserved after processing	[44]
Orange juice	UF	MD	• Polypropylene hollow fiber membrane modules • Feed flow rate – 200 L/h • Water flow rate – 100 L/h	• All antioxidant compounds were well preserved • The final product was rich in sensory and organoleptic properties	[41]
Orange juice	Nil	RO	• TMP – 20, 40 and 60 bar • Concentration factors – 2.3, 3.8 and 5.8 • Flow rate – 650 L/h	• TSS content were 16, 28 and 36°Brix respectively for three TMP • Ascorbic acid content were 53.9, 82.7 and 101.1 mg/100 g respectively for three TMP • Maximum ascorbic acid was obtained at highest TMP • RO juice had the best-preserved flavor	[24]
Pomegranate juice	UF	OD	• TMP – 0.4 bar • Flow rate – 500 mL/min	• Antioxidant activity, anthocyanins, and polyphenol contents were preserved • Achieved a 3.2 fold TSS concentration • All main physic-chemical parameters were preserved	[9]

TABLE 6.4 *(Continued)*

Filtered Product	Pre-Treatment	Membrane Treatment	Operating Parameters	Salient Results	References
Red grape juice	Nil	OD	• Temperature – 35 ± 1°C • Concentrations of feed – 5, 10 and 20°Brix • Concentration of stripping solution – 50% $CaCl_2$	• Antioxidant activity and phenolic index were well preserved • The concentration of final product, the concentration of initial juice and the time of OD were proportional to each other.	[27]

Another study on MD evaluated the effect of flow rate and concentration of draw solution and temperature of feed on permeate flux and nutritional characteristics of concentrated apple juice also proved that efficiency of MD for concentrating fruit juices [25]. In this study, feed temperature of 30°C and draw solution temperature, concentration, and flow rate of 14°C, 630 g/L and 3 mL/s, respectively were obtained as optimum parameters to achieve a concentrated apple juice of 35°Brix TSS content. However, the concentration using MD has not yet been commercialized successfully, and hence most of the studies are in laboratory scales only [33].

Like MD, OD is also widely used for concentrating juices. A comparative study of apple juice concentration using different techniques such as thermal evaporation, MD, and OD reported that the combination of MD and OD preserves all the nutritive and sensory property of the apple juice and thus can be a promising alternative to conventional evaporative concentration [34]. Cassano et al. [9] performed a study on the concentration of ultrafiltered pomegranate juice using a hollow fiber OD unit and achieved a concentrated final product with 3.2 fold TSS content and well preserved total phenols, anthocyanins, and antioxidant activity. Similar results were obtained while concentrating red grape [27] and blood orange juices [14] using OD membrane. In both the studies, the antioxidant activity and polyphenol content were unaffected. Also, the concentration of the final concentrated juice, concentration of the initial feed and the time taken for OD were proportional to each other.

RO is another widely accepted pressure-driven concentration technique, which generally concentrates a feed up to about 25–30°Brix [33]. It is a process where water is completely removed from the feed. The major advantages of using RO technology for the concentration process is its ease of operation, compact installation, low energy consumption, and the preservation of flavor, aroma, and nutritional compounds due to its low-temperature action [24]. Pap et al. [37] studied the capability of the RO membrane to concentrate blackcurrant juice pretreated with pectinase enzyme followed by centrifugation. A tubular RO membrane with more than 80% salt rejection is used for the study. The result showed that the RO membrane with 25C operating temperature and 60 bar TMP could be effectively used for the concentration of blackcurrant juice.

Souza et al. [47] tried to understand the synergistic effect of RO and osmotic evaporation on the concentration of juice from Camu-Camu with primary focus on the antioxidant activity, polyphenols, and ascorbic acid of the finished product. A concentration of 285 g/Kg soluble solids and 530 g/Kg soluble solids, respectively, were achieved after RO and osmotic evaporation. This study revealed after examining the quality parameters that the combination of RO and osmotic evaporation can be successfully utilized as a replacement for thermal processes to achieve a seven times concentration level.

Menchik et al. [32] on concentrating Greek-style yogurt by the combination of RO and FO reported that this combined non-thermal process could be effectively used for concentrating all temperature-sensitive beverages and foods. In this study, the TSS of the juice had increased from 6.6°Brix to more than 19°Brix after RO, which was further increased to more than 40°Brix after FO.

FO is a critical technology for concentration, which works on the osmotic pressure difference. Unlike other MPTs, FO requires very low electrical energy since it operates at low hydraulic pressure. The first reported application of FO technology for fruit juice concentration was conducted in 1966, where a flat sheet and tubular membrane developed using polymeric CA is used for concentrating grape juice up to 60°Brix TSS with the help of saturated NaCl as draw solution [45].

Recently, a study undergone by Sant'Anna et al. [44] to find the effect of FO and its transmembranal flux of salt and water on the concentration of jaboticaba juice. A complete factorial design was followed with a feed flow rate and NaCl concentration as independent variables. Also, the quality attributes of the FO concentrated juice such as TSS, pH, total polyphenols, monomeric anthocyanins, total anthocyanin, antioxidant activity, color intensity, non-enzymatic browning and sodium content were evaluated and compared with thermally concentrated juice. The result revealed that the NaCl concentration has a significant impact on salt and water transmembrane flux. However, the juice's flow rate only influenced the permeate flux of water. Furthermore, the sensory and nutritional results displayed that FO can be successfully used as an alternative to conventional thermal concentration technologies.

6.7.3 RECOVERY OF BIOACTIVE COMPOUNDS

Fresh foods, especially vegetables and fruits, contain numerous biologically active compounds such as anthocyanin, phenols, vitamins E, C, and A and many other components. The demand for food rich in bioactive compounds is increasing due to the growing health concern among the consumers. The consumption of these bioactive compounds can help to prevent diseases related to heart, cancer, decrease cholesterol levels and neurodegenerative

Membrane Processing in the Food Industry

disorders [27]. Conventional thermal processing for bioactive recovery from food can lead to the reduction of volatile components, color degradation and antioxidant activity reduction. In this context, MPT emerged as an alternative technology for bioactive compound recovery since it is non-thermal, more energy-efficient, ease of operation, selective, and minimizes nutritional and sensory losses [4]. Various studies on bioactive compound extraction using MPT are summarized in Table 6.5. Arend et al. [4] found out that NF can be effectively used to concentrate and recover polyphenols, anthocyanin, and other antioxidant compounds from strawberry juice without adversely affecting its color.

Similarly, Abd-Razak et al. [1] investigated the efficiency of UF membrane to separate phytosterol, a bioactive compound with antioxidant, anti-carcinogenic, and cholesterol-lowering properties, from orange juice. Rejection towards sugar, protein, antioxidant activity and phytosterol were analyzed for the membrane with MWCO of 10 kDa under a TMP range of 0.5–2 bar and CFV range of 0.5–1.5 m/s. Excellent efficiency for phytosterol separation (32%) with a yield of 40 mg/L were observed in the study. In another study, Conidi et al. [12] separated phenolic compounds and anthocyanins from pomegranate juice using combinations of UF, NF, and diafiltration (DF) and obtained a yield of 90.7% and 84.8% for anthocyanins and polyphenols respectively. Similarly, Machado et al. [30] successfully extracted carotenoid and polyphenol from the aqueous and alcoholic extract of pequi using NF technology. All these studies indicate that MPT can be effectively used to extract bioactive compounds.

6.7.4 COLD STERILIZATION OF LIQUID FOODS

Another significant application of MPT is the elimination of microorganisms using MF and UF membranes. Since these technologies do require heat, it is known as cold sterilization or cold pasteurization. Kotsanopoulos et al. [26] reported that many food industries, especially dairy industries, are already successfully utilizing MF technologies for removing microorganisms from their final product. De Oliveira et al. [13] studied the impact of MF on the microbial rejection while clarifying passion fruit juice and reported that the tubular ceramic MF membrane eliminated mesophilic bacteria from the group. Panigrahi et al. [35] also reported a similar result while studying on the MF of sugarcane juice. They found that the MF minimized 4 log scale yeast and mold, and 5 log scale total viable plate count. Another study on UF of sugarcane juice revealed that 6 log scale decline of total plate count were

TABLE 6.5   Examples for Recovery of Bioactive Compounds Using Membrane Technology

Filtered Product	Membrane Treatment	Bioactive Compounds Recovered	Salient Results	References
Strawberry juice	NF-MF	Polyphenols and anthocyanin	• Total phenolic content increased by 1.2 fold • Anthocyanin content increased by 1.5 fold • Antioxidant activity increased by 51%	[4]
Orange juice	UF	Phytosterols	• Good separation efficiency of 23% • Phytosterol yield – 40 mg/L	[1]
Pomegranate juice	UF, NF, and DF	Phenolic compounds and anthocyanins	• Final polyphenol yield – 84.8% • Final anthocyanin yield – 90.7%	[12]
Pequi	NF	Polyphenols and carotenoids	• For alcoholic extract – polyphenols (15%) and carotenoids (10%) • For aqueous extract – polyphenols (100%) and carotenoids (97%)	[30]
Thymus capitatus	NF	Carnosic acid and rosmarinic acid	• Successfully extracted carnosic and rosmarinic acids • Synthetic NF membrane performed better than synthetic UF and commercial NF membranes.	[2]

Membrane Processing in the Food Industry

observed and no yeast and mold were detected after UF [36]. These results show that MPT, especially MF and UF, can be successfully used as a cold sterilization technology for liquid foods.

6.7.5 DESALINATION

MPT can be used for the desalination process, i.e., to produce food with minimum salinity, potable water and to manufacture salt from seawater. MD is used primarily for this purpose. Similarly, MPT can be used for the desalination of effluents produced during the processing of seafood and fish. These effluents can contaminate the environment if disposed of untreated. However, it has a significant amount of salts and other high valued aroma compounds which can be recovered by ED followed by RO treatment [26].

6.7.6 RECOVERY OF AROMATIC COMPOUNDS

Adsorption, flash distillation and solvent extraction are the most commonly used conventional methods for recovery of aroma. Aroma compounds are susceptible to heat. Hence thermal technologies cannot be used to recover these compounds since a high temperature can deteriorate its quality. Furthermore, these technologies have very minimum yield and extraction efficiency. MPT, being a highly selective, non-thermal technology can be a possible replacement for the recovery of aroma. PV is the most common membrane technology used for aroma recovery since it lacks the use of chemical compounds during the extraction, which helps to minimize the contamination risk and preserve the functional properties, aroma, and flavor compounds. It has been already found successful in recovering aroma compounds from different fruits such as orange, grape, banana, passion fruit, pineapple, strawberry, bergamot, kiwifruit, and apple [6].

6.7.7 OTHER APPLICATIONS AND RECENT ADVANCES IN MPT

It is reported that MPT can be used successfully in sugar concentration process along with evaporation. Sievers et al. [46] used membrane testing units containing NF and RO membranes along with evaporation to concentrate biomass-derived sugar and found out that membrane concentration followed by evaporation can be successfully used with less energy consumption, more product recovery and quality. Onsekizoglu et al. [34] modified the surface characteristics of thin-film PA composite RO membrane with low-pressure nitrogen plasma. They observed that the hydrophobic membrane changed its nature to hydrophilic; in addition, a significant rise in water flux was also seen during the RO process. This modification helped to promote higher soluble solid contents and saved 30% of total processing time during the concentration process.

Recently, researchers have explored the use of MPT in various other applications, such as demineralization of dairy products and vinasse, production of bases, acids, and alcohols, separation of lactic acid, and dealcoholization. However, all these technologies are still in an emerging stage and not reached a position for commercialization.

6.8 ECONOMIC IMPORTANCE AND ENVIRONMENTAL IMPACTS OF MPT

Its substitution coefficient defines the energy savings of a MPT, i.e., the ratio among the saved thermal primary energy of the MPT compared to the traditional technique and its consumed electrical energy compared to the conventional method. The substitution coefficient is defined as follows:

$$\text{Substitution coefficient} = \frac{T1 - T2}{E2 - E1} \tag{1}$$

where; T1 is the primary thermal energy consumed by the conventional technique, in Mcal or MJ; T2 is the primary thermal energy consumed by the novel technique, in Mcal or MJ; E1 is the electrical energy consumed by the conventional technique, in kWh; E2 is the electrical energy consumed by the novel technique, in kWh.

Several researchers have studied the economic feasibility of different MPT and reported that the substitution coefficient values of this technology are very high [6]. Moreover, studies on other economic factors such as maintenance cost, membrane cost, initial investment and pre and post-operational cost also show that MPT has a better financial performance over conventional separation technologies [23].

Likewise, the impact of MPT on the environment is relatively less. This is because this process doesn't have involvement of heat, and also it doesn't use or produce any toxic/harmful chemicals during the process [18]. Thus, MPT can be a successful replacement for the existing technologies with

Membrane Processing in the Food Industry

better energy efficiency, less exploitation of natural resources, less capital cost, and less environmental impact.

6.9 SUMMARY

In this chapter, the current state of MPT has been reviewed in detail with an emphasis on its types and application. MPTs have a critical advantage over similar conventional technologies since it plays a crucial part in the sustainable growth of the industry by optimizing and minimizing the utilization of raw materials, minimizing environmental impact, reducing capital cost, and saving energy. As a result, the number of applications of MPTs is increasing in every sector of the food industry, i.e., in the dairy industry, beverage (alcoholic and non-alcoholic) industries, sugar industry, salt manufacturing industries (desalination), wastewater treatment and drinking water processing industries. The significant disadvantages of this technology that concerns the industries are its high equipment cost and its susceptibility to pore blocking. The later can be effectively solved by applying certain pretreatments such as enzyme treatment, centrifugation, ultrasonication, etc., before membrane filtration.

In broad, MF, and UF technologies are mainly used to separate suspended solids, colloids, and macromolecules from liquid feeds, and to eliminate or reduce its microbial count. NF, RO, MD can be used to concentrate liquid solution, separate bioactive compounds and other high value-added components from feed and PV can be used for aroma recovery. Also, PV, and MD also can produce non-alcoholic beverages. However, these two technology-based studies are mostly lab-scale experiments and studies on the industrial application is still scarce. Hence there is a lot of scope for future research.

Currently, the worldwide demand for MPT is increasing, since it has emerged as a promising alternative to traditional food processing operations as it has the capability to achieve superior quality final product with less energy, cost, and chemical requirement. It provides a variety of options for optimizing and designing innovative productions. The combination of membrane technologies is found more effective in reducing operational cost, environmental impact and energy consumption in food and agro-industries, separation of aqueous and gaseous mixtures, and biotechnical processes. Hence, the efficiency of MPTs can be significantly improved if new researchers can emphasis more on integrating different membrane operations.

KEYWORDS

- electrodialysis
- forward osmosis
- membrane processing
- microfiltration
- nanofiltration
- reverse osmosis
- ultrafiltration

REFERENCES

1. Abd-Razak, N. H., Chew, Y. M. J., & Bird, M. R., (2019). Membrane fouling during the fractionation of phytosterols isolated from orange juice. *Food and Bioproducts Processing, 113*, 10–21.
2. Achour, S., Khelifi, E., Attia, Y., Ferjani, E., & Hellal, A. N., (2012). Concentration of antioxidant polyphenols from *Thymus capitatus* extracts by membrane process technology. *Journal of Food Science, 77*(6), C703–C709.
3. Alkhudhiri, A., Darwish, N., & Hilal, N., (2012). Membrane distillation: A comprehensive review. *Desalination, 287*, 2–18.
4. Arend, G. D., Adorno, W. T., Rezzadori, K., Di Luccio, M., Chaves, V. C., Reginatto, F. H., & Petrus, J. C. C., (2017). Concentration of phenolic compounds from strawberry (*Fragaria* X *ananassa Duch*) juice by nanofiltration membrane. *Journal of Food Engineering, 201*, 36–41.
5. Aryanti, P. T. P., Subroto, E., Mangindaan, D., Widiasa, I. N., & Wenten, I. G., (2020). Semi-industrial high-temperature ceramic membrane clarification during starch hydrolysis. *Journal of Food Engineering, 274*, 109844.
6. Bhattacharjee, C., Saxena, V. K., & Dutta, S., (2017). Fruit juice processing using membrane technology: A review. *Innovative Food Science & Emerging Technologies, 43*, 136–153.
7. Biswas, P. P., Mondal, M., & De, S., (2016). Comparison between centrifugation and microfiltration as primary clarification of bottle gourd (*Lagenaria siceraria*) juice. *Journal of Food Processing and Preservation, 40*(2), 226–238.
8. Caro, J., (2016). Hierarchy in inorganic membranes. *Chemical Society Reviews, 45*(12), 3468–3478.
9. Cassano, A., Conidi, C., & Drioli, E., (2011). Clarification and concentration of pomegranate juice (*Punica granatum* L.) using membrane processes. *Journal of Food Engineering, 107*(3, 4), 366–373.
10. Castro-Muñoz, R., Boczkaj, G., Gontarek, E., Cassano, A., & Fíla, V., (2020). Membrane technologies assisting plant-based and agro-food by-products processing: A comprehensive review. *Trends in Food Science & Technology, 95*, 219–232.

Membrane Processing in the Food Industry

11. Castro-Muñoz, R., Conidi, C., & Cassano, A., (2019). Membrane-based technologies for meeting the recovery of biologically active compounds from foods and their by-products. *Critical Reviews in Food Science and Nutrition, 59*(18), 2927–2948.

12. Conidi, C., Cassano, A., Caiazzo, F., & Drioli, E., (2017). Separation and purification of phenolic compounds from pomegranate juice by ultrafiltration and nanofiltration membranes. *Journal of Food Engineering, 195*, 1–13.

13. De Oliveira, R. C., Docê, R. C., & De Barros, S. T. D., (2012). Clarification of passion fruit juice by microfiltration: Analyses of operating parameters, study of membrane fouling and juice quality. *Journal of Food Engineering, 111*(2), 432–439.

14. Destani, F., Cassano, A., Fazio, A., Vincken, J. P., & Gabriele, B., (2013). Recovery and concentration of phenolic compounds in blood orange juice by membrane operations. *Journal of Food Engineering, 117*(3), 263–271.

15. Díez, B., & Rosal, R., (2020). A critical review of membrane modification techniques for fouling and biofouling control in pressure-driven membrane processes. *Nanotechnology for Environmental Engineering, 5*(2), 1–21.

16. Domingues, R. C. C., Ramos, A. A., Cardoso, V. L., & Reis, M. H. M., (2014). Microfiltration of passion fruit juice using hollow fiber membranes and evaluation of fouling mechanisms. *Journal of Food Engineering, 121*, 73–79.

17. El Rayess, Y., & Mietton-Peuchot, M., (2016). Membrane technologies in wine industry: An overview. *Critical Reviews in Food Science and Nutrition, 56*(12), 2005–2020.

18. Field, R. W., Bekassy-Molnar, E., Lipnizki, F., & Vatai, G., (2017). *Engineering Aspects of Membrane Separation and Application in Food Processing* (1st edn., p. 390). Boca Raton: CRC Press.

19. Ghosh, P., Garg, S., Mohanty, I., Sahoo, D., & Pradhan, R. C., (2019). Comparison and storage study of ultra-filtered clarified jamun (*Syzygium cumini*) juice. *Journal of Food Science and Technology, 56*(4), 1877–1889.

20. Ghosh, P., Pradhan, R. C., & Mishra, S., (2018). Clarification of jamun juice by centrifugation and microfiltration: Analysis of quality parameters, operating conditions, and resistance. *Journal of Food Process Engineering, 41*(1), e12603.

21. Girard, B., & Fukumoto, L. R., (2000). Membrane processing of fruit juices and beverages: A review. *Critical Reviews in Biotechnology, 20*(2), 109–175.

22. Hu, K., & Dickson, J., (2015). *Membrane Processing for Dairy Ingredient Separation* (p. 296). Chicago: John Wiley & Sons.

23. Ilame, S. A., & Satyavir, V. S., (2015). Application of membrane separation in fruit and vegetable juice processing: A review. *Critical Reviews in Food Science and Nutrition, 55*(7), 964–987.

24. Jesus, D. F., Leite, M. F., Silva, L. F. M., Modesta, R. D., Matta, V. M., & Cabral, L. M. C., (2007). Orange (*Citrus sinensis*) juice concentration by reverse osmosis. *Journal of Food Engineering, 81*(2), 287–291.

25. Julian, H., Yaohanny, F., Devina, A., Purwadi, R., & Wenten, I. G., (2020). Apple juice concentration using submerged direct contact membrane distillation (SDCMD). *Journal of Food Engineering, 272*, 109807.

26. Kotsanopoulos, K. V., & Arvanitoyannis, I. S., (2015). Membrane processing technology in the food industry: Food processing, wastewater treatment, and effects on physical, microbiological, organoleptic, and nutritional properties of foods. *Critical Reviews in Food Science and Nutrition, 55*(9), 1147–1175.

27. Kujawski, W., Sobolewska, A., Jarzynka, K., Güell, C., Ferrando, M., & Warczok, J., (2013). Application of osmotic membrane distillation process in red grape juice concentration. *Journal of Food Engineering, 116*(4), 801–808.
28. Le, N. L., & Nunes, S. P., (2016). Materials and membrane technologies for water and energy sustainability. *Sustainable Materials and Technologies, 7*, 1–28.
29. Lohaus, T., Beck, J., Harhues, T., De Wit, P., Benes, N. E., & Wessling, M., (2020). Direct membrane heating for temperature induced fouling prevention. *Journal of Membrane Science, 612*, 118431.
30. Machado, M. T., Mello, B. C., & Hubinger, M. D., (2013). Study of alcoholic and aqueous extraction of pequi (*Caryocar brasiliense* Camb.) natural antioxidants and extracts concentration by nanofiltration. *Journal of Food Engineering, 117*(4), 450–457.
31. Martín, J., Díaz-Montaña, E. J., & Asuero, A. G., (2018). Recovery of anthocyanins using membrane technologies: A review. *Critical Reviews in Analytical Chemistry, 48*(3), 143–175.
32. Menchik, P., & Moraru, C. I., (2019). Nonthermal concentration of liquid foods by a combination of reverse osmosis and forward osmosis. Acid whey: A case study. *Journal of Food Engineering, 253*, 40–48.
33. Onsekizoglu, B. P., (2015). Potential of membrane distillation for production of high quality fruit juice concentrate. *Critical Reviews in Food Science and Nutrition, 55*(8), 1098–1113.
34. Onsekizoglu, P., Bahceci, K. S., & Acar, M. J., (2010). Clarification and the concentration of apple juice using membrane processes: A comparative quality assessment. *Journal of Membrane Science, 352*(1, 2), 160–165.
35. Panigrahi, C., Karmakar, S., Mondal, M., Mishra, H. N., & De, S., (2018). Modeling of permeate flux decline and permeation of sucrose during microfiltration of sugarcane juice using a hollow-fiber membrane module. *Innovative Food Science & Emerging Technologies, 49*, 92–105.
36. Panigrahi, C., Mondal, M., Karmakar, S., Mishra, H. N., & De, S., (2020). Shelf-life extension of sugarcane juice by cross-flow hollow fiber ultrafiltration. *Journal of Food Engineering, 274*, 109880.
37. Pap, N., Kertész, S., Pongrácz, E., Myllykoski, L., Keiski, R. L., Vatai, G., & Hodúr, C., (2009). Concentration of blackcurrant juice by reverse osmosis. *Desalination, 241*(1–3), 256–264.
38. PovrenoviÄ, D., VukosavljeviÄ, P., & StevanoviÄ, S., (2019). Recent developments in microfiltration and ultrafiltration of fruit juices. *Food and Bioproducts Processing, 106*, 147–161.
39. Pronk, W., Ding, A., Morgenroth, E., Derlon, N., Desmond, P., Burkhardt, M., & Fane, A. G., (2019). Gravity-driven membrane filtration for water and wastewater treatment: A review. *Water Research, 149*, 553–565.
40. Qin, G., Lü, X., Wei, W., Li, J., Cui, R., & Hu, S., (2015). Microfiltration of kiwifruit juice and fouling mechanism using fly-ash-based ceramic membranes. *Food and Bioproducts Processing, 96*, 278–284.
41. Quist-Jensen, C. A., Macedonio, F., Conidi, C., Cassano, A., Aljlil, S., Alharbi, O. A., & Drioli, E., (2016). Direct contact membrane distillation for the concentration of clarified orange juice. *Journal of Food Engineering, 187*, 37–43.
42. Rai, P., & De, S., (2009). Clarification of pectin-containing juice using ultrafiltration. *Current Science*, 1361–1371.

Membrane Processing in the Food Industry

43. Sagu, S. T., Karmakar, S., Nso, E. J., & De, S., (2014). Primary clarification of banana juice extract by centrifugation and microfiltration. *Separation Science and Technology, 49*(8), 1156–1169.

44. Sant'Anna, V., Gurak, P. D., Vargas, N. S. D., Da Silva, M. K., Marczak, L. D. F., & Tessaro, I. C., (2016). Jaboticaba (*Myrciaria jaboticaba*) juice concentration by forward osmosis. *Separation Science and Technology, 51*(10), 1708–1715.

45. Sant'Anna, V., Marczak, L. D. F., & Tessaro, I. C., (2012). Membrane concentration of liquid foods by forward osmosis: Process and quality view. *Journal of Food Engineering, 111*(3), 483–489.

46. Sievers, D. A., Stickel, J. J., Grundl, N. J., & Tao, L., (2017). Technical performance and economic evaluation of evaporative and membrane-based concentration for biomass-derived sugars. *Industrial & Engineering Chemistry Research, 56*(40), 11584–11592.

47. Souza, A. L., Pagani, M. M., Dornier, M., Gomes, F. S., Tonon, R. V., & Cabral, L. M., (2013). Concentration of camu–camu juice by the coupling of reverse osmosis and osmotic evaporation processes. *Journal of Food Engineering, 119*(1), 7–12.

48. Urošević, T., Povrenović, D., Vukosavljević, P., Urošević, I., & Stevanović, S., (2017). Recent developments in microfiltration and ultrafiltration of fruit juices. *Food and Bioproducts Processing, 106*, 147–161.

49. Wang, J., Cahyadi, A., Wu, B., Pee, W., Fane, A. G., & Chew, J. W., (2020). The roles of particles in enhancing membrane filtration: A review. *Journal of Membrane Science, 595*, 117570.

50. Yousefnezhad, B., Mirsaeedghazi, H., & Arabhosseini, A., (2017). Pretreatment of pomegranate and red beet juices by centrifugation before membrane clarification: A comparative study. *Journal of Food Processing and Preservation, 41*(2), e12765.

CHAPTER 7

POTENTIAL OF MEMBRANE TECHNOLOGY IN FOOD PROCESSING SYSTEMS

OLVIYA S. GONSALVES, RAHUL S. ZAMBARE, and
PARAG R. NEMADE

ABSTRACT

Membrane processes are used for efficient separation or fractionation of high-added-value compounds, recovery or concentration of nutraceuticals, clarification, de-alcoholization, dehydration, deacidification, biotransformation, and treatment of industrial food processing waste. Membrane technology gives flexibility in operation, and a membrane can be developed or modified as per the requirements of the feed and separation. Different membrane-based technologies such as microfiltration (MF), ultrafiltration (UF), nanofiltration (NF), forward osmosis (FO), membrane distillation (MD), electrodialysis (ED), pervaporation (PV), and membrane bioreactors (MBR) are analyzed. The applications of membrane processes in dairy processing, food, and starch processing, beverage processing, sugar manufacturing, oil processing, and other food product processing are discussed. Further, the role of membrane technology in the treatment of industrial food processing waste is also discussed. Advantages of membrane technologies such as ease of operation, desired purity achievement, mild processing conditions for heat-sensitive food compounds, reduced energy consumption, and other developments over conventional food processing methods are discussed. Also, the integration of membrane-based processes shows more benefits in enhancing product quality, process capacity, selectivity, and energy consumption. Fouling of

---

Advances in Food Process Engineering: Novel Processing, Preservation, and Decontamination of Foods.
Megh R. Goyal, N. Veena, & Ritesh B. Watharkar (Eds.)
© 2023 Apple Academic Press, Inc. Co-published with CRC Press (Taylor & Francis)

membrane is a major concern in pressure-driven membrane processes used in food processing. Membrane fouling in bio-based streams is observed to be a challenge due to its high fouling tendency and complex composition.

7.1 INTRODUCTION

Food industries are growing at a rate of about 5% every year [77], due to the increasing global population. In food industries, various methods have been used to process food products to achieve the desired quality and production. Among these methods, membrane-based processes demonstrated rapid growth due to its selectivity, ease of usage, and cost-effectiveness. Membrane technology has a wide scope in food technology. Membrane technology is a mild and gentle technology that keeps the organoleptic properties and characteristics of the product intact, which are affected by conventional technologies. Membrane technology reduces the number of unit operations involved in a process. It combines operations like microbiological sterilization, clarification, and stabilization. There are no external chemical agents like fining agents, stabilization agents involve in membrane technology, and therefore it does not affect the food taste and its characteristics. It also deals with the environmental hazard factor, as there is no hazardous waste like diatomaceous earth, generated during the process. The product treated by membrane technology is observed to have better quality than the conventional method. The ease of use and low energy consumption makes membrane technology a vital alternative for the conventional process used.

Different types of membrane-based processes are used in food industries, such as dairy, fruit juices, food and beverages, and sugar. Food and beverage industry is the second-largest industrial market of membranes, the largest being water and wastewater treatment. Membranes can be used for clarification, separation, or concentration of various products in food industries based on their application. Microfiltration (MF), ultrafiltration (UF), nanofiltration (NF), and reverse osmosis (RO) are the key pressure-driven membrane technologies used in the food industries. Along with these techniques, forward osmosis (FO), membrane distillation (MD), osmotic distillation (OD), electrodialysis (ED), and pervaporation (PV) are also employed in various food processing systems.

Types of membrane based on their material are divided into ceramic membranes and polymeric membranes. Ceramic membranes consist of porous support of α-alumina or silicon carbide covered with a porous inorganic membrane layer of aluminum oxide (Al_2O_3), titanium oxide (TiO_2) or zirconium

Potential of Membrane Technology

oxide (ZrO_3). They have strong mechanical stability, high temperature, chemical resistance, and a long and reliable life. These characteristics of ceramic membrane make their use in various food industries such as sugar, juices, wine, oil, and starch processing attractive. However, high prices, fragility to shock, and high weight are the biggest disadvantages of ceramic membranes. On the other hand, polymer membranes are relatively cheap, made from the polymerization of organic compounds. They are mostly made from thermoplastic polymers and are available in a wide range of pore sizes and molecular weight cut-off (MWCO). Polysulfone (PS), polytetrafluoroethylene (PTFE), polyvinylidene fluoride (PVDF), polyethersulfone (PES), nylon-based membranes (Figure 7.1) are widely used in food processing industries due to their compatibility, good mechanical, and thermal stability.

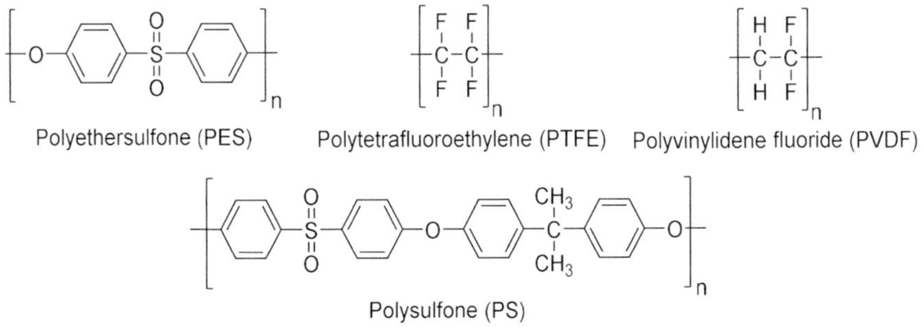

FIGURE 7.1 Molecular structure of polymeric membranes.

Perspective on current applications and future usage of the membrane processes in the food industry is discussed in this chapter.

7.2 MEMBRANE PROCESSES

Membrane separation processes have broad applications in food processing industries (Table 7.1). They are subdivided by the origin of the driving force, phases of feed and permeate, and pore size as well as MWCO.

7.2.1 MICROFILTRATION (MF)

Microfiltration (MF) is a pressure-driven process employed to separate the suspended solids in the range of 0.1–10 μm (MWCO > 200 kDa) under an

180   Advances in Food Process Engineering

applied pressure gradient of 0.5–3 bar across the membrane. In the food industries, MF is primarily used for clarification and removal of spores and bacteria.

TABLE 7.1 Membrane Processes with its Application in Food Industry

Membrane Technology	Food Application	Processes	Process Replaced
MF/UF	Dairy	Removal of bacteria and spores	High temperature treatment/ pasteurization
	Dairy	Fractionation of milk protein (cheese making)	Adjustment of processing parameters
	Dairy	Concentration of milk (cheese making)	Adjustment of processing parameters; thermal evaporation
	Dairy	Fractionation of whey protein	Spray drying; waste/animal feed
	Juices	Clarification of fruit juices	Addition of fining agents/ conventional filtration
	Sugar	Clarification of sugar juices	External chemical agents/ flocculation/sludge separation
	Wine	Clarification of wine	Addition of fining agents/ conventional filtration
	Beer	Recovery of beer/surplus yeast from tank bottom	Settling tanks/drum filters/yeast press
	Beer	Clarification of beer	Kieselguhr filtration
NF/RO/OD/FO	Dairy	Concentration of milk	Thermal evaporation
	Dairy	Isolation of lactose from whey, demineralization of whey	Thermal evaporation; spray drying; waste/animal feed
	Juices	Concentration of fruit juices	Thermal evaporation
	Sugar	Concentration of sugar juice	Thermal evaporation
	Wine	Dealcoholization of wine	Distillation/vacuum distillation
	Beer	Dealcoholization of beer	Distillation/vacuum distillation
Electrodialysis	Dairy	Demineralization of whey	Waste/animal feed/ion-exchange
	Sugar	Demineralization of sugar juice	Ion exchange
	Wine	Wine tartaric stabilization	Cold stabilization
Pervaporation	Juices	Recovery of aroma	Solvent extraction/steam distillation
	Beverages	Recovery of aroma	Solvent extraction/steam distillation

Potential of Membrane Technology 181

7.2.2 ULTRAFILTRATION (UF)

Ultrafiltration (UF) is a pressure-driven process, which is used for the separation of macromolecules like protein, sugar, dextrans, and polysaccharides. This process separates molecules in the range of 0.001–0.1 µm and MWCO ranges 1–200 kDa. A hydraulic pressure range of about 2–10 bar is employed across the membrane to overcome the resistance of the membrane.

7.2.3 NANOFILTRATION (NF)

Nanofiltration (NF) is a pressure-driven process used to retain glucose, low molecular weight organic molecules in the range of 0.1–1 kDa, and multivalent ions. It works in the hydraulic pressure range of 5–20 bar. It is employed for the demineralization of salts and in the concentration process.

7.2.4 REVERSE OSMOSIS (RO)

Reverse osmosis (RO) is a pressure-driven process used for the separation of salts and organic molecules. It works at a high hydraulic pressure range of 20–80 bar. It requires high pressure to overcome the concentration gradient across the membrane. It is primarily used for the removal of monovalent ions, concentration, and purification of various products in the food industries.

7.2.5 ELECTRODIALYSIS (ED)

Electric potential gradient acts as the driving force for ED. An electric field applied across ion exchange membrane, selectively permeable to ions is used for separation of ions from aqueous solution. ED is used for demineralization in the food industries.

7.2.6 PERVAPORATION (PV)

Pervaporation (PV) is a selective membrane technique based on the sorption diffusion mechanism. A non-porous permselective membrane separates the liquid feed mixture by partial vaporization by lowering the partial pressure of the solute at the permeate side than the corresponding equilibrium partial pressure in the feed. PV is usually applied for the recovery of aroma in food industries.

7.2.7 MEMBRANE DISTILLATION (MD)

Membrane distillation (MD) is a non-isothermal membrane process; it includes transportation of water vapors through hydrophobic microporous membranes, as vapor pressure acts as a driving force, which is provided by the difference in the temperature or concentration of solute across the membrane. MD is employed for the concentration of fruit juices in the food industries.

7.2.8 OSMOTIC DISTILLATION (OD)

Osmotic distillation (OD), also known as osmotic evaporation, is a non-pressure driven and isothermal process. The process involves a hydrophobic membrane separated by the feed solution and a hypertonic solution, which is usually brine. The driving force for the process is the difference in concentration across the membrane, which facilitates the mass transfer. OD is used in the food industries for the concentration of juices.

7.2.9 FORWARD OSMOSIS (FO)

Forward osmosis (FO) is an osmotically driven membrane separation process; the water molecule from a lower solute concentration moves towards higher solute concentration through a semipermeable membrane. It is an energy efficient process and relies on natural osmotic phenomena, predominantly for dewatering applications.

The membranes as mentioned above processes are widely used in food processing as a new or alternative technique to conventional techniques. Sometimes the techniques are also hyphenated to create hybrid processes such as ED followed by ultrafiltration (EDUF). The efficiency and lifespan of these membrane processes used in the food industries depend upon the type of solute, the concentration of feed solution, membrane material, and surface properties of the membrane, which control the fouling of the membrane.

Potential of Membrane Technology

7.3 TERMINOLOGY USED

7.3.1 FLUX

The permeate flux evaluates the productivity of a membrane. Permeate flux is the rate of mass transport across the membrane per unit area per unit time ($L\,m^{-2}\,h^{-1}$).

$$J = \frac{\Delta Vp}{A\,\Delta t} \tag{1}$$

where; J is the permeate flux ($Lm^{-2}h^{-1}$ or kg/m²h); ΔV_p is the total volume of the permeate (L); A is the effective membrane area (m²); and Δt is the filtration time (h).

7.3.2 PERMEANCE

It is the characteristic parameter to quantify the permeance of the solvent through the membrane. It is the ratio of permeate flux J to the transmembrane pressure (TMP) across the membrane ΔP.

$$L_p = \frac{J}{\Delta P} \tag{2}$$

where; L_p is the permeance of the membrane ($L\,m^{-2}\,h^{-1}\,bar^{-1}$); J is the flux of the membrane; and ΔP is the TMP.

7.3.3 RETENTION FACTOR

$$R = 1 - \frac{C_{permeate}}{C_{feed}} \tag{3}$$

where; R is the retention factor; $C_{permeate}$ is the concentration of the permeate solution; and C_{feed} is the concentration of the feed solution.

The selectivity of the membrane is described on the basis of its retention factor (R).

7.3.4 MOLECULAR WEIGHT CUT-OFF (MWCO)

Membranes retaining ability is commonly described through the MWCO. It is the molecular weight of a solute that corresponds to 90% rejection by a given membrane [44]. It does not consider the operating parameters and the feed interaction with the membrane affecting its effective pore size.

7.3.5 TRANSMEMBRANE PRESSURE (TMP)

The pressure difference across the membrane from feed to the permeate side is the driving force for the pressure-driven membrane process. In a dead-end filtration, the TMP is the difference in the pressure of the feed solution P_f and the pressure on the permeate side of the membrane P_p.

$$TMP = P_f - P_p \qquad (4)$$

In a tangential/crossflow filtration, the feed flows parallel to the membrane. In the feed side, the pressure decreases in the direction of the flow, i.e., the outlet pressure P_r is lower than the inlet pressure P_f, whereas, on the permeate side, the pressure is constant P_p. The TMP for tangential flow filtration is:

$$TMP = \left(\frac{P_f + P_r}{2}\right) - P_p \qquad (5)$$

7.3.6 CROSSFLOW VELOCITY (CFV)

Crossflow velocity (CFV) (*v*) is the linear velocity of the flow tangential to the membrane surface (Figure 7.2). CFV is expressed in m/s and has an effect on the fouling rate and concentration polarization (CP) at the membrane surface.

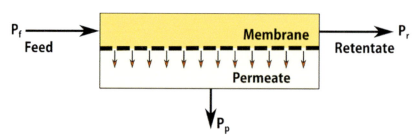

FIGURE 7.2 Schematic representation of tangential flow filtration.

7.3.7 FOULING OF MEMBRANES

Membrane technology has a wide application in the food industry, owing to its ease in operation and low energy consumption. It is majorly used in the

Potential of Membrane Technology

dairy industry, sugar industry, fruit juice industry and beverage industry. MF and UF are used for the clarification process of various products in different industries. NF and RO act as an efficient alternative to conventional methods for the concentration process. The new emerging technologies like ED, OD, and MD also found their application in the food industry. Membrane fouling is a major drawback associated with membrane technology. It affects the performance and the selectivity of the membrane, hurting the membrane processes.

Membrane fouling affects the performance of the membrane, such as selectivity and permeates flux (Figure 7.3(a)). Fouling is defined as the deposition or adsorption of solute/particles on the membrane surface or in the pores of the membrane. Many factors are responsible for the fouling of a membrane, such as concentration/composition of the feed, membrane material and module, membrane porosity, and pore size distribution.

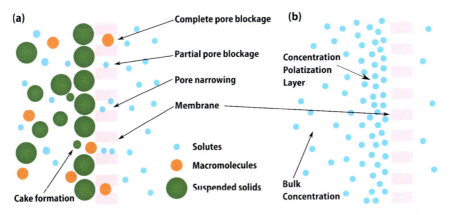

FIGURE 7.3 Schematic representation of: (a) fouling; and (b) concentration polarization.

Membrane process operating parameters like TMP, flow velocity, pH, and temperature affect the permeability and membrane fouling. Fouling is associated with different mechanisms such as pore blocking–partial, complete, or internal, chemical interaction, and cake formation. In case, when the particle size of the feed is smaller than the pore size, it diffuses through the membrane and is absorbed in the membrane pore. This deposition reduces the pore diameter of the membrane, prevents the passage of the solute, is known as internal pore blocking, and affects the performance of the membrane. In the matter of feed having particle size comparable with the pore size, the pores of the membrane are affected, resulting in the membrane fouling. When the particle size of the feed is larger than the pore size of the membrane, it leads to cake formation on membrane due to particle interaction and binding to each other, reducing its active area, affecting the flux and its performance.

Fouling can be a result of a mechanism or integration of two or more mechanisms collectively affecting the performance and permeability of the membrane. Fouling control is an important issue that needs to be addressed to maintain the acceptable flux of the process. Critical flux is an important concept for understanding the fouling control of the membrane. If the membrane separation process is operated below critical flux, no fouling is observed. Critical flux depends on hydrodynamic conditions, pore size, and composition of the feed. Reducing the concentration of rejected solute near the membrane surface by local mixing with the help of a mechanical or magnetic stirrer, or turbulence assist in reducing the adsorption of the solute on the membrane. It also improves the mass transfer across the membrane.

Backflushing or pulsing is another approach towards fouling control. Increasing the affinity of the membrane material to the solvent, in most cases water, decreases the tendency for fouling. Commonly employed strategies to minimize fouling by organic solutes are to increase hydrophilicity of the membrane or its surface by physical or chemical modifications. Backflushing removes the cake formed on the feed side by reversing the flow through the membrane. It displaces the foulant on the membrane and regenerates the flux. It is mainly used when the feed has high solid content leading to cake formation. Membrane performance that can be recovered after employment of fouling control strategies such as backflushing, flushing, etc., is due to control of reversible fouling. The irreversible fouling causes permanent damage to the membrane and cannot be recovered. If the membrane performance is severely degraded, chemical cleaning methods may be employed.

Chemical cleaning protocols involve either cleaning with only distilled water, NaOH solution, NaOH solution with free chlorine, an alkaline cleaning agent (hydroxides and carbonates), acids (nitric and phosphoric), enzymes (Proteases and lipases), or surfactants (anionic, non-ionic, etc.). Chemical cleaning involves the use of aggressive chemicals and may cause damage to the membranes. Therefore, chemical cleaning is used with discretion and unfrequently. After chemical cleaning the membranes may also be rejuvenated by deposition of thin layers of hydrophilic polymers.

Concentration polarization (CP) is the resistance developed due to the aggregation of rejected particles and solutes on the surface of the membrane (Figure 7.3(b)). This concentration of the solute contributes to additional resistance to flow towards the membrane reducing the flux of the membrane.

Potential of Membrane Technology

The CP has a significant effect on the performance and permeability of the membrane and can be reversed by cleaning the membranes at various intervals. CP is particularly severe in the case of MF, UF, and ED. With an increase in the flux with increasing the TMP, the CP also increases. CP cannot be eliminated but can be minimized by increasing turbulence of the fluid on the feed side by using turbulence promoters.

7.4 DAIRY PROCESSING

The new developments in membrane technology in the food industry are encouraged by the dairy industry. Membrane technology is commonly used in the dairy industry for processes, such as milk concentration, microbiological stabilization of milk, fractionation of skimmed milk, whey fractionation and processing, and effluent treatment. A schematic of application of membranes is given in Figure 7.4. Milk is a complex mixture of suspended particles like caseins, emulsions of fat globules, and soluble molecules like lactose and minerals having different charges and varying in their sizes. Therefore, membrane technology is suitable for fractionation, separation, and concentration of milk constituents for various applications. The physiochemical properties of milk and dairy product depend largely on the concentration, temperature, and pH, which are also important parameters of membrane technology.

7.4.1 REMOVAL OF BACTERIA AND SPORES

Milk is subjected to high-temperature treatments like pasteurization to kill pathogens and to increase the shelf life, which gives a typical "cooked" flavor. MF can be an effective alternative to heat treatments for the removal of spores and bacteria, giving prolong life and less intense pasteurization without affecting its organoleptic and sensory characteristics. The MF for skimmed milk was performed through the average 1.4 μm pore size ceramic membrane at a uniform TMP of 0.5 bar and CFV of about 6–9 m/s and about 50°C. Pathogens are remained in the retentate, which is mixed with cream separated earlier and the mixture undergoes high heat treatment at 130°C and is then added back in the permeate [67]. In this process, only 10% of the total milk volume undergoes heat treatment. It was found that MF at 6°C followed by high-temperature short-time (HTST) pasteurization gave a near-complete removal of vegetative microflora and a minor microbial growth was observed after 92 days at 6°C [75].

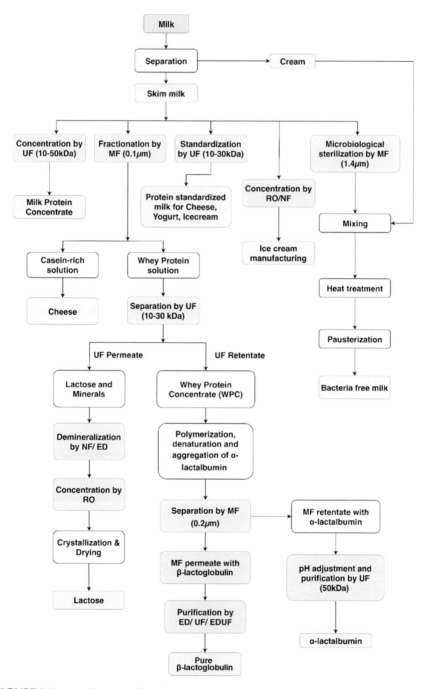

FIGURE 7.4 Applications of membrane processes in the dairy industry.

Potential of Membrane Technology

7.4.2 *CONCENTRATION AND FRACTIONATION OF MILK PROTEIN FOR MILK PROTEIN STANDARDIZATION*

The quality of the milk is not uniform and keeps changing throughout the year along with location, which leads to inconsistency in cheese yield. Standardization of the milk by UF allows manipulating the protein content without the addition of casein, milk powder, or whey protein concentrate (WPC) to improve the consistency in the milk quality. With an increase in the protein content, appearance, and the viscosity of the milk is improved, attracting the consumer. UF is also employed for the concentration of milk with DF in a continuous process at adjusted pH values and temperature. Water is the major constituent of milk, and dairy products like ice cream require all solids and only 30% water. The concentration of milk that is conventionally performed by thermal evaporation can be replaced by RO. RO is a non-thermal process and does not denature the protein, as in the case of heat-induced techniques. Generally, cellulose acetate (CA) membranes are used for RO and operated at 10–30°C with 25–45 bar pressure. Studies with polyamide (PA) membranes have also been reported due to the working limitation of the CA membrane to pH 2–8 and about 35°C [21]. NF membranes are also used for the concentration of milk. However, along with water, the NF permeate also contains minerals and salts. FO also has an application in the concentration of milk and whey. PA and CA are few membrane materials with NaCl, $MgCl_2$ and $CaCl_2$ as osmotic agents used in FO. The total solid concentration by FO can reach up to 40%, which is higher than those achieved by RO and NF without affecting the product quality.

MF for fractionation of milk protein is the most promising application in the dairy industry. Ceramic MF membrane with pore size 0.1 µm separates the micellar casein and the whey protein at uniform TMP. The permeate solution comprises mainly of whey protein and lactose. The whey protein, which is the by-product of cheese making industry, is further processed and valorized. The retentate is rich in casein protein and is used for cheese making. The volume of the retentate is much lower than the original milk and hence requires less quantity of rennet for coagulation to produce cheese.

7.4.3 *CHEESE-MAKING PROCESS*

The cheese-making industry has been revolutionized by membrane technology over the years. Standardization and concentration of casein rich cheese milk have improved the consistency and quality of the cheese products. The use of UF concentrated milk for cheese making, increase in total solids, increasing the cheese yield. The volume of the concentrated milk is low; hence, the use of the rennet enzyme for coagulation is lower. The standard cheese milk has a uniform composition, which improves the quality of cheese. The concentrated UF milk has a good potential for coagulation and generates a lower amount of wastewater.

The pre-concentrated UF milk with a 1.2–2 concentration factor is used for the production of Cheddar, Cottage, and Mozzarella cheese. The APV-Siro Curd process and similar other processes are used for the production of Cheddar, Queso Fresco, Feta, Camembert, and Brie, with partial concentrated UF milk with 2–6 concentration factor. The total concentration of milk by UF to final solid cheese production gives maximum yield and no whey generation. Feta, quark, cheese cream, Ricotta, and Mascarpone are cheese produced by this approach. The UF permeate is concentrated by RO, which consists of major lactose. The RO permeate after pasteurization, or UV treatment can be circulated for various processes in the plant, reducing the water cost.

7.4.4 WHEY PROCESSING

For the longest time, whey was a by-product of the cheesemaking industry and was used either as animal feed or left in the sewage. Whey is processed using membrane technology to yield WPC or Whey protein isolates (WPI), which has high nutritional and functional properties with high economic value. WPC is used as additives in dairy products, bakery, meat, beverages, and infant formula products. Fractionation of the whey proteins is carried out using UF. Whey undergoes MF pretreatment first to remove the bacteria, spores, and fat from the solution. UF separates the proteins from lactose and minerals to form the WPC. The UF membrane can be polymeric (PS, PES, CA, PVDF, etc.), as well as ceramic (zirconium or aluminum oxide) membrane ranging from 10–30 kDa in MWCO [76]. There is a reduction in selectivity and efficiency in UF observed due to membrane fouling.

Whey protein consists of α-lactalbumin, β-lactoglobulin, lactoferrin, and other minor proteins that have a high nutritional value. The WPC, which is the UF retentate, is further fractionated into α-lactalbumin and β-lactoglobulin. The α-lactalbumin is polymerized by adjusting the pH 4–5 and heat at 55°C. The α-lactalbumin denatures and aggregates, as calcium is soluble in the solution, along with other immunoglobulin (IG) and lipids. MF then separates the soluble β-lactoglobulin as permeate, which is purified by using UF and ED for demineralization. The MF retentate with α-lactalbumin is solubilized by adjusting the pH and is purified by UF with MWCO of 50 kDa.

Although UF and MF play a major role in the dairy industry, yet there are limitations associated with membrane fouling and separation of proteins with similar sizes. The electrostatic charge of proteins plays a vital role in its structure, binding, and related biological functions. Charged membranes can be used to separate the proteins with an electrostatic charge. Charged UF membrane with defined pore size has low fouling due to electrostatic repulsion amongst the membrane and the foulant. ED combined with UF (ED/UF) is also employed for the separation of protein. In this, the electrostatic field is applied across the ion exchange, and UF membranes and the separation depends on molecular size and electrical charge [1].

The demineralization of whey is one more application of membrane technology and is carried out using NF and ED depending on the extent of mineralization. NF is employed for partial demineralization typically 35%; the monovalent ions, organic acids, and some amount of lactose permeate through the membrane. NF performs demineralization and concentration of whey in combination. For a higher degree of demineralization of whey, ED is the most cost-effective technique in comparison to ion exchange.

7.5 SUGAR PROCESSING

The sugar industry includes one of the most energy-intensive processes, and membrane separation technology has a great scope in the sugar industry. Membrane technology has potential in both cane sugar as well as the sweet beet sugar industry. Membrane technology has an edge over conventional techniques as it is a gentle treatment, does not require external chemicals, and is energy efficient. MF and UF are used for clarification of juice, NF/RO is used for concentration of the juice before crystallization, and ED is applied for the demineralization of sugar juice.

7.5.1 CLARIFICATION OF SUGAR JUICE

The juice extracted from cane or beet needs to be clarified to remove suspended and colloidal particles, and non-sugar impurities. Conventional clarifiers are heavy equipment and require external chemical agents. Conventional techniques are insufficient in the removal of suspended and colloidal particles, and non-sugar impurities, high-energy consumption, disposal, and environment-related problems due to chemical treatments. Membrane technology is an efficient alternative for conventional techniques like the addition of chemicals like lime, sulfur dioxide, phosphoric acid for flocculation, and sludge separation. The sugar juice clarified by membrane technology like MF and UF, gives lower viscosity, reduced color, and excellent quality of clarified juice. The MF/UF reduces the proteins and polysaccharides, which reduces the viscosity, color of the juice and increases the production capacity with better performance in the crystallization process.

Typically, either a flat sheet or spiral wound polymeric membranes or ceramic membranes in tubular filtration systems are used in UF with either having their advantages and disadvantages. As the incoming feed of sugarcane or beet juice is at high temperature, temperature resistant membranes are used for this application. In polymeric membranes, PS, PES, PVDF, and PTFE are used in for clarification in the sugar industry, as they are temperature resistant. Ceramic membranes with alumina or ZrO_3 as an active layer are used in the sugar industry due to their robust nature. It is observed that the MF (0.1–0.2 μm) membrane gave a better flux in comparison to UF (0.05 μm, 15–50 kDa), but UF performance was superior in removing the non-sucrose impurity, removal of turbidity, and color. The flux of the membrane depends on the operating parameters like TMP, temperature, and crossflow as it affects membrane fouling. Pretreatment like carbonation, pre-filtration of raw juice, evaporation affects the permeate flux of the membrane [70]. The addition of soda ash as a pretreatment followed by crossflow UF, showed about 98% removal of turbidity, an increase in crystal growth rate, and 99.3% color reduction after crystallization [74].

7.5.2 DEMINERALIZATION AND CONCENTRATION OF SUGAR JUICE/SYRUP

After clarification, the sugar syrup contains some minerals, which are removed conventionally by ion exchange. The ion exchange resin needs to be regenerated after treatment, which is not the case with ED. The syrup contains melassogenic ions, which bind to sugar molecules and reduce the yield. ED is applied to remove the melassogenic ions to increase the yield and reduce the sugar loss to molasses. ED demineralizes as much as 90% of the sugar syrup.

The concentration of sugar syrup up to 90°Brix is done before the crystallization of sugar. Evaporation of sugar syrup is the traditional approach,

Potential of Membrane Technology

which is an energy-consuming process. RO is applied for concentration of the sugar syrup, as there is no phase change; heat damage is avoided and consumes a lower amount of energy. It is found that RO can achieve a maximum of only 50% concentration after multistage hence is used as a pre-concentration step. RO is performed at a temperature of 80°C; hence, a non-acetate membrane is used.

7.5.3 CLARIFICATION AND CONCENTRATION OF FRUIT JUICES

Raw fruit juice consists of solid and colloidal particles, proteins, enzymes, sugar, salts, and flavor as well as aroma compounds. Clarification of the raw juice is necessary for its longer durability and storage, keeping its flavor and aroma intact. Enzymatic treatment with enzymes like pectinase and amylase reduces pectin and polysaccharide molecules from the juice, which leads to aggregation and sedimentation of particles. The addition of fining agents like gelatin and bentonite enhances flocculation and precipitation in the juices, reducing the haziness and viscosity of the juice. The suspended solids and colloidal particles and proteins are removed by conventional filtration.

Complete removal of the additives, fining agents is one of the major concerns as it affects the taste of the juice. Later on, it is subjected to thermal evaporation for the concentration of the juice. Along with the high-energy requirement, retaining the aroma, flavor, appearance, and mouthfeel are some of the concerns associated with the conventional thermal processing of fruit juices. Membrane technologies, like MF, UF, NF, RO, OD, PV as well as MD are effective alternatives for conventional techniques like the addition of fining agents and thermal evaporation. Membrane technology concentrates the juice at low temperatures with lower energy consumption as well as works excellent for heat-sensitive compounds in juices. It also eliminates the concern related to the removal of fining agents as well as conserves the flavor, aroma, and the mouthfeel of fresh juices. It does not involve any external chemical agents, hence maintains the nutritional profile of the fruit juice.

7.5.4 CLARIFICATION OF FRUIT JUICES

Generally, clarification is required to remove the haziness and reduce the viscosity of the juice. Clear fruit juice is one of the deciding factors at the customer ends. MF and UF are the major techniques employed for clarification of juices. Polymeric as well as ceramic membranes are used for the clarification process. MF and UF remove the high molecular weight compounds, the solid and colloidal particles, separate the concentrated fibrous pulp and reduce the microbial content of the juice, which causes spoilage. MF and UF also enhance the total antioxidant and the phenolic contents of the clarified juice. They eliminate the use of fining agents, which addresses the issue related to the taste of the juice, also enhances the yield and quality of clarified juice at a low cost. The presence of pectin and polysaccharides block the pore of the membranes reducing the performance of the membranes. Enzymatic treatment before membrane filtration reduces the pectin content and viscosity of the juices. De-pectinized juices increase the permeate flux of the membranes [3, 69]. Recirculation of the low concentrate enzyme treatment increases the permeate flux of the membranes [28]. Enzymatic treatment, along with UF and MF, gives high yield and high-quality clarified juice. The list of juices, membranes used, and operating parameters are mentioned in Table 7.2.

7.5.5 CONCENTRATION OF FRUIT JUICES

The concentration of fruit juices involves a reduction in the water content and an increase in the soluble solid content of up to 65–75% (°Brix) of the weight. The concentration of juices reduces the storage and handling cost and increases the durability of the juices due to low microbial activity as the water content is low. Conventionally the concentrate of juice is obtained by thermal evaporation of the clarified juice at about 90°C to remove water. In this process, there is a loss of aroma and antioxidants, color degradation, and cooked taste to the concentrated juice. Membrane technology such as RO, NF, MD, OD, and PV is implemented for the concentration of fruit juices. Membrane technology works at low temperatures and has a lower energy requirement in comparison to thermal evaporation. There is no phase change and operates at low temperatures. Therefore, it prevents the degradation of heat-sensitive compounds, maintaining the nutritional profile intact.

RO is a great alternative to thermal evaporation technique as it works at a lower temperature, requires less energy consumption. The aroma and flavors of the concentrated juice are maintained without causing any degradation of thermo-sensitive compounds. The operating pressure is based on the osmotic pressure of the juice, which is a factor of the concentration of the clarified juice. RO is insufficient to reach the standard product equivalent to thermal

Potential of Membrane Technology 195

TABLE 7.2 Membranes Used in Fruit Juice Clarification and Processing

Fruit	Membrane Material	Process	Flux (L m^{-2} h^{-1})	TMP (bar)	Temperature (°C)	Clarity (%)	Achievement	References
Apple	Polymeric PS, PES, regenerated CA	UF	21.6[a]	2.5	25	ND	63.2% TPC was maintained in permeate along with 52% malic acid	[38]
	Polymeric Al_2O_3/TiO_2 on PS/PEI	UF	8.3	5.4	ND	99.9[b]	Good quality clarified juice was obtained with a turbidity of 0.02 NTU	[68]
Guava	Polymeric PES	UF	8.6[a]	2	ND	97	Reduced turbidity up to 97%	[59]
Key lime	Ceramic metakaolin-based	UF	178.7	7	ND	90.4	Almost 90.4% clarity was obtained with a color of 0.049	[41]
Kiwi fruit	Ceramic metakaolin-based	UF	ND	7	ND	90.1	Almost 90.1% clarity was obtained with a color of 0.071.	[41]
	Ceramic fly-ash	MF	20	1.5	ND	90	Clarity was increased to 90% and color decreased to 0.24 from 3.1 for feed	[64]
	Polymeric modified PEEK	UF	16.25	1.6	25	ND	Total removal of suspended particles and bioactive compounds were preserved	[16]
Lemon juice	Polymeric PVDF	UF	17.2	0.87	25	ND	Increases the yield by increasing the volume of clarified juice	[13]
	Polymeric polyolefin	MF	8.7	2	27	94.7	Increase in flux observed due to enzymatic treatment	[22]
	Polymeric PVDF/PMMA/PVP	MF	32.5	0.4–1	20	ND	Total solids were completely removed	[34]

TABLE 7.2 (Continued)

Fruit	Membrane Material	Process	Flux (L m⁻² h⁻¹)	TMP (bar)	Temperature (°C)	Clarity (%)	Achievement	References
Mousambi	Ceramic kaolin based	MF	156.5	2.07	25	98.9	A maximum of 98.9% clarity was observed with negligible AIS value	[32]
	Polymeric CA	MF	8.5	1.38	30	97.85	Improvement in color and clarity	[65]
Mulberry	Polymeric PVDF	UF	0.55ᵃ	3	25	ND	Increased total antioxidant activity and improved color	[48]
Orange juice	Polymeric PVDF	UF	23	0.87	25	ND	Increases the yield by increasing the volume of clarified juice	[13]
	Ceramic Kaolin based quartz boric acid	MF	34.4	1.4	ND	65.35	Decrease in the color of the juice to 0.227 and increased clarity to 65.34. No suspended solids found	[58]
Passion fruit	Polymeric polyetherimide	MF	15.2	1	RT	95	Obtained turbidity of 1.06 NTU	[27]
	Ceramic Al_2O_3/TiO_2	MF	39ᵃ	1	25	ND	Clean and clarified permeate observed	[25]
Pineapple	Polymeric PS	MF	52.9	0.7	20	ND	Complete removal of suspended solids, recovered vitamin C up to 94.3% and phenolic content up to 93.4%	[47]
	Polymeric PES	MF	38.3	1.5	25	ND	90% increase in luminosity and turbidity	[12]
Pomegranate	Polymeric modified PEEK PS	UF	40.9	1.15	25	ND	24.1% rejection for flavonoids and 25.1% for phenolic compounds	[15]

TABLE 7.2 *(Continued)*

Fruit	Membrane Material	Process	Flux (L m⁻² h⁻¹)	TMP (bar)	Temperature (°C)	Clarity (%)	Achievement	References
	Polymeric PVDF	UF	3.66	3	25	ND	Reduction in phenolic groups, improved clarity	[6]
Red beet Juice	Polymeric mixed cellulose ester	MF	ND	0.1–1	ND	90.91	Decrease in turbidity, color, total solids, and phenolic content	[4]
Red raspberry	Polymeric CA/CNF	MF	108[a]	7.5	25	78	39% decrease in TS, No changes in TSS and phenolics	[8]
Strawberry	Polymeric CA/CNF	MF	84[a]	7.5	25	71	41% decrease in TS, 34% decrease in TSS, 19% in phenolics	[8]
Watermelon	Ceramic metakaolin based	UF	121.9	7	ND	95.5	Almost 95.5% clarity was obtained with a color of 0.064.	[41]

[a]$kg/(m^2.h.bar)$.
[b]*Calculated based on value mentioned.*

ND: not defined.

evaporation. Due to this drawback, the RO is used in an integrated process along with other membrane technologies like OD or MD, where RO is a part of the pre-concentration treatment [2, 7]. NF is another membrane process that is engaged in the concentration of fruit juices. One of the advantages of NF over RO is it requires low energy in comparison to RO and also works at low pressure avoiding any damages to the juice constituents due to high pressure.

MD is a membrane process, which involves a microporous hydrophobic membrane and operates at atmospheric pressure and temperature lower than the boiling point of the solutes. MD can be incorporated for the concentration of juices as it works at lower temperatures and gives the desired °Brix value. One of the shortcomings of MD is the loss due to the evaporation of volatile compounds contributing to the aroma and flavor of the juices.

OD or FO is an efficient alternative for conventional thermal evaporation. OD has the potential to attend higher total soluble solids (TSS) content while operating at lower temperatures and pressure. It is an isothermal process, which involves a macroporous hydrophobic membrane separated by aqueous solution differing in the concentration. One of the aqueous solutions is the dilute clarified juice solution, and the other is the stripping solution, which is a concentrate brine solution. The most common osmotic agents used are $MgSO_4$, $CaCl_2$, and K_2HPO_4. OD has the upper hand in comparison to RO and MD, because it overcomes the drawbacks of RO related to the limitation with low TSS, and that of MD with loss of volatile compounds when operated at room temperature.

Generally, the compounds contributing to the aroma of the juices are volatile and are lost during the concentration of the fruit juices due to high temperatures. The lost aroma compounds due to evaporation can be fed back in the final product with the help of the PV technique to restore the sensory quality of the fruit juice. PV involves a selective membrane, which separates the feed mixture based on partial vaporization. The volatile organic aroma compounds permeate through the non-porous selective membrane and are condensed on the permeate side by cold traps.

The integrated system has great potential for manufacturing an industrial standard product with lower energy consumption. In the literature, many integrated systems have been studied that gives high-quality product with lower energy consumption. Few examples have been mentioned below. For pomegranate juice, UF was employed for clarification of the juice. The clarified juice was then treated for RO, followed by OD to give a concentrated

Potential of Membrane Technology

pomegranate juice of 65°Brix [7]. Kiwi fruit juice concentrate of 60Brix was prepared by UF, followed by OD [14]. Bélafi-Bakó and Koroknai studied the concentration of apple juice wherein they coupled OD with MD [9]. The integrated membrane process system involves an enzymatic pretreatment, followed by clarification by UF or MF. The pre-concentration of the clarified juice was done by RO; followed by concentration with OD, or MD, along with aroma recovery using PV to give the standard product.

7.6 WINE PROCESSING

Wine clarification and stabilization are two of the major and important processes of winemaking. The result of alcoholic and malolactic fermentation is a turbid and unstable crude wine. It consists of solids and colloidal particles, grape-pulp remains, macromolecules like polysaccharides and polyphenols, and microorganisms that contribute to the haze in the wine, which is not acceptable at the consumer end. Traditionally, clarification is performed by the addition of fining agents (proteins such as gelatin and egg albumin, and bentonites), centrifugation, and dead-end filtration through diatomaceous earth materials. Disposal of diatomaceous earth materials is a problem as well as requires several successive filtrations to obtain the desired limpidity and reduced microorganisms. Membrane processes are a valid alternative for replacing the traditional techniques.

7.6.1 CLARIFICATION OF WINE

Crossflow MF and UF have gained popularity in winemaking as they can replace filtration through diatomaceous earth material. It replaces several steps in traditional technique by combining clarification, microbiological stabilization, and sterilization with a unit operation, which reduces the product loss. The cost of the process is also lowered due lower number of steps and reduced energy requirement. One of the major limitations of MF and UF in the wine-making industry is the membrane fouling due to large macromolecules (polyphenols and polysaccharides) and particles (cell debris and microorganisms like yeast and bacteria). The adsorbed polyphenols and polysaccharides block the pore of the membrane and reduce the flux and permeability of the membrane. It also retains some necessary components of the wine affecting its organoleptic characteristics.

The performance of clarification of wine greatly depends on the membrane selection. Membrane material plays its parts in membrane fouling as the interaction and adsorption of polyphenols and polysaccharides with the membrane [73]. Membrane pore size and pore size distribution have a huge effect on membrane fouling, which decides the membrane performance. The operating parameters, like TMP and concentration factor, also affect membrane performance and fouling [29, 62].

7.6.2 WINE TARTARIC STABILIZATION

ED is an effective technique for wine tartaric acid stabilization. Potassium hydrogen tartrate is the natural constituent of grapes. Instability of the tartaric salts is a major issue as a there is decrease in its solubility in the presence of ethanol and at low temperatures, which forms crystals in the bottle. Cold stabilization is used to deal with the problem where the wine is cooled about its freezing point for several days to precipitate the tartaric salts. In some cases, they also add potassium hydrogen tartrate crystals to initiate seeding. Limitation of cold stabilization is that in the process of precipitating tartaric salts, it also precipitates other components of wine like polyphenols and polysaccharides affecting the sensory properties of the wine [35, 36]. In ED, under an electric field, the positively charged ions K^+, Ca^+ migrate towards the cathode, and the tartaric ions migrate towards the anode, which passes through ion exchange membranes. ED overcomes the limitation of cold stabilization, as it maintains the organoleptic properties of wine as it does not alter the polyphenol and polysaccharide composition of wine, energy consumption is low and requires lesser duration. With ED, the degree of potassium hydrogen tartrate removal can be controlled by varying the operating time [71]. An additional advantage of ED is that it also removes other ions like calcium and malate ions [36].

7.6.3 DEALCOHOLIZATION OF WINE

The alcoholic content of wine is regulated by the law and affects the organoleptic characteristics of the wine. There are two approaches involved in the dealcoholization of wine. One is reducing the sugar concentration in must, which leads to an increase in alcohol concentration after fermentation. Another approach is the removal of the alcohol content before bottling the product. Viticulture practices, microbiological techniques, post-fermentation

Potential of Membrane Technology 201

techniques, adsorption on zeolites, and membrane processes are some techniques applied for the dealcoholization of wine. Membrane processes operate by reducing the alcohol content after the fermentation process. RO is applied to reduce alcohol in wine, but along with alcohol, water also permeates out, which is reinstated back into the wine. The addition of water to wine is legally prohibited in some countries. The capacity of RO is also restricted due to the low permeate flux of the membrane. NF overcomes the limitation of low flux associated with RO. NF has a higher flux compared to RO. On comparison based on the organoleptic property, the product after NF and RO are comparable [55].

OD is used in the partial dealcoholization of wine as it operates at low temperature, atmospheric pressure, low energy consumption, low loss of volatile compounds, and no by-products formed. It selectively permeates the ethanol through the membrane. A high degree of dealcoholization affects the sensory properties of the wine.

A very recent development is liquid emulsion membranes, a double emulsion with innermost, and the outermost phase usually aqueous and the intermediate phase as organic. The organic phase forms the liquid membrane, which is stabilized by a surfactant which plays the most important role in liquid emulsion liquid. It works at low temperatures and pressure. It can remove alcohol up to 92% with multistage operation [19].

Dealcoholization is also demonstrated by reducing the sugars in the must by combining MF or UF with RO or NF. In 2004, the reduction of sugars was performed by the Redux® process, which involves UF, followed by NF. The UF performs clarification of the must, and the 'clear' must is concentrated through NF [66].

7.7 BEER PROCESSING

In the beverage industry, beer commands a large market share and has a strong economic position in the food industry. Beer is prepared by mashing of the cereal, fermentation of wort by yeast, and subsequent clarification. It is essential to remove the microorganisms like yeast and bacteria, to meet the standard quality of the beer and sterilize for stability. Beer production procedure involves clarification, microbiological stabilization, and dealcoholized as per requirement. Membrane technologies like MF, UF are employed for clarification and microbiological stabilization; RO and OD are used for the dealcoholization of alcohol.

7.7.1 RECOVERY OF BEER/SURPLUS YEAST FROM THE TANK BOTTOM

The yeast settles at the base of the tank after fermentation along with 1.5–2% of the total volume of beer. For recovery of this beer, crossflow MF is employed with pore size between 0.5 µm and 0.8 µm. Ceramic, as well as polymeric membranes are used, but the ceramic membrane has an advantage due to its robust properties. After filtration, the yeast slurry of about 20% dry weight (DW) is obtained. A recent development is a plate and frame arrangement with polymer membrane (0.45 µm) working at low pressure [49, 61].

7.7.2 CLARIFICATION OF BEER

Beer undergoes clarification to eliminate the haze and colloidal particles to give a bright, clear, and sparkling beer. In a traditional approach, clarification is associated with kieselguhr filtration. Disposal of diatomaceous earth materials is an issue and has a negative environmental impact. MF and UF techniques are applied for clarification of beer. Along with colloidal particles and haze, these techniques also remove the bacteria and spores giving microbiological stabilization to the beer. They combine the process of clarification, microbiological stabilization, and sterilization in one operation, giving an upper hand with low energy consumption and better quality of beer. The major challenges with MF and UF are fouling and low permeate flux. The operating parameters, such as TMP and CFV, affect the fouling of the membrane [72]. It was observed that pre-enzymatic treatment with Brewers Clarex®, followed by crossflow MF performed at 30°C, gave a higher flux and colloidal stability to the product [23].

7.7.3 DEALCOHOLIZATION OF BEER

There is an increasing demand for non-alcoholic or low alcoholic beer to abide by the laws and perceive a healthier lifestyle. Biological and physical processes are employed to prepare an alcohol-free beer or low alcoholic beer. In the biological process, they limit the fermentation by yeast to the desired alcohol content but affect the typical taste of the product. The physical process can be thermal or membrane processes. The thermal process like vacuum evaporation and distillation are used to reduce the alcohol content but affect the flavor of the product. Membrane technologies such as RO and OD are employed for lowering the alcohol content from the beer.

Potential of Membrane Technology 203

Dialysis (DI) is used for the dealcoholization of beer. In DI, the ethanol is separated from the membrane, which acts as molecular sieves due to concentration difference. The separation is controlled by the pore size and the surface properties of the membrane. The filtration process is performed at a low temperature and hence does not have a thermal impact on the product. The low molecular weight volatile compounds associated to the flavor and aroma of the product are lost during this filtration.

During RO, the beer is permeated through a semipermeable membrane with TMP higher than the osmotic pressure. Ethanol and water permeate to the other side, maintaining the flavor and aroma almost intact. Through RO, beer is processed at lower temperatures without affecting the flavor and aroma in contrary to distillation. Higher operating pressure give higher permeate flux accompanied by higher ethanol rejection and higher aroma rejection [18].

OD is an isothermal distillation engaged for the dealcoholization of beer. It works at low temperatures and pressure, restricting the damage due to heat. Polypropylene membrane is used commonly in OD, working at ambient temperature with water as a stripping agent [54]. There is literature for the use of MD for dealcoholization, applying pressure to permeate alcohol through non-porous membranes and vacuum to condensate the alcohol on the permeate side [63]. PV is also used for the dealcoholization of beer along with aroma recovery [17].

7.7.4 AROMA RECOVERY IN BEVERAGES

During beer and wine processing, dealcoholization is carried out at the expense of volatile compounds associated with the aroma. PV, a membrane technique extracts the aroma compounds from the wine and beer earlier to the dealcoholization process and is reincorporated in the dealcoholized product. In PV, vacuum is introduced on the permeate side enabling the transport of the aroma compounds. The vapor travels through the selective membrane towards the permeate side, where the vapor is condensed by cooling. The flux of the membrane depends on the membrane material, as well as operating conditions. Hydrophobic membranes are generally used to extract the organic volatile aroma compounds. A linear increase in flux observed along with the increase in temperature, but the selectivity of the membrane decreases. A good selection of membrane material along with optimized operating conditions, will lead to higher productivities of PV for aroma recovery.

7.8 ENZYME PROCESSING

Enzymes are commonly used in food and juice processing. Enzymes may need to be isolated from their source or have to be recovered from the process. Enzymes have a fragile structure, which needs to be maintained for its biological activity. Enzymes are macromolecules and UF is used for purification and concentration of enzymes at low temperature and pressure. Generally, polymeric membranes (PS, CA, PVDF) in spiral wound and hollow fiber module are used to maintain the enzyme activity. Despite good operating parameters, enzyme activity is also influenced by pH, initial concentration, and composition. Hence, the choice of the membrane should meet these requirements. PES membrane with MWCO of 100–300 kDa was used for purification of tyrosinase through tangential flow filtration [46]. Ceramic membranes have also been used due to better thermal, mechanical, and chemical stability than polymeric membranes. Crossflow UF ceramic membrane of Al_2O_3 coated with TiO_2 was used for the primary concentration of the enzyme endopeptidase [45].

7.9 WASTE TREATMENT

Food industries consume a significant quantity of water for food processing. The cost of water is increasing due to a shortage of sources, and ever-increasing demand for water. This water shortage effects on productivity and profit of industries. The generation of wastes from food industries has been increasing due to tremendous food product demand at each level. The treatment of food waste, inhibition, volume minimization is crucial to avoid its detrimental impact in the environment upon untreated disposal. Recovery of water and valuable components from waste streams in food industries provides a solution to maintain environmental sustainability and the industry's benefits. However, waste from various food industries contains complex constituents and large volumes, which face difficulties and challenges in the treatment. Conventional methods are time-consuming, expensive, and require excessive efforts to make high-quality water/ recovered product. Also, the selective recovery of contaminants from wastewater is important before mixed into other streams to avoid complex mixtures. The use of membrane technology reduces the several steps during treatment, gives enhanced recovery of water reducing wastewater volume, and saves energy needed to treat wastewater from food processing compared to conventional methods.

Potential of Membrane Technology 205

7.9.1 DAIRY WASTE TREATMENT

Dairy effluents generally have high organic loads due to milk components, which differ in the pH due to cleaning agents used, high levels of phosphorus, and nitrogen resulting in high biological oxygen demand (BOD) and chemical oxygen demand (COD). Whey is one of the wastes from the dairy industry, which is now valorized. Dairy industries use huge amount of water for cleaning and sanitation, generating large volumes of wastewater. There are two options available for reuse of water, one where there will be direct contact with the food (cleaning/rinsing the equipment; diafiltration (DF) water for UF or NF; ion exchange resin rinsing or elution), and the other is no contact with the food (cooling process, boiler water, steam generation, and sanitation). The option of reuse determines the hazard and the type of water treatment needed.

The evaporate condensate from milk concentration was treated with ceramic UF membranes to remove the microorganisms and protein materials. The permeate from UF was later treated by RO to give high-quality water. Pretreatment by UF before RO was employed to lower the fouling of the RO membrane that leads to a decrease in the permeate flux. Dairy raised wastewater is treated through a single staged RO or a two-staged NF/RO. The treatment was sufficient to reuse the water for heating, cooling, cleaning, or in boiler feed. A two-stage RO/RO process produces high-quality water; the quality of which depends on the integrity of the membrane on continuous use. The quality of the water needs to be monitored frequently to ensure water safety and quality. Dairy wastewater was treated by the PES UF membrane, showed a retention rate of 99% for turbidity and BOD, 80% suspended solids, 95% proteins, and the TDS was reduced to 55% [10]. Membrane bioreactor (MBR) integrated with NF showed great results with overall 99.99% efficiency for removal of COD and 93.1% for total solids. The treated water could be reused for steam generation, cooling, and cleaning the external areas [5].

7.9.2 BREWING INDUSTRY WASTES

Water is the major component of beer and cider, and the production process consumes a high level of water. Environmental concerns and water shortages are a concern worldwide and need action on water issues. Currently, a final packaged beer requires around three-four volumes of water. The usage of water volumes needs to be reduced and can be attained by recycling and reusing the water. Membrane technology is an energy-efficient and cost-effective technology for water treatment and recovery.

In 2004, Braeken et al. investigated NF for the treatment of brewery waste, where COD was reduced to an average of about 100%, and around 55–70% removal of salts. The use of NF was only applicable to biologically treated wastewater and required pretreatment processes [11]. Polyethylene (PE) terephthalate RO membrane was used by Madaeni and Mansourpanah for biologically treated wastewater reduced COD 100% with high flux value [53]. A sono-electrocoagulation process integrated with RO was employed for the treatment of brewery wastewater that gave 100% color removal and 90.8% COD removal [26].

MBR is a critical technology in the treatment of brewing wastewater. An anaerobic MBR with 0.04 μm UF membrane removed 98% COD with a critical flux of about 8 L m^{-2} h^{-1} [20]. Other studies using MBR for brewery effluent treatment were reported with COD removal above 90%, indicating MBR can be a good option for the treatment of brewery effluent [24, 33, 37, 56]. Recently, a photosynthetic bacteria-MBR was employed for continuous-mode long term brewery wastewater treatment, which gave COD of 80 mg/L and recovery of biomass [50, 51].

7.9.3 SUGAR INDUSTRY WASTES

In sugar and other food industry, sugar (sucrose, glucose, starch, fructose, etc.), is the main constituent of the product. Nevertheless, along with final products, sugar is also present in the effluent streams of those industries creating an environmental impact on waste generation. Product recovery addresses the economic as well as environmental aspects of waste generation. Molasses from the sugar factory is a raw material in the distillery for the production of ethyl alcohol [42]. A ceramic tight UF membrane was used for the decoloration of molasses [39, 40]. NF membrane was employed for the recovery of sucrose and reduced sugar like fructose and glucose from cane molasses [52]. Bagasse is also a by-product of the sugar industry, which is used for biogas production, electricity production as well as pulp for paper production [43].

Enzymatic hydrolysis of starch can be carried out to generate dextrin using a membrane reactor in one step using amylase enzyme [60]. In membrane reactors, the enzymes are not immobilized, and the membrane is used for separation. The major limitation associated with membrane reactors is the substantial decrease in the permeate flux due

Potential of Membrane Technology 207

to concentration polarization and fouling of the membrane. A Kubota™ MBR was used for sugar mill effluent treatment and showed a good removal of COD [57].

7.9.4 OLIVE MILL WASTEWATER

Every year, more than 30 million m^3 of wastewater is generated during olive oil extraction processing. Olive oil wastewater effluents have extremely high content of solids and organic compounds along with low pH and biodegradability, which causes a harmful impact on the environment. Olive wastewater treatment uses various physical and biological treatment such as coagulation, oxidation, aerobic or anaerobic treatment, electrolysis, and evaporation. However, these processes showed limitations associated to low efficiency and solid waste/sludge disposal. Membrane technology has attained importance for olive wastewater treatment due to its selective separation and ease of operation. New emerging membrane technologies like OD and osmotic membrane distillation (OMD) displayed potential for recovery of polyphenols from olive oil wastewater effluent and its treatment without any additional pretreatment. There is a study that shows that the PTFE membranes display realistic permeate fluxes (2.9–4.2 L h^{-1} m^2), along with higher retention values for phenolic compounds in the feed side and high fouling resistance [31].

Further, commercial PTFE membranes having different pore sizes by using direct contact membrane distillation (DCMD) process showed a concentration of natural polyphenols and produced pure water from olive mill wastewater effluent. There was no major effect on the phenolic content and its anti-oxidant properties at high temperatures in DCMD [30]. MD using PTFE membranes are shown promising alternative technologies for the treatment and concentration of olive mill wastewater.

7.10 PERSPECTIVES OF MEMBRANE TECHNOLOGY

Food processing is crucial in food industries to turn fresh food into the food product. Food processing contains one or a combination of various stages, including mixing, fermentation, purification, pasteurization, sterilization, concentration, washing, freezing, drying, and packaging. Among the use of different methods and unit operations, membrane-based processes have tremendous potential for food processing due to their selective separation, ease of operation, easy to scale up and cost-effectiveness. Based on the

208   Advances in Food Process Engineering

end application, desired membranes can be developed to use in different arrangement for the separation of product, recovery or concentration of nutraceuticals, volume reduction, and recovery of material without affecting its original properties and characteristics. Along with excellent quality products and low energy consumption, membrane processes do not need external chemical agents as well as generate the low amount of waste compared to conventional food processes. Further, food product in membrane processes can be handled at lower operating temperature and pressure compared to conventional methodologies practiced for food processing.

In food industries, membrane-based processes show their potential to combine several steps like purification, clarification, concentration, micro-biological sterilization, and stabilization into one process. Various food industries like dairy, sugar, beverages, fruit juice uses membrane processes to make value added product. Almost all the membrane-based processes such as MF, UF, NF, RO, FO, MD, OD, PV, ED can be used as an individual or in combination for food processing and food waste treatment.

Fouling of the membrane is apparent to be a concern in the food processing which reduces efficiency and increases the operation cost of the membrane processes. The formation of concentration polarization on fouled membrane affects the performance and the stability of the membrane that require frequent membrane cleaning or replacement of the membrane module, increasing the cost of the process. A self-cleaning membrane with anti-fouling property and turbulence fluid flow in the system can assist in addressing the issue of concentration polarization and fouling of the membrane. Moreover, the type or composition of the feed affects the performance of the membrane.

A pretreatment of the feed would be a positive impact on the efficiency of the membrane process. Scaling is also an issue in demineralization and dewatering applications especially in OD, FO, NF, and RO processing. Another challenge is to develop better materials for membranes to overcome limitations associated with fouling, adhesion, and production of finely porous membranes with tight pore size distribution and high pore density to overcome flux limitations with high degree of perm-selectivity with long operating life.

7.11   SUMMARY

Membrane based processes have shown great potential in food processing for the purification or recovery of solutes from different streams. The chapter has deliberated different applications of membrane used in the food industries such as dairy, sugar, fruit juice, beverages as well as food waste treatment. Membrane technology overpowers the conventional and the traditional technology as they are more energy-efficient, time saving, have a low operating cost, and give a stabilized, and excellent quality product. Membrane technologies like MF and UF have shown a broad application in dairy, sugar, fruit juice, and beverage industries mainly associated with clarification of crude product, and microbiological sterilization. In dairy industries, it shows an excellent potential in fractionation of whey protein and milk protein especially in the cheesemaking industry. NF, RO, and OD have demonstrated their application in the concentration process in the food industries; without affecting the flavor and characteristics of the product contrary to the evaporation and thermal process used in the conventional methodology. New emerging technologies like ED and MD have displayed good performance in demineralization process and concentration process. PV showed a noteworthy scope in the recovery of aroma in the food industry. Further, membrane technology showed its important application for the treatment of wastewater generated from the food industries. Capital cost, fouling of the membrane, membrane life and handling of different feed compositions are some of the concerns of membrane processes in food industries.

KEYWORDS

- **beverages**
- **dairy**
- **food waste**
- **forward osmosis**
- **fouling**
- **fruit juice**
- **membrane applications**
- **sugar**

REFERENCES

1. Aguero, R., Bringas, E., Román, M. F., Ortiz, I., & Ibáñez, R., (2017). Membrane processes for whey proteins separation and purification: A review. *Current Organic Chemistry, 21*(17), 1740–1752.

2. Aguiar, I. B., Miranda, N. G. M., Gomes, F. S., Santos, M. C. S., Freitas, D. G. C., Tonon, R. V., & Cabral, L. M. C., (2012). Physicochemical and sensory properties of apple juice concentrated by reverse osmosis and osmotic evaporation. *Innovative Food Science & Emerging Technologies, 16*, 137–142.

3. Alvarez, S., Alvarez, R., Riera, F. A., & Coca, J., (1998). Influence of depectinization on apple juice ultrafiltration. *Colloids and Surfaces A: Physicochemical and Engineering Aspects, 138*(2, 3), 377–382.

4. Amirasgari, N., & Mirsaeedghazi, H., (2015). Microfiltration of red beet juice using mixed cellulose ester membrane. *Journal of Food Processing and Preservation, 39*(6), 614–623.

5. Andrade, L. H., Motta, G. E., & Amaral, M. C. S., (2013). Treatment of dairy wastewater with a membrane bioreactor. *Brazilian Journal of Chemical Engineering, 30*(4), 759–770.

6. Bagci, P. O., (2014). Effective clarification of pomegranate juice: A comparative study of pretreatment methods and their influence on ultrafiltration flux. *Journal of Food Engineering, 141*, 58–64.

7. Bagci, P. O., Akbas, M., Gulec, H. A., & Bagci, U., (2019). Coupling reverse osmosis and osmotic distillation for clarified pomegranate juice concentration: Use of plasma modified reverse osmosis membranes for improved performance. *Innovative Food Science & Emerging Technologies, 52*, 213–220.

8. Battirola, L. C., Andrade, P. F., Marson, G. V., Hubinger, M. D., & Gonçalves, M. C., (2017). Cellulose acetate/cellulose nanofiber membranes for whey and fruit juice microfiltration. *Cellulose, 24*, 5593–5604.

9. Bélafi-Bakó, K., & Koroknai, B., (2006). Enhanced water flux in fruit juice concentration: Coupled operation of osmotic evaporation and membrane distillation. *Journal of Membrane Science, 269*(1, 2), 187–193.

10. Bennani, C. F., Ousji, B., & Ennigrou, D. J., (2015). Reclamation of dairy wastewater using ultrafiltration process. *Desalination and Water Treatment, 55*(2), 297–303.

11. Braeken, L., Van, D. B. B., & Vandecasteele, C., (2004). Regeneration of brewery wastewater using nanofiltration. *Water Research, 38*(13), 3075–3082.

12. Carvalho, L. M. J., & Silva, C. A. B., (2010). Clarification of pineapple juice by microfiltration. *Food Science and Technology, 30*(3), 828–832.

13. Cassano, A., Conidi, C., & Tasselli, F., (2015). Clarification of pomegranate juice (*Punica granatum L.*) by hollow fiber membranes: Analyses of membrane fouling and performance. *Journal of Chemical Technology and Biotechnology, 90*(5), 859–866.

14. Cassano, A., Conidi, C., Tasselli, F., & Drioli, E., (2017). Quality of kiwifruit juice clarified by modified poly(ether ether ketone) hollow fiber membranes. *Journal of Membrane Science & Research, 3*(4), 313–319.

15. Cassano, A., Drioli, E., Galaverna, G., Marchelli, R., Di Silvestro, G., & Cagnasso, P., (2003). Clarification and concentration of citrus juices by integrated membrane processes. *Journal of Food Engineering, 57*(2), 153–163.

16. Cassano, A., Jiao, B., & Drioli, E., (2004). Production of concentrated kiwifruit juice by integrated membrane process. *Food Research International, 37*(2), 139–148.

17. Castro-Muñoz, R., (2019). Pervaporation-based membrane processes for the production of non-alcoholic beverages. *Journal of Food Science and Technology, 56*, 2333–2344.

18. Catarino, M., Mendes, A., Madeira, L. M., & Ferreira, A., (2007). Alcohol removal from beer by reverse osmosis. *Separation Science and Technology, 42*(13), 3011–3027.

Potential of Membrane Technology

19. Chanukya, B. S., & Rastogi, N. K., (2013). Extraction of alcohol from wine and color extracts using liquid emulsion membrane. *Separation and Purification Technology, 105*, 41–47.

20. Chen, H., Chang, S., Guo, Q., Hong, Y., & Wu, P., (2016). Brewery wastewater treatment using an anaerobic membrane bioreactor. *Biochemical Engineering Journal, 105*(Part B), 321–331.

21. Cheryan, M., Veeranjaneyulu, B., & Schlicher, L. R., (1990). Reverse osmosis of milk with thin-film composite membranes. *Journal of Membrane Science, 48*(1), 103–114.

22. Chopde, S. A., (2019). Optimization of membrane separation process for comparative study of clarification process of enzyme-treated and untreated lemon juice. *International Journal on Emerging Trends in Technology, 6*(1), 12042–12044.

23. Cimini, A., & Moresi, M., (2018). Combined enzymatic and crossflow microfiltration process to assure the colloidal stability of beer. *LWT, 90*, 132–137.

24. Cornelissen, E. R., Janse, W., & Koning, J., (2002). Wastewater treatment with the internal MEMBIOR. *Desalination, 146*(1–3), 463–466.

25. De Oliveira, R. C., Docê, R. C., & De Barros, S. T. D., (2012). Clarification of passion fruit juice by microfiltration: Analyses of operating parameters, study of membrane fouling and juice quality. *Journal of Food Engineering, 111*(2), 432–439.

26. Dizge, N., Akarsu, C., Ozay, Y., Gulsen, H. E., Adiguzel, S. K., & Mazmanci, M. A., (2018). Sono-assisted electrocoagulation and cross-flow membrane processes for brewery wastewater treatment. *Journal of Water Process Engineering, 21*, 52–60.

27. Domingues, R. C. C., Ramos, A. A., Cardoso, V. L., & Reis, M. H. M., (2014). Micro-filtration of passion fruit juice using hollow fiber membranes and evaluation of fouling mechanisms. *Journal of Food Engineering, 121*, 73–79.

28. Echavarría, V. A. P., Pagán, J., & Ibarz, A., (2011). Effect of previous enzymatic recirculation treatment through a tubular ceramic membrane on ultrafiltration of model solution and apple juice. *Journal of Food Engineering, 102*(4), 334–339.

29. El Rayess, Y., Albasi, C., Bacchin, P., Taillandier, P., Raynal, J., Mietton-Peuchot, M., & Devatine, A., (2011). Cross-flow microfiltration applied to oenology: A review. *Journal of Membrane Science, 382*(1, 2), 1–19.

30. El-Abbassi, A., Khayet, M., Kiai, H., Hafidi, A., & García-Payo, M. C., (2013). Treatment of crude olive mill wastewaters by osmotic distillation and osmotic membrane distillation. *Separation and Purification Technology, 104*, 327–332.

31. El-Abbassi, A., Kiai, H., Hafidi, A., García-Payo, M. C., & Khayet, M., (2012). Treatment of olive mill wastewater by membrane distillation using polytetrafluoroethylene membranes. *Separation and Purification Technology, 98*, 55–61.

32. Emani, S., Uppaluri, R., & Purkait, M. K., (2013). Preparation and characterization of low cost ceramic membranes for mosambi juice clarification. *Desalination, 317*, 32–40.

33. Fakhru'l-Razi, A., (1994). Ultrafiltration membrane separation for anaerobic wastewater treatment. *Water Science & Technology, 30*(12), 321–327.

34. Firmán, L. R., Pagliero, C., Ochoa, N. A., & Marchese, J., (2015). PVDF/PMMA membranes for lemon juice clarification: Fouling analysis. *Desalination Water Treatment, 55*(5), 1167–1176.

35. Gómez, B. J., Palacios, M. V. M., Szekely, G. P., Veas, L. R., & Pérez, R. L., (2003). Comparison of electrodialysis and cold treatment on an industrial scale for tartrate stabilization of sherry wines. *Journal of Food Engineering, 58*(4), 373–378.

36. Gonçalves, F., Fernandes, C., Cameira, D. S. P., & De Pinho, M. N., (2003). Wine tartaric stabilization by electrodialysis and its assessment by the saturation temperature. *Journal of Food Engineering, 59*(2), 229–235.
37. Guglielmi, G., Andreottola, G., Foladori, P., & Ziglio, G., (2009). Membrane bioreactors for winery wastewater treatment: Case-studies at full scale. *Water Science & Technology, 60*(5), 1201–1207.
38. Gulec, H. A., Bagci, P. O., & Bagci, U., (2017). Clarification of apple juice using polymeric ultrafiltration membranes: A comparative evaluation of membrane fouling and juice quality. *Food Bioprocess Technology, 10*(5), 875–885.
39. Guo, S., Luo, J., Wu, Y., Qi, B., Chen, X., & Wan, Y., (2018). Decoloration of sugarcane molasses by tight ultrafiltration: Filtration behavior and fouling control. *Separation and Purification Technology, 204*, 66–74.
40. Hamachi, M., Gupta, B. B., & Ben, A. R., (2003). Ultrafiltration: A means for decolorization of cane sugar solution. *Separation and Purification Technology, 30*(3), 229–239.
41. Hubadillah, S. K., Othman, M. H. D., Harun, Z., Jamalludin, M. R., Zahar, M. I. I. M., Ismail, A. F., Rahman, M. A., & Jaafar, J., (2019). High strength and antifouling metakaolin-based ceramic membrane for juice clarification. *Journal of Australian Ceramic Society, 55*(2), 529–540.
42. Kawa-Rygielska, J., Pietrzak, W., Regiec, P., & Stencel, P., (2013). Utilization of concentrate after membrane filtration of sugar beet thin juice for ethanol production. *Bioresource Technology, 133*, 134–141.
43. Kiatkittipong, W., Wongsuchoto, P., & Pavasant, P., (2009). Life cycle assessment of bagasse waste management options. *Waste Management, 29*(5), 1628–1633.
44. Koros, W. J., Ma, Y. H., & Shimidzu, T., (1996). Terminology for membranes and membrane processes (IUPAC recommendations 1996). *Pure & Applied Chemistry, 68*(7), 1479–1489.
45. Krstić, D. M., Antov, M. G., Peričin, D. M., Höflinger, W., & Tekić, M. N., (2007). The possibility for improvement of ceramic membrane ultrafiltration of an enzyme solution. *Biochemical Engineering Journal, 33*(1), 10–15.
46. Labus, K., Wiśniewski, Ł., Cieńska, M., & Bryjak, J., (2020). Effectivity of tyrosinase purification by membrane techniques versus fractionation by salting out. *Chemical Papers, 74*, 2267–2275.
47. Laorko, A., Li, Z., Tongchitpakdee, S., Chantachum, S., & Youravong, W., (2010). Effect of membrane property and operating conditions on phytochemical properties and permeate flux during clarification of pineapple juice. *Journal of Food Engineering, 100*(3), 514–521.
48. Li, F., Yan, H., Li, W., Zhao, J., & Ming, J., (2019). A comparative study of the effects of ultrafiltration membranes and storage on phytochemical and color properties of mulberry juice. *Journal of Food Science, 84*(12), 3565–3572.
49. Lipnizki, F., (2010). Basic aspects and applications of membrane processes in agro-food and bulk biotech industries. In: Drioli, E., & Giorno, L., (eds.), *Comprehensive Membrane Science and Engineering* (1st edn., Vol. 4, pp. 165–194). Oxford–United Kingdom: Elsevier.
50. Lu, H., Peng, M., Zhang, G., Li, B., & Li, Y., (2019). Brewery wastewater treatment and resource recovery through long term continuous-mode operation in pilot photosynthetic bacteria-membrane bioreactor. *Science of the Total Environment, 646*, 196–205.

Potential of Membrane Technology

51. Lu, H., Zhang, G., Dai, X., Schideman, L., Zhang, Y., Li, B., & Wang, H., (2013). A novel wastewater treatment and biomass cultivation system combining photosynthetic bacteria and membrane bioreactor technology. *Desalination, 322*, 176–181.

52. Luo, J., Guo, S., Wu, Y., & Yinhua, W., (2018). Separation of sucrose and reducing sugar in cane molasses by nanofiltration. *Food Bioprocess Technology, 11*, 913–925.

53. Madaeni, S. S., & Mansourpanah, Y., (2006). Screening membranes for COD removal from dilute wastewater. *Desalination, 197*(1–3), 23–32.

54. Mangindaan, D., Khoiruddin, K., & Wenten, I. G., (2018). Beverage dealcoholization processes: Past, present, and future. *Trends Food Science & Technology, 71*, 36–45.

55. Mietton-Peuchot, M., (2010). New applications for membrane technologies in enology; Chapter 6. In: Peinemann, K. V., Nunes, S. P., & Giorno, L., (eds.), *Membranes for Food Applications* (Vol. 3, pp. 119–127). Weinheim: Wiley-VCH Verlag GmbH & Co.

56. Nagano, A., Arikawa, E., & Kobayashi, H., (1992). The treatment of liquor wastewater containing high-strength suspended solids by membrane bioreactor system. *Water Science & Technology, 26*(3, 4), 887–895.

57. Naidoo, D., Ramdhani, N., & Bux, F., (2008). Microbial community analysis of a full-scale membrane bioreactor treating industrial wastewater. *Water Science & Technology, 58*(8), 1589–1594.

58. Nandi, B. K., Das, B., & Uppaluri, R., (2012). Clarification of orange juice using ceramic membrane and evaluation of fouling mechanism. *Journal of Food Process Engineering, 35*(3), 403–423.

59. Omar, J. M., Nor, M. Z. M., Basri, M. S. M., & Che Pa, N. F., (2020). Clarification of guava juice by an ultrafiltration process: Analysis on the operating pressure, membrane fouling and juice qualities. *Food Research, 4*(Suppl. 1), 85–92.

60. Paolucci-Jeanjean, D., Belleville, M. P., Rios, G. M., & Zakhia, N., (2000). The effect of enzyme concentration and space-time on the performance of a continuous recycle membrane reactor for one-step starch hydrolysis. *Biochemical Engineering Journal, 5*(1), 17–22.

61. Peinemann, K. V., Nunes, S. P., & Giorno, L., (2010). *Membranes for Food Applications* (Vol. 3, p. 264). Weinheim: Wiley-VCH Verlag GmbH & Co.

62. Pinho, M. N. D., (2010). Membrane processes in must and wine industries; Chapter 5. In: Peinemann, K. V., Nunes, S. P., & Giorno, L., (eds.), *Membranes for Food Applications* (Vol. 3, pp. 105–118). Weinheim: Wiley-VCH Verlag GmbH & Co.

63. Purwasasmita, M., Kurnia, D., Mandias, F. C., Khoiruddin, & Wenten, I. G., (2015). Beer dealcoholization using non-porous membrane distillation. *Food and Bioproducts Processing, 94*, 180–186.

64. Qin, G., Lü, X., Wei, W., Li, J., Cui, R., & Hu, S., (2015). Microfiltration of kiwifruit juice and fouling mechanism using fly-ash-based ceramic membranes. *Food and Bioproducts Processing, 96*, 278–284.

65. Rai, P., Majumdar, G. C., Sharma, G., Das, G. S., & De, S., (2006). Effect of various cutoff membranes on permeate flux and quality during filtration of mosambi (*Citrus sinensis* (L.) Osbeck) juice. *Food and Bioproducts Processing, 84*(3), 213–219.

66. Rayess, Y. E., & Mietton-Peuchot, M., (2016). Membrane technologies in wine industry: An overview. *Critical Reviews in Food Science and Nutrition, 56*(12), 2005–2020.

67. Ribeiro, H., Janssen, J., Kobayashi, I., & Nakajima, M., (2010). Membrane emulsification for food application; Chapter 7. In: Peinemann, K. V., Nunes, S. P., & Giorno, L., (eds.),

Membranes for Food Applications (Vol. 3, pp. 129–166). Weinheim: Wiley-VCH Verlag GmbH & Co.,

68. Severcan, S., Uzal, N., & Kahraman, K., (2020). Clarification of apple juice using new generation nanocomposite membranes fabricated with TiO_2 and Al_2O_3 nanoparticles. *Food and Bioprocess Technology, 13*(3), 391–403.

69. Sharma, H. P., Patel, H., & Sugandha (2017). Enzymatic added extraction and clarification of fruit juices: A review. *Critical Reviews in Food Science and Nutrition, 57*(6), 1215–1227.

70. Shi, C., Rackemann, D. W., Moghaddam, L., Wei, B., Li, K., Lu, H., Xie, C., Hang, F., & Doherty, W. O. S., (2019). Ceramic membrane filtration of factory sugarcane juice: Effect of pretreatment on permeate flux, juice quality and fouling. *Journal of Food Engineering, 243*, 101–113.

71. Soares, P. A. M. H., Geraldes, V., Fernandes, C., Dos Santos, P. C., & De Pinho, M., (2009). Wine tartaric stabilization by electrodialysis: Prediction of required deionization degree. *American Journal of Enology and Viticulture, 60*(2), 183–188.

72. Thomassen, J. K., Faraday, D. B. F., Underwood, B. O., & Cleaver, J. A. S., (2005). The effect of varying transmembrane pressure and crossflow velocity on the microfiltration fouling of a model beer. *Separation and Purification Technology, 41*(1), 91–100.

73. Ulbricht, M., Ansorge, W., Danielzik, I., König, M., & Schuster, O., (2009). Fouling in microfiltration of wine: The influence of the membrane polymer on adsorption of polyphenols and polysaccharides. *Separation and Purification Technology, 68*(3), 335–342.

74. Vu, T., LeBlanc, J., & Chou, C. C., (2020). Clarification of sugarcane juice by ultrafiltration membrane: Toward the direct production of refined cane sugar. *Journal of Food Engineering, 264*, 109682.

75. Wang, D., Fritsch, J., & Moraru, C. I., (2019). Shelf life and quality of skim milk processed by cold microfiltration with a 1.4-μm pore size membrane, with or without heat treatment. *Journal of Dairy Science, 102*(10), 8798–8806.

76. Wen-qiong, W., Yun-chao, W., Xiao-feng, Z., Rui-xia, G., & Mao-Lin, L., (2019). Whey protein-membrane processing methods and membrane fouling mechanism analysis. *Food Chemistry, 289*, 468–481.

77. Wynberg, R., (2016). Access and benefit-sharing: Key points for policy-makers. *The Food and Beverage Industry and Agriculture* (p. 13). The ABS Capacity Development Initiative, Eschborn, Germany.

PART III
MATHEMATICAL MODELLING IN FOOD PROCESSING

CHAPTER 8

MATHEMATICAL MODELING IN FERMENTATION PROCESSES

ANKIT PALIWAL and HIMJYOTI DUTTA

ABSTRACT

Fermentation is any microorganism-regulated food processing method. It targets at enhancing functional properties like texture, flavor, aroma, and nutritional status through bioconversion of the food components and liberation of specific metabolites. The primary products of food fermentation are alcohols and acids. However, industrially, fermentation processes are also designed for production and subsequent isolation of important commercial enzymes and metabolites. Any fermentation process is regulated by microbial activity, growth, viability, formation, and interaction of products and changes in acidity and pH of the fermenting medium. The process parameters like substrate availability, growth limiting factors/inhibitor, availability of oxygen, and a number of environmental conditions like temperature, and pH are responsible for the proliferation and activity of the fermenting species. In industry, each of these conditions needs to be optimized for best control of the process. The process in itself is hard to predict and quantify without considering all the factors involved in microbial growth and bioconversion. A number of mathematical modeling approaches have been taken up to study and optimize different fermentation processes. These include simpler microbial growth kinetics for real-time production quantification to more complicated models involving response surface methodology and artificial intelligence-based neural network. Such models help in understanding of the fermentation processes and their large-scale application. This chapter discusses all mathematical approaches for fermentation process optimization

---

Advances in Food Process Engineering: Novel Processing, Preservation, and Decontamination of Foods.
Megh R. Goyal, N. Veena, & Ritesh B. Watharkar (Eds.)
© 2023 Apple Academic Press, Inc. Co-published with CRC Press (Taylor & Francis)

218   Advances in Food Process Engineering

and their development in research, considering the readership of students, researchers, and industrial process developers.

8.1 INTRODUCTION

Fermentation is among the oldest methods used for conversion, processing, and preservation of foods. Long before the understanding of the process, fermentation was used for conversion and preservation of milk, vegetables, and meat in an ethno-artistic manner. Gradually, the traditional practices were developed for handling, processing, and preservation of food. These processes not only help in preservation but also improve quality and acceptability of the final product. In most of the situations, the main objective of fermentation is to improve functional properties like, aroma, flavor, texture, and nutritive value of the food product.

Being a biological process, fermentation is an inter-related complex process, which might involve single strain or mixed strain microbial cultures for fermentation. Although a starter has been used traditionally as the seed for a fermentation process, a pure microbial culture can also be utilized for fermentation. In recent times, active enzymes are also utilized for commercial fermentation processes. Due to the unpredictable nature of microbial growth and dependency on various environmental, physical, and chemical parameters, it is not easy to maintain desired consistency in product quality. Apart from the carbon source, factors like time, temperature, and pH have the major effects on microbial growth and metabolism of the fermenting microbes, and hence on the overall fermentation process. A study on the effects of change in these environmental parameters on final product formation, which is mostly acid or alcohol, could help in optimization of the process and maximization of the final intended product. Modeling of the process would help in planning and standardization of the process.

Even for the set parameters, the required time for a single batch process could differ as it not only depends upon substrate concentration and temperature but also on the initial microbial load of the starter culture, growth phase of microbes is present in culture and their viability. It is difficult to summarize all process parameters in a simple growth-associated or temperature-dependent model. Each fermentation process requires a unique model for optimization of process and standardization of final product.

This chapter discusses different mathematical and computational models for measuring and regulating fermentation procedures by researchers.

8.2 BASICS OF FERMENTATION PROCESSES

By its nature, fermentation is a metabolic process, during which sugar is consumed anaerobically for generation of energy in the form of adenosine triphosphate (ATP) using glycolysis. The glycolysis pathway ends with the production of pyruvate (pyruvic acid). During fermentation, this pyruvate is reduced in the form of acid or alcohol through a series of reactions for regenerating nicotinamide adenine dinucleotide (NAD^+), which in turn is required for continuation of glycolysis. On the basis of the final product, fermentation has been classified into three different categories: lactic acid fermentation, alcoholic fermentation, and acetic acid fermentation.

Lactic acid fermentation is the most common method of food preservation, where the food is preserved by a reduction in pH due to the formation of the acid. The fermentation is carried out by a consortium of bacteria cumulatively known as the lactic acid bacteria (LAB) and few yeast species (*Saccharomyces boulardii*). It is generally used for preservation of milk, fruits, vegetables, meat, and cereals. Of the two classes of lactic acid fermentation, homo-lactic fermentation yields two lactic acid molecules per glucose molecule metabolized. Hetero-lactic fermentation also called the phospoketolase pathway produces ethanol and carbon dioxide (CO_2) along with a lactic acid molecule. In alcoholic fermentation, mostly yeasts and some bacteria converts sugars into alcohol and CO_2. This requires anaerobic condition. However, under development or induction of aerobic atmosphere, acetic acid bacteria ferment sugar and alcohol in the broth into acetic acid.

Therefore, acetic acid is also sometimes called a byproduct of alcoholic fermentation due to occurrence in the same batch of fermentation, although different metabolic pathways as follows (Figure 8.1):

- **Lactate:** $C_6H_{12}O_6 + 2ADP + 2P_i \rightarrow 2C_3H_6O_3 + 2ATP + 2H_2O.$
- **Acetate:** $C_6H_{12}O_6 + 4ADP + 4P_i + 4NAD^+ \rightarrow 2C_2H_4O_2 + 4ATP + 4NADH + 4H^+ + 2CO_2 + 2H_2O.$
- **Ethanol$_1$:** $C_6H_{12}O_6 + 2ADP + 2P_i \rightarrow 2C_2H_6O + 2ATP + 2CO_2 + 2H_2O.$
- **Ethanol$_2$:** $C_6H_{12}O_6 + 2ADP + 2P_i + 2NAD^+ + 2NADPH \rightarrow 2C_2H_6O + 2ATP + 2NADH + 2NADP^+ + 2CO_2 + 2H_2O.$
- **Acetone:** $C_6H_{12}O_6 + 3ADP + 3P_i + 4NAD^+ \rightarrow C_3H_6O + 3ATP + 4NADH + 4H^+ + 3CO_2 + 2H_2O.$
- **Butyrate:** $C_6H_{12}O_6 + 3ADP + 3P_i + 2NAD^+ \rightarrow C_4H_8O_2 + 3ATP + 2NADH + 2H^+ + 2CO_2 + 3H_2O.$

- **Butanol$_1$**: $C_6H_{12}O_6 + 2ADP + 2P_i \rightarrow C_4H_{10}O + 2ATP + 2CO_2 + 3H_2O$.
- **Butanol$_2$**: $C_6H_{12}O_6 + 2ADP + 2P_i + NAD^+ + NADPH \rightarrow C_4H_{10}O + 2ATP + NADH + NADP^+ + 2CO_2 + 3H_2O$.

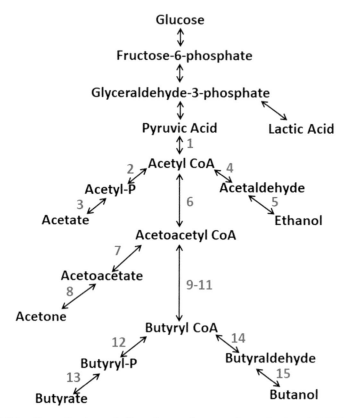

FIGURE 8.1 Generalized metabolic pathway of acetone–butanol–ethanol (ABE) fermentation with enzymes involved, such as: (1) pyruvate ferrodoxin oxidoreductase; (2) phosphotransacetylase; (3) acetate kinase; (4) acetaldehyde dehydrogenase; (5) ethanol dehydrogenase; (6) thiolase; (7) CoA transferase; (8) acetoacetate decarboxylase; (9) 3-hydroxybutyryl-CoA dehydrogenase; (10) crotonase; (11) butyryl-CoA dehydrogenase; (12) phosphotransbutyrylase; (13) butyrate kinase; (14) butyraldehyde dehydrogenase; (15) butanol dehydrogenase.

8.3 ROLE OF FERMENTATION IN FOOD PROCESSING

Food fermentation can be defined as a controlled microbial growth and enzymatic conversions of major and minor food components. With the help of metabolic activity of microbes, fermentation helps in bioconversion,

Mathematical Modeling in Fermentation Processes 221

transformation, stabilization, and preservation of food materials. Static state fermentation models are therefore necessarily designed for industrial application [2]. Although fermentation was initially developed for the preservation of perishable agriculture produces, later it was developed to be applied for enhancing functional, organoleptic, and nutritional characteristics of food products. Fermentation reportedly affects the bioavailability of minerals and phytonutrients, enhances the availability of proteins, peptides, and amino acids, vitamins, and improves digestibility of food [21]. Most of the time, fermentation also removes anti-nutritional factors like tannins, phytates, oxalic acid, and various other inhibitors [18].

8.4 PRINCIPLE OF FERMENTATION KINETICS

Kinetic relates to the rate of a reaction and can be defined as progressive change in quantity of substrate, formation of product and the intermediate steps involved in it. Mathematical models have proven as one of the most utilized and effective tools for optimizing any process, including fermentation. Any mathematical model could be defined as a set of propositions with reference to mathematical relations between measured or measurable quantities associated with the system and process. The complexity or simplicity of a mathematical model depends on the number of inputs, popularly termed as variables.

A bioreactor model is generally used to describe a complete bioprocess system. This includes two sub-models. The first is the balance model, which explains the heat and mass transfer within the system, and the second kinetic model explains microbial growth and bio-conversion of the substrate. The majority of the developed bioreactor models focus majorly on the second aspect of kinetics. This is mainly due to the heterogeneity of the system making the detail of mass balance and heat transfer very complex. Moreover, these aspects cannot be explained without considering the growth kinetics. For simplification of the whole situation, most of these models are based upon specific conditions like specific substrate conversion, ingredient incorporation, product formation or specific environmental conditions like temperature, pH, and atmospheric pressure. Hence, most of fermentation models are based upon growth kinetics. These models are based upon physical, chemical, or biological principles, which form the basis of a model. It is followed by developing the hypothesis and correlation between measurable data, generating the final kinetic model.

Numerous kinetic models or profiles have been proposed for explanation of the fermentation processes. A few are linear curve, exponential curve, logistic curve, and fast-acceleration/slow-deceleration curve models. They are primarily based upon nature of fermentation and the relationship between microbial growth and product formation. Fermentation, being a complex process involving microorganisms and a high level of complexity between reactions, where several interrelated reactions occur simultaneously, it is not easy to obtain strict reproducibility of the results after each trial. A small variation in process or substrate can cause substantial differences in fermentation process output. However, a good and appropriate model can provide the best assessment for optimization of the process with maximization of product.

According to Gaden [5], production of product or metabolite produces of fermentation could be divided in three categories. The first category directly relates to the substrate utilization, like ethanol. The second category products like citric acid indirectly relates to the substrate utilization. The third category product formation is non-associated with substrate utilization, like penicillin formation.

Maxon [14] provided slightly different classification. According to Maxon, the first category products' formation and growth are synonymous, for example, microbial biomass production. In the second category, product formation is associated with growth, for example, ethanol and the last category product formation are not associated with microbial growth, like penicillin [3]. For simplification, the fermentation product is expressed in the term of yield coefficient. Yield coefficient is the overall efficiency of the conversion of substrate to cell mass or specific product as follows:

$$Y_{X/S} = \frac{\partial X}{\partial S}$$

$$Y_{P/S} = \frac{\partial P}{\partial S} \tag{1}$$

Most of the models start at a single point with quantification of the rate of change of a metabolite, which could be the product or the substrate [16].

Rate of change of metabolite = [Population size] × [Stoichiometric matrix] × {[Flux vector] – [Dilution rate]} × [Metabolite concentration] (2)

Different components in Eqn. (2) could be explained differently with specific assumptions. The population size could be explained as either dry cell mass or optical cell density or cell growth rate. Stoichiometric matrix includes metabolic network of different reactions which includes different metabolites (both substrate and product). A positive notion expresses product formation while negative represents depletion of substrate. It is the one of the most important parts of the equation which depends upon information regarding biological reaction, its quantification, and experimental data. The flux vector represents the reaction rate, which could show the influence of enzyme concentration, limiting substrate and inhibitors concentration on reaction rate. Dilution rate expresses the movement of metabolites during continuous and plug flow process.

On the basis of operation mode, fermentation processes can be divided into three categories (Figure 8.2): batch fermentation, continuous fermentation, and tubular or fed-batch fermentation.

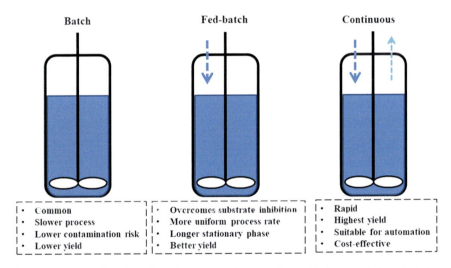

FIGURE 8.2 Batch, fed-batch, and continuous fermentation and their general characteristics.

Batch fermentation occurs in a closed system and contains a limited amount of nutrients on which inoculate grows. In such a system, cells grow until the required components get exhausted from the system or else the environmental conditions become unsuitable for growth due to formation of any inhibitor molecule. The batch operation could be divided in four distinct phases. During the lag phase, the microbial inoculate adapts to the environment. In the exponential phase, cells multiply and grow exponentially in population. This is followed by the stationary phase during which microbial growth rate remains equal to microbial death rate and finally comes the deceleration/death phase when microbial death rate overtake the growth rate resulting in ending of the microbial population. Mathematically, growth during the exponential phase can be described by two methods, i.e., cell biomass and cell numbers.

$$\frac{dx}{dt} = \mu x \tag{3}$$

$$x_t = x_0 e^{\mu t}$$

Specific growth rate is given by Monod equation:

$$\mu = \frac{\mu_{max} \cdot S}{K_s + S} \tag{4}$$

In a continuous fermentation operation, fresh media is continuously added to the system and also the fermented broth containing microbial biomass and other metabolites are continuously removed from the system. The main object of continuous fermentation is to prolong the exponential growth phase as is ideal for growth associated product formation. During the operation, the flow rate or the dilution rate is maintained to keep the fermentation conditions in steady-state. Dilution rate refines the volume which replaces within the fermentation system.

$$D = \frac{F}{V} \tag{5}$$

$$\frac{dx}{dt} = \mu x - Dx$$

For steady state during the operation:

$$dx/dt = 0$$

$$\mu = D \tag{6}$$

Monod equation for continuous fermentation is given as follows:

$$D = \frac{\mu_{max} \cdot S}{K_s + S} \tag{7}$$

During the fed-batch operation, growth limiting substrate is continuously added to the system to keep the fermentation in steady-state. During the operation, the volume of the fermenter is increased continuously and the biomass concentration and product concentration are kept at constant levels. The volume at time t is given by:

Mathematical Modeling in Fermentation Processes

$$V_t = V_0 + F.t \tag{8}$$

The dilution rate is given as:

$$D = \frac{F}{\left(V_0 + F.t\right)} \tag{9}$$

Most common mathematical models work on a basic relationship, which take into account specific synthesis pathways, biomass production, and possible effect of inhibitions, temperature, pH, etc. Two approaches can be utilized for explanation of a mathematical model of fermentation process, namely: structured model and unstructured model.

Structured model uses metabolic network structure information for accessing metabolic activity of the cell. It considers the basic aspects of cell structure, its function and composition. Being the complex model, generally structured model assumes a quasi-steady-state assumption to solve the equation using flux balance analysis or elementary mode analysis [24]. Flux balance analysis considers metabolic pathways for steady-state balance during the reaction. Generally, it uses stoichiometric equations for substrate utilization, and conversion into primary product and biomass generation [12]. Elementary mode analysis converts steady-state metabolic pathways into irreducible linear equations [13].

Unstructured models do not consider any physiological characteristics of the cell and consider biomass and metabolic routes as common variable. It makes unstructured model less sensitive to certain changes in metabolic pathways and reaction rate. The main parameters for unstructured models include operation mode, substrate characterization and concentration, substrate utilization rate, biomass, and product formation rate, initial substrate concentration and inhibition.

ABE fermentation has been extensively studied and could be used as a base for understanding of different fermentation model. Lactic acid synthesis is another fermentation system, where numerous mathematical models have been developed for the expression of the biological process. Although both structured and unstructured model has been utilized, unstructured models are predominantly used for lactic acid production owing to their nature [8]. Lactic acid fermentation can be described in three different ways like in the terms of substrate concentration, product concentration or biomass concentration. Apart from these, there are different environmental factors which can affect the metabolic rate and hence fermentation kinetics [8]. Due to the complexity involved in the representation of such factors, unstructured models have been often preferred. They can be utilized for representing particular metabolic pathway and set of reaction occurring during the process.

Effect of enzyme and yeast during bread making and dough leavening has been also vastly studied. The presence of sucrose as raw material has been found to improve *Z. mobilis* ability in leavening of dough in a better way due to better microbial growth, more gaseous production, and better retention in the structure. The results are comparable with the effect of *S. cerevisiae* on dough leavening [19]. Yeast level and fermentation temperature also affect the production of ochratoxin A, a foodborne mycotoxin during bread manufacturing. It was reported that the increase of fermentation temperature by 10°C along with yeast level from 1% to 2% could reduce ochratoxin A by 59% in the final baked bread [17].

Doehlert approach has been utilized for optimization of sorghum fermentation conditions, mainly time and temperature for ting (a fermented product) production [4]. Adebo et al. [1] used a second-order polynomial equation for determination of optimum fermentation time and temperature for maximum content of phenolic, tannin, flavonoid, and antioxidant activity with good microbial growth and high organic acid production with a reduced pH during formation of ting.

8.5 MODELLING OF A FERMENTATION PROCESS

8.5.1 PRIMARY MODELS

The change in substrate concentration or microbial population or product concentration under specific environmental circumstances can be explained by primary models. Primary models use elementary mathematics, where one factor is considered at a time for establishing the relationship. Monod model, Logistic model, Rosso model, Baranyi model, and Gompertz functions are the most common primary models. Most commonly derived primary models for cell growth, substrate consumption and production of metabolites are based on simple differential equation. Commonly, cell growth rate is explained using sigmoidal equation, modified logistic equation with consideration of specific death rate or self-inhibition due to nutrient limitation (sugar and amino acids) or end-product inhibition (alcohol and acids) of pH.

Gompertz model, being sigmoidal in nature, has been used extensively for microbial growth. It has been used to represent the relationship between microbial population and specific growth rate, considering time as a factor [11]. The elementary exponential equation could be represented as:

Mathematical Modeling in Fermentation Processes

$$\frac{dN}{dt} = \mu N \tag{10}$$

where; N is the microbial population.

Maintenance energy model is used for expression of substrate consumption during the fermentation process. Model for production of metabolites is developed assuming growth-associated product [14, 19, 27].

Cell growth:

$$\frac{dX}{dt} = \left[\mu_{max} \left(1 - \frac{X}{X_{max}} \right)^n - \propto \right] X, \text{ For } t > 1$$

$$\frac{dX}{dt} = \left[\mu_{max} \gamma_i \right] X, \text{ For } \geq \lambda \tag{11}$$

Sugar consumption:

$$\frac{dM}{dt} = -\frac{1}{Y_{X/M}} \frac{dX}{dt} - m_M X \tag{12}$$

$$\frac{dF}{dt} = -\frac{1}{Y_{X/F}} \frac{dX}{dt} - m_F X \tag{13}$$

Lactic acid production:

$$\frac{dL}{dt} = -\frac{1}{Y_{L/S}} (dM/dt + dF/dt) \tag{14}$$

Acetic acid production:

$$\frac{dA}{dt} = -\frac{1}{Y_{A/S}} (dM/dt + dF/dt) \tag{15}$$

Bacteriocin production:

$$\frac{dB}{dt} = \frac{k_b dX}{dt} - k_{inact} XB \text{ for } X > X \tag{16}$$

where; X is the biomass concentration (in grams of cell dry mass (CDM) per liter); X_{max} is the maximum attainable biomass concentration (in grams of CDM per liter); μ_{max} is the maximum specific growth rate (per hour); 't' is the time (in hours); λ is the duration of the lag phase (in hours); 'n' is the inhibition exponent; $\propto$ is the specific rate of death (per hour); γ_i is the complex inhibitory function; 'M' is the residual maltose concentration (in grams of maltose per liter); 'F' is the residual fructose concentration (in grams of fructose per liter); m_M is the maintenance coefficient of maltose (in grams of maltose per gram of CDM per hour); m_F is the maintenance coefficient of fructose (in grams of fructose per gram of CDM per hour); $Y_{X/M}$ is the cell yield coefficient based on maltose (in grams of CDM per gram of maltose); $Y_{X/F}$ is the cell yield coefficient based on fructose (in grams of CDM per gram of fructose); 'L' is the amount of lactic acid produced (in grams of lactic acid per liter); $Y_{L/S}$ is the yield coefficient for lactic acid production based on fructose and maltose (in grams of lactic acid per gram of substrate); 'A' is the amount of acetic acid produced (in grams of acetic acid per liter); $Y_{A/S}$ is the yield coefficient for acetic acid production based on fructose and maltose (in grams of acetic acid per gram of substrate); 'B' is the soluble bacteriocin activity (in arbitrary units (AU) per liter); k_B is the specific bacteriocin production (in AU per gram of CDM); k_{inact} is the apparent rate of bacteriocin inactivation (in liters per gram of CDM per hour); and X' is the biomass concentration which is needed to produce detectable amounts of bacteriocin.

Rosso model proposed the break-in cell growth rate between lag phase and exponential phase [23]. For representation of this, the model was utilized to different time zones, lag times and generation times. The final equation of Rosso model could be represented as:

$$\ln\left(\frac{N}{N_o}\right) = \ln\left(\frac{N_{max}}{N_o}\right) \times \exp\left(-\exp\left(-\frac{\mu_{max}e}{\ln\left(\frac{N_{max}}{N_o}\right)}(\lambda - t) + 1\right)\right) \quad (17)$$

$$\text{For } t < \lambda, \ln(N) = \ln(N_o) \quad (18)$$

$$\text{For } t \geq \lambda \ln(N) = \ln(N_{max}) - \ln\left[1 + \left(\frac{N_{max}}{N_o} - 1\right)\exp\left(-\mu_{max}(t - \lambda)\right)\right], \quad (19)$$

where; N_o is the inoculum bacterial concentration; N_{max} is the maximum bacterial concentration.

8.5.2 SECONDARY MODELS

Secondary models are formulated to show the interdependency of environmental conditions on fermentation parameters. Square root model, response surface models and ɤ-concept models are the most common secondary models used to explain dependency of pH, temperature, and other environmental factors on fermentation parameters like specific growth rate, product formation rate, etc.

Mathematical Modeling in Fermentation Processes

The square root model was proposed by Ratkowsky et al. [22] for calculating microbial growth rate under refrigerated state. In place of Arrhenius law, the linear relationship was proposed between the square root of specific growth rate and temperature [22]. The model also utilized the minimum temperature required for microbial growth for calculation of maximum specific growth rate. The equation also utilized a coefficient which could be described in a different manner.

$$\sqrt{\mu_{max}} = b(T - T_{min}) \tag{20}$$

where; μ_{max} is the maximum growth rate; T_{min} is the minimum temperature for bacteria growth; and 'b' is a coefficient.

Due to the complex nature of fermentation, generally there are other factors too which affect the fermentation process. Root square model cannot consider factors other than temperature in an effective way. The separate quadratic and cubic equations are utilized to represent the effect of those factors. Response surface model could be utilized for calculation of different growth associated and non-growth associated factors for a given temperature, pH range or other factors. Initial data showing the effect of temperature, pH, and other factors on specific growth rate might be required for the development of response surface model.

ɤ-concept models like Zwietering model and Rosso's cardinal model utilize a new optimum growth factor. Here it is assumed that all external factors could be represented in the form of ɤ-function and multiplication of this ɤ-function with optimum growth factor will result in specific growth rate. A simple ɤ-concept model could be represented as:

$$\mu = \mu_{opt} \cdot \gamma \tag{21}$$

Zwietering et al. [29] proposed this ɤ-factor as growth factor. For optimum conditions, the value of ɤ-factor is equal to 1, while for all other scenarios it ranges between 0 and 1 [29]. It is the ratio of specific growth rate to optimum growth rate, which depends upon the product of function of environmental factors like temperature, water activity (a_w), pH, and oxygen availability. Here it is assumed that all these factors are independent of each other.

$$\gamma = \frac{\mu}{\mu_{opt}} = g_T(T)g_{pH}(pH)g_{a_w}(a_w)g_{O_2}(O_2) \tag{22}$$

where; μ is the actual growth rate; μ_{opt} is the growth rate at optimal conditions; γ is the actual growth factor; gF(F) is a function of environmental factor F (T, temperature, pH, a_w, O_2).

230 Advances in Food Process Engineering

Rosso's cardinal model is based upon a similar approach, and it assumes the effect of environmental factors on maximum specific growth rate. Similar to τ-concept model, environmental factors like temperature, a_w, and pH act independently on microbial growth and affect maximum specific growth rate. For calculation of the effect of environmental factors, minimum, optimum, and maximum values for microbial growth of each are considered.

$$\mu_{max} = \mu_{opt} CM_2(T) CM_1(pH) CM_2(a_w) \tag{23}$$

where; μ_{max} is the maximum growth rate; CM_n is the function for environmental factor with minimum, optimum, and maximum values.

8.6 ASSESSMENT OF FERMENTATION KINETICS

8.6.1 EFFECTS OF TEMPERATURE AND pH

Temperature and pH are the most predominant factors which affect microbial growth and thus production rate during fermentation. Several mathematical models have been developed to describe effect of pH and temperature on specific growth rate, which represent metabolic activity of microorganism. Change in pH or acidity could easily be monitored during fermentation process. Although generally both pH and acidity are co-related to each other, it is also possible that with increase in acidity may not result in a change in process pH due to buffering effect.

One of the initial models given by Wijtes et al. [25], which was based upon modified Ratkowsky's square root model, showed the effect of temperature and pH on maximum specific growth rate.

$$\mu_{max} = b_1 \left\{ \left(pH - pH_{min}\right) \left[1 - e^{c_1(pH - pH_{max})}\right] \left(T - T_{min}\right)\right\}^2 \tag{24}$$

where; pH_{max} is the pH above which no growth occurs; pH_{min} is the pH below which no growth occurs; 'T' is the temperature (in °C); T_{min} is the temperature below which no growth occurs; and b_1 (per hour per square C) and c_1 (dimensionless) are biologically meaningless parameters.

The above equations were used by Rosso et al. [23], for developing relationship between the maximum specific growth rate and pH assuming constant temperature during fermentation:

$$\mu_{max} = b_2 \left\{ \left(pH - pH_{min}\right) \left[1 - e^{c_1(pH - pH_{max})}\right]\right\}^2 \tag{25}$$

where; b_2 is (per hour per square °C) biologically meaningless parameter.

Mathematical Modeling in Fermentation Processes

In the same study, Rosso et al. [23] used the equation given by Zwietering et al. [29]. They modified it to relate maximum specific growth rate to optimum, minimum, and maximum pH and temperature, respectively.

$$\mu_{max} = \mu_{opt}\gamma(T)\gamma(pH)$$

$$\gamma(pH) = \left\{\frac{\left(pH - pH_{min}\right)\left[1 - e^{c_2\left(pH - pH_{max}\right)}\right]}{\left(pH_{opt} - pH_{min}\right)\left[1 - e^{c_1\left(pH_{opt} - pH_{max}\right)}\right]}\right\}^2$$

$$\gamma(T) = \left\{\frac{\left(T - T_{min}\right)\left[1 - e^{c_2\left(T - T_{max}\right)}\right]}{\left(T_{opt} - T_{min}\right)\left[1 - e^{c_1\left(T_{opt} - T_{max}\right)}\right]}\right\}^2 \qquad (26)$$

where; pH_{opt} is the pH at which the μ_{max} is optimal; T_{opt} is the temperature at which the μ_{max} is optimal; T_{max} is the temperature above which no growth occurs; μ_{opt} (per hour) is the μ_{max} under optimal conditions (pH_{opt}, T_{opt}); c_2 (dimensionless) and c_3 (degrees Celsius $\geq$ 1) are regression coefficients.

Messens et al. [15] used the same model to find the influence of temperature and pH on growth and bacteriocin production by *Lactobacillus amylovorus* DCE 471. The model assumed no interaction between optimized temperature and pH for maximizing specific growth rate. Zhang et al. [27] used the cardinal model to study the influence of temperature and pH on growth and production of bacteriocin by *Pediococcus acidilactici* PA003 [27]. Gordeeva et al. [9] proposed the following model for studying the effect of pH on specific growth rate on product formation rate of lactic acid.

$$\mu_{max} = \frac{\mu_m}{1 + \left(k_{\mu 1} / \left[H^+\right]\right) + k_{\mu 2}\left[H^+\right]}$$

$$K_P = \frac{K_{Pm}}{1 + \left(k_{P1} / \left[H^+\right]\right) + k_{P2}\left[H^+\right]} \qquad (27)$$

where; K_P is constant of product inhibition (g per liter); $[H^+]$ is H ion concentration; μ_m is constant (per hour); and $k_{\mu 1}$, $k_{\mu 2}$, K_{Pm}, k_{P1}, k_{P2} are dimensionless constants.

Guerra [10] used three-dimensional Lotka–Volterra equation to establish a relationship between pH drop, nisin production and specific growth rate during batch fermentation of *Lactococcus lactis* CECT 539. The author first used logistic equations for determination of the absolute value of the rate of pH change, growth, and nisin production. Then, using those absolute values, proposed the following equations, which is based on a three-dimensional Lotka–Volterra equation to show the effect of pH on cell growth rate and nisin production rate:

$$\frac{dX}{dt} = \alpha_x.X - \beta_x.X.BT - \delta_x.X.pH$$

$$\frac{dBT}{dt} = -\alpha_{BT}.BT.X + \beta_{BT}.BT.pH - \delta_{BT}.BT \tag{28}$$

where; BT is nisin concentration (BU/ml); α_x is intrinsic growth rate (h–1); β_x is efficiency of nitrogen source utilization (ml/BU/h) for growth rather than for bacteriocin production; δ_x is constant (h–1) that represents the effect of the time course of pH on the growth; α_{BT} is efficiency of nitrogen source utilization (L/g/h) for bacteriocin production rather than for growth; β_{BT} is constant (h–1) that represents the effect of the time course of pH on nisin production; and δ_{BT} is coefficient adsorption of bacteriocin (h–1) onto producer cells.

To show the influence of pH and temperature on cell biomass growth, the primary model based on the Gompertz model was developed by Gibson et al. [7]. The influence as a function of time under specific environmental conditions was expressed as following logistic model [26, 28].

$$x(t) = C + \frac{A}{1 + e^{-B(t-D)}} \tag{29}$$

where; $x(t)$ is log10 (CFU/ml) of cell concentration at time 't'; C is the lower asymptote in units of log10 (CFU/ml); 'A' is equal to log10 (x_{max}/x_0); x_0 is the initial population density; x_{max} is the maximum population density; 'B' is the maximum relative growth rate (μ_{max}); and 'D' is the time in hours at which the absolute growth rate is maximum.

Cell biomass production is generally expressed in the term of CDM. Zhou et al. [28] used the Luedeking-Piret model, for estimation of CDM and bacteriocin production by batch fermentation using BC-25 strain of *L. plantarum*.

$$CDM = \frac{x_m}{1 + exp\left[2 + \frac{4.v_m}{X_m}(L-t)\right]}$$

$$AU = \frac{Y_{A/X}.X_m}{1 + \left(\frac{X_m}{X_0} - 1\right).e^{\frac{-4v_m.t}{X}}} - Y_{A/X} \cdot X_0 + \frac{K \cdot X_m^2}{4 \cdot v_m} \ln\left[\frac{X_0 \cdot \left(e^{\frac{-4v_m.t}{X}} - 1\right) + X_m}{X_m}\right] \tag{30}$$

Mathematical Modeling in Fermentation Processes 233

where; CDM is the biomass (g/L); X_m is the maximum biomass concentration (g/L); v_m is the CDM maximum growth rate (/h); 'L' is the lag time (h); 't' is the time (h); 'AU' is the bacteriocin activity (AU/ml); X_0 is the initial biomass (g/l); $Y_{A/X}$ is the growth-associated constant for bacteriocin production (AU/g); and 'K' is the non-growth-associated constant for bacteriocin production [(AU/g)/h].

For showing the combined effect of temperature and pH on cell growth during bacteriocin production, Zhou et al. [28] used response surface model, which also considered growth associated and non-growth associated constants.

$$\ln(\mu_{max}),(Y_{A/X} \text{ and } K) = C_0 + C_1.T + C_2.T^2 + C_3.pH + C_4.pH^2 + C_5.T.pH \quad (31)$$

where; 'T' is temperature in degree Celsius (°C); and C_n are constants.

8.6.2 EFFECTS OF OTHER FACTORS

Secondary models could be utilized to show the effect of other factors like salt, organic acids, and alcohol on the growth of microorganisms. Response surface model and ɤ-concept models are generally used for developing model to show the effect. Ganzle et al. [6] used ɤ-concept model for assessment of the effect of salt, alcohol, lactate, and acetate on microbial growth rate of *C. milleri* and *L. sanfranciscensis*.

$$\mu = \mu_{opt} \frac{I(I - I_{max})}{I(I - I_{max}) - I(I - I_{opt})^2} \quad (32)$$

where; 'I' is ionic strength; I_{max} and I_{opt} are maximum and optimum values of ionic strength.

The same model was also used by Neysens et al. [20] for calculating the effect of added salt or amylovorin L471 production on the maximum specific growth rate of *Lactobacillus amylovorus* DCE 471. The assessment of these factors on growth was done independent of each another.

8.7 SUMMARY

Although the kinetics of industrial fermentation processes are difficult to be optimized due to the wide versatility of process and product variables, yet number of the developed mathematical models have been discussed in this chapter, and these have markedly supported the bioprocessing industries to automatize the process protocols. Relevant research studies carried out to date suggest that multifunctional approaches in designing a suitable model for any specific fermentation process is the solution and hence is also the major challenge. Computational control of fermentation kinetics requires insight into the metabolite production, identification, and determination of reaction rates and developed precision for a defined controlled system.

KEYWORDS

- **adenosine triphosphate**
- **cell dry mass**
- **fermentation**
- **lactic acid bacteria**
- **mathematical modeling**
- **metabolite**
- **nicotinamide adenine dinucleotide**

REFERENCES

1. Adebo, O. A., Njobeh, P. B., Mulaba-Bafubiandi, A. F., Adebiyi, J. A., Desobgo, Z. S. C., & Kayitesi, E., (2018). Optimization of fermentation conditions for ting production using response surface methodology. *Journal of Food Processing and Preservation, 42*(1), e13381.
2. Bilal, M., & Iqbal, H. M. N., (2019). Sustainable bioconversion of food waste into high-value products by immobilized enzymes to meet bio-economy challenges and opportunities: A review. *Food Research International, 123*, 226–240.
3. Deindoerfer, F. H., (1960). Fermentation kinetics and model processes. *Advances in Applied Microbiology, 2*, 321–334.
4. Desobgo, S. C., Nso, J. E., & Tenin, D., (2011). Use of the response surface methodology for optimizing the action of mashing enzymes on wort reducing sugars of the madjeru sorghum cultivar. *African Journal of Food Science, 5*, 91–99.
5. Gaden, Jr. E. L., (1959). Fermentation process kinetics. *Journal of Biochemical and Microbiological Technology and Engineering, 1*(4), 413–429.
6. Ganzle, M. G., Ehmann, M., & Hammes, W. P., (1998). Modeling of growth of *Lactobacillus sanfranciscensis* and *Candida milleri* in response to process parameters of sourdough fermentation. *Applied and Environmental Microbiology, 64*(7), 2616–2623.
7. Gibson, A. M., Bratchell, N., & Roberts, T. A., (1988). Predicting microbial growth: Growth Responses of *Salmonellae* in a laboratory medium as affected by pH, sodium

Mathematical Modeling in Fermentation Processes 235

chloride and storage temperature. *International Journal of Food Microbiology, 6*(2), 155–178.

8. Gordeev, L. S., Koznov, A. V., Skichko, A. S., & Gordeeva, Y. L., (2017). Unstructured mathematical models of lactic acid biosynthesis kinetics: A review. *Theoretical Foundations of Chemical Engineering, 51*, 175–190.

9. Gordeeva, Y. L., Rudakovskaya, E. G., Gordeeva, E. L., & Borodkin, A. G., (2017). Mathematical modeling of biotechnological process of lactic acid production by batch fermentation: A review. *Theoretical Foundations of Chemical Engineering, 51*, 282–298.

10. Guerra, N. P., (2014). Modeling the batch bacteriocin production system by lactic acid bacteria by using modified three-dimensional Lotka-Volterra equations. *Biochemical Engineering Journal, 88*, 115–130.

11. Isabelle, L., & André, L., (2006). Quantitative prediction of microbial behavior during food processing using an integrated modeling approach: A review. *International Journal of Refrigeration, 29*(6), 968–984.

12. Kauffman, K. J., Prakash, P., & Edwards, J. S., (2003). Advances in flux balance analysis. *Current Opinion in Biotechnology, 14*(5), 491–496.

13. Kumar, M., Saini, S., & Gayen, K., (2014). Elementary mode analysis reveals that *Clostridium acetobutylicum* modulates its metabolic strategy under external stress. *Molecular BioSystems, 10*(8), 2090–2105.

14. Maxon, W. D., (1959). Aeration-agitation studies on the novobiocin fermentation. *Journal of Biochemical and Microbiological Technology and Engineering, 1*(3), 311–324.

15. Messens, W., Neysens, P., Vansieleghem, W., Vanderhoeven, J., & De Vuyst, L., (2002). Modeling growth and bacteriocin production by *Lactobacillus amylovorus* DCE 471 in response to temperature and pH values used for sourdough fermentations. *Applied and Environmental Microbiology, 68*(3), 1431–1435.

16. Millat, T., & Winzer, K., (2017). Mathematical modeling of clostridial acetone-butanol-ethanol fermentation. *Applied Microbiology and Biotechnology, 101*(6), 2251–2271.

17. Mozaffary, P., Milani, J. M., & Heshmati, A., (2019). The influence of yeast level and fermentation temperature on ochratoxin a decrement during bread making. *Food Science & Nutrition, 7*(6), 2144–2150.

18. Mukherjee, R., Chakraborty, R., & Dutta, A., (2016). Role of fermentation in improving nutritional quality of soybean meal: A review. *Asian-Australasian Journal of Animal Sciences, 29*(11), 1523–1529.

19. Musatti, A., Cappa, C., Mapelli, C., Alamprese, C., & Rollini, M., (2020). *Zymomonas mobilis* in bread dough: Characterization of dough leavening performance in presence of sucrose. *Foods, 9*(1), 89.

20. Neysens, P., & De Vuyst, L., (2005). Kinetics and modeling of sourdough lactic acid bacteria. *Trends in Food Science & Technology, 16*(1–3), 95–103.

21. Nkhata, S. G., Ayua, E., Kamau, E. H., & Shingiro, J. B., (2018). Fermentation and germination improve the nutritional value of cereals and legumes through activation of endogenous enzymes. *Food Science & Nutrition, 6*(8), 2446–2458.

22. Ratkowsky, D. A., Lowry, R. K., Mcmeekin, T. A., Stokes, A. N., & Chandler, R. E., (1983). Model for bacterial culture growth rate throughout the entire biokinetic temperature range. *Journal of Bacteriology, 154*(3), 1222–1226.

23. Rosso, L., Lobry, J. R., Bajard, S., & Flandrois, J. P., (1995). Convenient model to describe the combined effects of temperature and pH on microbial growth. *Applied and Environmental Microbiology, 61*(2), 610–616.

24. Segrè, D., Vitkup, D., & Church, G. M., (2002). Analysis of optimality in natural and perturbed metabolic networks. *Proceedings of the National Academy of Sciences, 99*(23), 15112–15117.
25. Wijtes, T., McClure, P. J., Zwietering, M. H., & Roberts, T. A., (1993). Modeling Bacterial growth of *listeria monocytogenes* as a function of water activity, pH, and temperature. *International Journal of Food Microbiology, 18*(2), 139–149.
26. Yang, E., Fan, L., Yan, J., Jiang, Y., Doucette, C., Fillmore, S., & Walker, B., (2018). Influence of culture media, pH, and temperature on growth and bacteriocin production of bacteriocinogenic lactic acid bacteria. *AMB Express, 8*, Article No. 10.
27. Zhang, J., Zhang, Y., Liu, S. N., Han, Y., & Zhou, Z. J., (2012). Modeling growth and bacteriocin production by *Pediococcus acidilactici* PA003 as a function of temperature and pH value. *Applied Biochemistry and Biotechnology, 166*(6), 1388–1400.
28. Zhou, K., Zeng, Y. T., Han, X. F., & Liu, S. L., (2015). Modeling growth and bacteriocin production by *Lactobacillus plantarum* BC-25 in response to temperature and pH in batch fermentation. *Applied Biochemistry and Biotechnology, 176*(6), 1627–1637.
29. Zwietering, M. H., Wijtzes, T., Rombouts, F. M., & Van't Riet, K., (1993). A decision support system for prediction of microbial spoilage in foods. *Journal of Industrial Microbiology, 12*(3–5), 324–329.

CHAPTER 9

MODELING OF HYDRATION KINETICS IN GRAINS

ANKIT PALIWAL, ASHISH M. MOHITE, and NEHA SHARMA

ABSTRACT

During the processing of cereal grains and seeds, soaking is generally considered as an initial processing step for cleaning and conditioning. The process of soaking is a slow in nature, which involves absorption of water and then migration of water particles or moisture within the seed. Water movement in the seed and intra-endosperm migration of moisture, is governed by the molecular diffusion, which is subjected to various factors like physical structure, soaking water temperature, duration, grain to water ratio, and others. During the process, apart from water migration, the solid molecules and compounds also leached out due to solubilization in the existing aqueous atmosphere, which exist in the cereal grains. The mathematical models are frequently employed for understanding and designing of industrial process and for their optimization. The simplicity and easy usability of these models gives them an advantage over other advanced statistical tools and programming. Mathematical models for hydration kinetics could be categorized in two ways as downward concave shape (DCS) hydration behavior and sigmoidal hydration behavior, and as empirical model and phenomenological model. While empirical models are simpler and are used extensively for studies of water hydration kinetics, but they ignore the basic physical structure of the cereal grain. This omission makes them much less usable for large size or complex grains, while studying the hydration kinetics. A phenomenological model, which is more complex than the previous one considers phenomena like migration via

---

Advances in Food Process Engineering: Novel Processing, Preservation, and Decontamination of Foods.
Megh R. Goyal, N. Veena, & Ritesh B. Watharkar (Eds.)
© 2023 Apple Academic Press, Inc. Co-published with CRC Press (Taylor & Francis)

convection, and concentration of water inside the cereal grain apart from basic diffusion. Phenomenological model could be categorized in different way. For example, it could be grouped on the basis of the lumped parameters like dispersion of water inside the subject or on the basis of the distributed parameters like water gradient inside the subject. A better understanding of hydration model can help in optimization and scaling of soaking process on large industrial scale.

9.1 INTRODUCTION

Agricultural produce is conserved mostly by keeping their water activity at a low level, which is achieved by the different methods. Cereals, seeds, and legumes are the most common example of this methodology as they are stored in a dry state. They are harvested in a dry state and kept in the same way, or dried further if required for the long-term storage. While cereal grains are the main source of starch, legumes are rich in protein and seeds have substantially high amount of both protein and lipids. Generally, grains are considered as nutritional with rich source of fibers, proteins, and minerals.

Although dryness increases the shelf-life of these grains, but they are needed to be hydrated before using again. Among various methods of processing of the grains, steeping gained more attention and accepted for its effectiveness. It is first basic as well as most important step in the processing of dried products. It could be defined as pre-treatment also before different unit operations like cooking, parboiling, extraction, germination, fermentation, etc., which define product characteristics. A large number of studies have been performed which confirm the benefit and importance of hydration (or rehydration) processes.

The process of water absorption during the hydration could be expressed conveniently using the mathematical models. Numerous models have been explored and developed for explanation of water hydration and sorption kinetics. In general, mathematical models are used preferentially to describe any kind of process design and its optimization. Although various advanced statistical tools and programs are available, the simplicity and ease of use of these mathematical models make them as preferred tool [1, 20]. Various mathematical models have been developed and proposed for better understanding of water sorption kinetics of grains like rice [2, 3], barley [8], corn [10], soybean [1], chickpea [24], millets [7, 8], seeds [7] and various other beans [10, 11].

Modeling of Hydration Kinetics in Grains 239

For better understanding of assumptions, these models could be divided into two categories: empirical model and phenomenological model:

1. **Empirical Models:** Being simpler, are utilized most extensively for hydration kinetics studies. Many of these empirical models emphasis primarily up on the water uptake rate [5]. Peleg's expression represent one of the simplest non-exponential empirical model, which utilized two-parameter sorption equation [23]. Due to the linear nature of the curve, it cannot represent the initial lag phase during the hydration, which resulted in many cases primarily because of the coating layer development. For understanding and effective comprehension of such incident, the sigmoidal model has been suggested, where the water hydration rate increase after a lag phase [10, 12].

2. **Phenomenological Model:** It takes into account all combined effects of other factors like the amount of water present in the cereal grain and mass transfer due to convection apart from basic diffusion process. Phenomenological model is categorized based on different considerations. For example, the lumped parameters-based models which are built upon the distribution of water in the cereal grain or the distributed parameters-based models which are built upon the concentration gradient of water in the cereal grain [2, 5, 14, 15]. Omoto model is one of the most recognized lumped parameter-based phenomenological model, which considers uniform water concentration inside the grain during the hydration process [13].

This chapter explores the application and use of mathematical models for the hydration kinetics of grains.

9.2 APPLICATION OF HYDRATION PROCESS IN FOOD PROCESSING

The main objective of the hydration process is to increase the moisture content of grain via submerging them in the water. It has been reported as one of the major step which improves their nutritional and physicochemical properties. Steeping is also a necessary step before other operations like germination, fermentation, extractions, cooking, and malting. Hydration activates various enzymes which subsequently soften the cell wall of the grain.

Hydration also reduces the cooking time for the grain, as it increases the heat transfer rate inside the grain. It enhances the solubilization of galactan and polygalacturonan, which subsequently causes improved solubility of polysaccharide and reduced time for cooking. Hydration also enhances the denaturation of protein and gelatinization of starch during the cooking.

During the steeping process, the cell wall of grain gets soften and shows improvement in the extraction process. Starch is extracted from the steeped cereal grains using the wet milling process. Softened cell wall facilitates wet grinding and also starch purification. Steeping also helps in extraction of anti-nutritional components from beans and millets. During the steeping process, anti-nutritional elements like alkaloids, phytic acids, phenolic compounds, trypsin inhibitors, and amylase, are extracted from the grains.

For the fermentation process, the soaking of grain is prolonged up to 72 hours to 120 hours. The process of fermentation saturates the grain with maximum water absorption capacity and provides an environment for microbial activity. Soy sauce is one of the popular examples, which is produced by fermented soybean. Fermented grains are also used for development of various probiotic and LAB-based products.

Hydration of grains is a prerequisite for germination, which biologically help in development of new plant. Water uptake during hydration help in activation of enzymes which stimulate the germination process. Germination is also the precondition for the malting process. Malted grains develop characteristic aroma, flavor, and color due to various enzymatic activity during germination and also due to malting itself. Malted grains like barley and sorghum are most commonly used for the development of alcoholic beverages. Cocoa seeds are also malted for chocolate production.

9.3 PRINCIPLE OF HYDRATION KINETICS

Being the preliminary step of various processes, an optimum hydration is required for the success of the later processes. The absorption of water is a natural process with seed germination being the final goal of the process. This natural process could be divided in three stages:

- First stage involves passive water absorption via physical means. During this stage moisture content reaches to the level, required for activating the metabolic activity of seed. Water absorption in seed is counteracted by turgor pressure applied my living seeds.
- At stage two, activated enzymes start breaking down the reserve molecules in embryo to activate germination process. There is no further water uptake during this stage.

Modeling of Hydration Kinetics in Grains

- At stage three, germination starts with cell reproduction and growth in tissue. During this stage again moisture level increases inside the grain. For utilization and processing of grain, the main focus remains on the first stage of hydration-germination.

The grain hydration is the mass transfer process, where water movement could be attributed to the driving force developed mainly due to difference in the water activity. Due to which the water drifted from higher concentration bulk, i.e., soaking water in the surrounding to the lower concentration zone, i.e., grain via diffusion. Due to the complex heterogeneous structure of grain, different tissues and cells formulate channels of various sizes and composition with different levels of permeability. Water movement inside these channels is also happens due to capillary action, although diffusion plays the major part. Due to heterogeneous structure of grain, water transport is also not homogenous process and thus kinetics behavior varies according to grain composition and structure. Although diffusion is main factor behind the transportation of water, but it is not the only factor and there could be many other structural factors which affect the water movement in significant manner.

In the case of cereal grains due to water permeability of bran, water is transported across all the surface of grain equally via diffusion. The absorption of the diffused water is mainly due to presence of the starch in the cereal grain. Because starch hydration is a slow process, the grain hydration might take longer time. In such a situation the hydration kinetics follows the downward concave shape (DCS) curve, and hydration rate reduces as grain moisture content move towards equilibrium moisture content (EMC).

Some grains like corn have empty spaces inside the kernel with a porous surface where water is absorbed by capillary action. Water absorption through capillary action is much faster compared to diffusion alone. Such grain exhibits faster initial hydration rate. When the void spaces inside the kernel are occupied by the water, it is absorbed by diffusion alone which resulted in the slow hydration curve. Depending upon the impact of capillary action, the DCS curve could have 1 step, which is mainly due to diffusion or 2 steps, which start with higher rate of hydration as a result of capillary action, and followed by lower rate of hydration because of diffusion alone. Mathematical models which only focus on diffusion cannot describe the hydration behavior of the 2 step DCS curve.

On the other hand, the seed coat of legume grain is practically impermeable for water. Also, structure of legume grain is more complex compared to cereal grains. Due to the impermeable nature of legume seed coat, water transportation inside the grain is only possible from hilum. As the water enters from hilum, it is dispersed between seed coat and cotyledon and makes seed coat permeable for water. Now water enters from both, hilum, and hydrated seed coat mainly via diffusion. Depending upon the structure of legume grain, water is distributed in cotyledon by capillary action and diffusion till it attends to the EMC. Thus, in preference to DCS curve, hydration kinetics of most of the legumes exhibit sigmoidal curve. The sigmoidal curve could be divided in three stages:

- First lag phase is caused by impermeability of seed coat;
- Second exponential phase is due to diffusion of water from hydrated seed coat and hilum both; and
- In the last or third phase where internal moisture content approach towards the equilibrium moisture.

Few legume grains have permeable seed coat and they display DCS of hydration kinetics instead of sigmoidal as hydration start from the second stage.

9.4 MATHEMATICAL MODELLING OF A HYDRATION PROCESS

By fitting hydration kinetics data into a suitable model, the amount of moisture present in the grain could be calculated as the function of time (Figure 9.1).

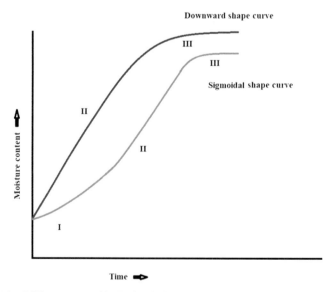

FIGURE 9.1 Different type of hydration behaviors.

Modeling of Hydration Kinetics in Grains

These models could utilize kinetics data for the calculation of the rate of hydration, EMC, and the lag phase of the sigmoidal type curve. Due to different behavior of grain display during hydration, various models have been proposed. These models could be divided either on the basis of hydration curve such as: (i) models for DCS hydration behavior; and (ii) models for sigmoidal hydration behavior; or on the basis of consideration of the physical structure of grain such as: (i) empirical models; and (ii) phenomenological models. The most accepted models applied for studying grain hydration kinetics have been discussed in subsections.

9.4.1 MATHEMATICAL MODELS FOR DCS HYDRATION BEHAVIOR

For a better understanding of DCS hydration behavior sufficient models have been proposed, where data from such grain hydration could be fitted. Empirical models are more general in this approach, while phenomenological models are more specific as they depend upon grain structure and composition and are suitable for only specific conditions.

9.4.1.1 EMPIRICAL MODELS

The first empirical model used for understanding of hydration kinetics was based upon Fick's second law of diffusion. It has many solutions, and it serves as a basic initial step in the development of different models. It is based upon diffusion process purely which describe the unsteady state of water diffusion. The basic equation is used for determination of diffusivity under isobaric and isothermal conditions also described as effective diffusivity. The effective diffusivity is presumed to be a constant property of the material. The empirical solution model assumes homogenous composition of grain with defined regular geometry with negligible change during the hydration process and water migration via diffusion only.

$$\frac{\partial C_A}{\partial t} = D_{AB} \nabla^2 C_A \tag{1}$$

where; D_{AB} is effective diffusivity, which is assumed to be constant property.

One more constraint with Fick's second law is large number of terms in final solution, which depends on shape of material and hydration conditions. The first-order model is another common hydration kinetics model, which assumes hydration kinetics as first-order kinetics reaction. According to model rate of hydration and the difference between the moisture content at a specific time and the EMC are directly proportional to each other. Model assumes a homogenous grain structure with uniform hydration throughout the grain with the same hydration rate. The model could be represented as the following:

$$\frac{M_t - M_s}{M_o - M_s} = \exp\left(-k_L t\right) \tag{2}$$

where; 't' is steeping time (hours); M_o is the original moisture content on a dry basis; M_t is moisture content on a dry basis at time t; M_s is EMC on a dry basis; and k_L is hydration rate.

With the prior knowledge of moisture content before the start of the hydration process and periodic moisture content at different time intervals throughout the process of hydration, the proposed model (first-order) assists in the estimation of hydration rate and EMC.

Peleg model is the utmost frequently applied empirical model for the explanation of both water absorption as well as water desorption kinetics. Model is widely used to describe hydration kinetics in both types of grains (i.e., cereal, and non-cereal). The main equation proposed by Peleg is the sorption equation based upon two parameters and could be expressed as:

$$M_t = M_0 \pm \frac{t}{K_1 + K_2 t} \tag{3}$$

where; K_1 and K_2 are the Peleg's constants. K_1 indicates the adsorption rate at the commencement of hydration process and known as Peleg rate constant (1/h%), while K_2 expresses maximum water holding capacity, when the sample attends the equilibrium state with adjoining liquid state at the end of water absorption process and is known as Peleg capacity constant [16]. The above equation could be rearranged in the form of a linear equation ($y = mx + c$) as:

$$\frac{t}{M_t - M_o} = K_1 + K_2 t \tag{4}$$

Hence based upon the above equation, a plot between $t/(M_t-M_o)$ versus steeping time t, could be plotted, which gives the values of Peleg's constants K_1 and K_2.

$$\frac{1}{M_t - M_o} = K_1\left(\frac{1}{t}\right) + K_2 \tag{5}$$

Singh and Kulshreshtha considered the EMC of the bean while proposing the following equation for the water absorption of pigeon peas and soybean [19].

$$\frac{M_t - M_o}{M_e - M_o} = \frac{Kt}{Kt + 1} \tag{6}$$

$$\text{Or } M_t = M_0 \pm \frac{K_1' K_2' t}{K_1' t + 1} \tag{7}$$

$$\frac{1}{M_t - M_o} = \frac{1}{K(M_e - M_o)t} + \frac{1}{M_e - M_o} \tag{8}$$

$$\text{Or } \frac{1}{M_t - M_o} = \left(\frac{1}{K_1' K_2'}\right)\left(\frac{1}{t}\right) + \left(\frac{1}{K_2'}\right) \tag{9}$$

The above expression by Singh and Kulshreshtha model [19] is very similar to Peleg's model. While the former model is most suitable for the homogenous materials, it cannot be fitted accurately for materials, which are heterogeneous in nature like grains. For most of the cereal grains, the composition of the outer layer of bran differs considerably from the area of endosperm and hence their water absorption characteristics. To overcome this situation, Paquet-Durand et al. [15] proposed application of Peleg model in 2-steps. According to the authors, the model should be applied individually for the outer layer (contain bran) and the inner area (contain endosperm). The summation of both water absorption equation should express the total water sorption of the grain. The modified Peleg model according to Paquet-Durand and coworkers [15] could be written as:

$$M_t = \frac{t}{K_{1b} + K_{2b}t} + \frac{t}{K_{1e} + K_{2e}t} \tag{10}$$

where; K_{1b} & K_{2b} and K_{1e} & K_{2e} are described as modified Peleg's constants, respectively for the outer layer of bran and for the endosperm.

In the above equation, the original moisture content at the start of the hydration process is neglected in comparison to absorbed moisture content during the process.

Drying could be assumed as reverse of hydration or water absorption as both exhibit DCS curve. Hence, Page model, which is widely used in the study of drying kinetics, could be utilized for study of hydration kinetics. Simpson proposed Page model with slight modification for hydration kinetics as [18]:

$$\frac{M_t - M_s}{M_e - M_s} = \exp\left(-k_p t^n\right) \tag{11}$$

where; k_p is the hydration rate of Page's model; and n is Page fitting parameter.

Simpson et al. [18] used fractional calculus for better understanding Page's parameter. While hydration rate could be associated with geometry of subject and coefficient of diffusion, the Page fitting parameter could be associated elementary structure of food and type of diffusion.

Cunha et al. [3] proposed Weibull distribution model for fitting hydration kinetics data. The proposed model has two defined parameters: (i) α is a scale parameter, whose reciprocal is directly proportional to process rate; and (ii) β is the shape parameter.

$$\frac{M_t - M_o}{M_e - M_o} = 1 - \exp\left(-\left(\frac{t}{\alpha}\right)^\beta\right) \tag{12}$$

9.4.1.2 PHENOMENOLOGICAL MODELS

The basic flaw of the empirical models is that they do not consider the rudimentary stages during the mass transfer and offer no statistics related to the water molecule moving mechanism. It ignores fundamental grain geometry and variation arising due to that. It also does not consider physical deviations taking place throughout the grain hydration process. Distributed parameter phenomenological model was proposed to answer these deficiencies, which took the elementary stage of mass transfer under consideration. The first phenomenological model was derived from Fick's Second Law of Diffusion. Water diffusion as per Fick's Second Law of Diffusion can be given by the following equation for spherical grain:

$$\frac{\partial M_t}{\partial t} = D\left(\frac{\partial^2 M_t}{\partial r^2} + \frac{2}{r}\frac{\partial M_t}{\partial r}\right) \tag{13}$$

Crank [2] proposed an analytical solution for above equation with spherical coordinates as:

$$MR = \frac{M_t - M_s}{M_o - M_s} = \frac{6}{\pi^2} + \sum_{n=1}^{\infty} \frac{1}{n^2} \exp\left(\frac{-D_e n^2 \pi^2}{r^2} t\right) \tag{14}$$

Another model based upon differential mass balance in the grain was proposed by Hsu [4], which considers a constant spherical diameter with no physical change during the hydration process and known concentration in the

Modeling of Hydration Kinetics in Grains

solid-liquid interface. Hsu also considered the effect of the temperature on water molecule diffusion inside the grain. As per Hsu model, diffusion takes place only in radial direction for spherical grain like soybean and diffusivity varies exponentially with the moisture present in the grain [4]:

$$X = (1 - \exp^{-\beta t})X_{eq} + X_o\exp^{-\beta t} \tag{15}$$

$$\text{For } r = R, t > 0: \ D = D_o\exp^{k'X} \tag{16}$$

where; X is moisture content in the surface of grain; 'R' is radius (m); D_o is the pre-exponential coefficient (m²/s); k' (kg_{solid}/kg_{H2O}) and β (s⁻¹) are the model fitting parameters.

Hsu model is too specific on the shape of the grain and could not be utilized of different shapes like rice kernel or wheat kernel.

Omoto et al. [13] proposed another model based upon phenomenological characteristics, which includes theoretical postulation along with rudimentary phase of mass transfer. Omoto suggested the below-mentioned mass transfer balance equation, with the assumption of consistent original water concentration throughout the grain. The left side of the equation signifies the variation of water mass throughout the grain in context of time, and the right side of the equation expresses the flux of water mass flow.

$$\frac{d(\rho_A V)}{dt} = N_A A \tag{17}$$

where; N_A is water mass flow (g cm⁻² h⁻¹); 'A' represents an external area of the grain; ρ_A is the water concentration (g/cm³) and 'V' is grain volume (cm³).

The basic equation for convective mass flow can be given as:

$$N_A = K_s\left(\rho_{eq} - \rho_A\right) \tag{18}$$

where; K_s is the mass transfer coefficient (cm h⁻¹); and ρ_{eq} is the water concentration in the grain at the point of equilibrium (g cm⁻³).

The Omoto model for mass transfer could be written as follows with the assumption that the grain has uniform spherical geometry with the constant volume:

$$\frac{d(\rho_A V)}{dt} = \frac{3K_s}{R\left(\rho_{eq} - \rho_A\right)} \tag{19}$$

where; R is the grain radius (cm).

The Nicolin–Jorge provided the simplified solution of Omoto model with considering mass transfer coefficient (K_s) as constant [11]:

$$\rho_A(t) = \rho_{eq} - \left(\rho_{eq} - \rho_{Ao}\right)\exp\left(-\frac{3K_s t}{R}\right) \tag{20}$$

Nicolin–Jorge model [11] assumes grain in regular shape with defined geometry like spherical, cylindrical or parallelepiped. They further proposed a linear dependency between overall mass transfer coefficient and water concentration inside the grain and proposed the following model:

$$\frac{d\rho_A}{dt} = \frac{3\left(a + b\rho_A\right)\left(\rho_{eq} - \rho_A\right)}{R} \tag{21}$$

$$\rho_A(t) = \frac{-a + \rho_{eq}K_1\exp\left(K_2 t\right)}{b + K_1\exp\left(K_2 t\right)} \tag{22a}$$

$$K_1 = \frac{a + b\,\rho_{Ao}}{\rho_{eq} - \rho_{A0}} \text{ and } K_2 = \frac{3\left(a + b\,\rho_{eq}\right)}{R} \tag{22b}$$

where; 'a' and 'b' are constants for linear overall mass transfer coefficient; K_1 and K_2 could be defined by Eqn. (22b).

9.4.2 MATHEMATICAL MODELS FOR SIGMOIDAL CURVE SHAPE HYDRATION BEHAVIOR

Contrary to DCS behavior, very limited models are available for sigmoidal curve shape behavior of hydration kinetics as very few grains exhibit such curve. Kaptso model and Ibarz-Augusto model are the only two recognized empirical models for fitting hydration kinetics data for sigmoidal curve shape hydration behavior.

Kaptso et al. [5] proposed the first model for sigmoidal behavior of hydration kinetics. Apart from saturation moisture content and constant rate of rehydration, the model includes a third parameter which expresses the steeping time duration required for grain to attain the half of EMC to explain the sigmoidal behavior of the curve [5]. Model finds good fit for explanation of hydration kinetics of majority of legume grains.

$$M_t = \frac{M_s}{1 + \exp\left[-k_k \cdot (t - \tau)\right]} \tag{23}$$

where; k_k is hydration rate (min^{-1}); and τ steeping time (hours) duration required for grain to attain the half of EMC.

Ibarz-Augusto model assumed sigmoidal behavior of hydration kinetics as second-order autocatalytic reaction. In contrast to Kaptso model, it is based upon only two parameters, saturation moisture content and combined kinetic parameter (k_{IA}) related to lag phase time and hydration rate. Due to limitation of parameters and independency from lag phase time this model is less suitable than Kaptso model.

$$M_t = \frac{M_s}{1 + \frac{(M_s - M_o)}{M_o} \exp[-k_{IA}.t.M_s]} \tag{24}$$

9.5 SUMMARY

Hydration is one of the most important steps of grain processing for direct consumption as well as industrial processing. Research down in the field of grain hydration kinetics has established two different kind of hydration curve, i.e., DCS, and the sigmoidal shape behavior curve. DCS behavior curve is more common among the grains whereas sigmoidal shape behavior curve is exhibited by few legume grains only. Various empirical and phenomenological models have been proposed for DCS curve. Although empirical models are easier to fit and show a broad spectrum of application, phenomenological models are much more accurate though they have very limited application. Very few mathematical models have been developed for sigmoidal shape hydration behavior. Those developed for sigmoidal shape are only empirical models, and there is great scope of development of phenomenological models for such behaviors. Due to complex chemistry and structure of the grain, limiting hydration process to diffusion only will not serve the purpose. As there are many things other than mere diffusion involved in the process. However, most of mathematical models consider diffusion as either only reason or as the main force. Therefore, there is need for more detailed model which could address capillary flow, molecule movement in the grain, volumetric expansion, and other changes occurring during hydration other than diffusion. There is a need for studying the relationship between macroscopic structure and composition of grain and its impact on water absorption rate. Further new approaches like the use of hydric potential theory and the use of modeling based on fractional calculus has been proposed for the development of better models.

KEYWORDS

- downward concave shape
- empirical model
- equilibrium moisture content
- grains
- hydration kinetics
- mathematical model
- phenomenological model

REFERENCES

1. Coutinho, M. R., Omoto, E. S., Dos Santos, C. W. A., Andrade, C. M. G., & Jorge, L. M. M., (2010). Evaluation of two mathematical models applied to soybean hydration. *International Journal of Food Engineering, 6*(6), Article 7.
2. Crank, J., (1979). *The Mathematics of Diffusion* (2nd edn., p. 414). New York–USA: Oxford University Press.
3. Cunha, L. M., Oliveira, F. A. R., & Oliveira, J. C., (1998). Optimal experimental design for estimating the kinetic parameters of processes described by the Weibull probability distribution function. *Journal of Food Engineering, 37*(2), 175–191.
4. Hsu, K. H., (1983). A diffusion model with a concentration-dependent diffusion coefficient for describing water movement in legumes during soaking. *Journal of Food Science, 48*(2), 618–622.
5. Kaptso, K. G., Njintang, Y. N., Komnek, A. E., Hounhouigan, J., Scher, J., & Mbofung, C. M. F., (2008). Physical properties and rehydration kinetics of two varieties of cowpea (*Vigna unguiculata*) and Bambara groundnuts (*Voandzeia subterranea*) seeds. *Journal of Food Engineering, 86*(1), 91–99.
6. Kashiri, M., Kashaninejad, M., & Aghajani, N., (2010). Modeling water absorption of sorghum during soaking. *Latin American Applied Research, 40*, 383–388.
7. Matias, J. R., Ribeiro, R. C., Aragão, C. A., Araújo, G. G. L., & Dantas, B. F., (2015). Physiological changes in osmo and hydroprimed cucumber seeds germinated in biosaline water. *Journal of Seed Science, 37*(1), 7–15.
8. Montanuci, F. D., Perussello, C. A., De Matos, J. L. M., & Jorge, R. M. M., (2014). Experimental analysis and finite element simulation of the hydration process of barley grains. *Journal of Food Engineering, 131*, 44–49.
9. Nicolin, D. J., Balbinoti, T. C. V., Jorge, R. M. M., & Jorge, L. M. M., (2017). Generalization of a lumped parameters model using fractional derivatives applied to rice hydration. *Journal of Food Process Engineering, 41*(1), e12641.
10. Nicolin, D. J., Marques, B. C., Balbinoti, T. C. V., Jorge, R. M. M., & Jorge, L. M. M., (2017). Modeling rice and corn hydration kinetic by Nicolin-Jorge model. *Journal of Food Process Engineering, 40*(6), e12588.

Modeling of Hydration Kinetics in Grains

11. Nicolin, D. J., Neto, R. M., Paraíso, P. R., Jorge, R. M. M., & Jorge, L. M. M., (2015). Analytical solution and experimental validation of a model for hydration of soybeans with variable mass transfer coefficient. *Journal of Food Engineering, 149*, 17–23.

12. Oliveira, A. L., Colnaghi, B. G., Silva, E. Z. D., Gouvêa, I. R., Vieira, R. L., & Augusto, P. E. D., (2013). Modeling the effect of temperature on the hydration kinetics of adzuki beans (*Vigna angularis*). *Journal of Food Engineering, 118*(4), 417–420.

13. Omoto, E. S., Andrade, C. M. G., Jorge, R. M. M., Coutinho, M. R., Paraíso, P. R., & Jorge, L. M. M., (2009). Mathematical modeling and analysis of pea grains hydration. *Food Science and Technology* (*Campinas*), *29*(1), 12–18.

14. Paliwal, A., & Sharma, N., (2019). Mathematical modeling of water absorption kinetics using empirical and phenomenological models for millets. *International Journal of Engineering and Advanced Technology, 8*(6), 1016–1021.

15. Paquet-Durand, O., Zettel, V., Kohlus, R., & Hitzmann, B., (2015). Optimal design of experiments and measurements of the water sorption process of wheat grains using a modified peleg model. *Journal of Food Engineering, 165*, 166–171.

16. Peleg, M., (1988). An empirical model for the description of moisture sorption curves. *Journal of Food Science, 53*(4), 1216–1217.

17. Shafaei, S. M., Masoumi, A. A., & Roshan, H., (2016). Analysis of water absorption of bean and chickpea during soaking using Peleg model. *Journal of the Saudi Society of Agricultural Sciences, 15*(2), 135–144.

18. Simpson, R., Ramírez, C., Nuñez, H., Jaques, A., & Almonacid, S., (2017). Understanding the success of page's model and related empirical equations in fitting experimental data of diffusion phenomena in food matrices. *Trends in Food Science & Technology, 62*, 194–201.

19. Singh, B. P. N., & Kulshrestha, S. P., (1987). Kinetics of water sorption by soybean and pigeonpea grains. *Journal of Food Science, 52*(6), 1538–1541.

20. Soares, M. A. B., Jorge, L. M. D. M., & Montanuci, F. D., (2016). Drying kinetics of barley grains and effects on the germination index. *Food Science and Technology* (*Campinas*), *36*(4), 638–645.

21. Sopade, P. A., Ajisegiri, E. S., & Badau, M. H., (1992). The use of Peleg's equation to model water absorption in some cereal grains during soaking. *Journal of Food Engineering, 15*(4), 269–283.

22. Sopade, P. A., Xun, P. Y., Halley, P. J., & Hardin, M., (2007). Equivalence of the Peleg, Pilosof, and Singh-Kulshrestha models for water absorption in food. *Journal of Food Engineering, 78*(2), 730–734.

23. Tunde-Akintunde, T. Y., (2010). Water absorption characteristics of Nigerian acha (*Digitaria exilis*). *International Journal of Food Engineering, 6*(5), Article 11.

24. Turhan, M., Sayar, S., & Gunasekaran, S., (2002). Application of Peleg model to study water absorption in chickpea during soaking. *Journal of Food Engineering, 53*(2), 153–159.

PART IV
DECONTAMINATION METHODS FOR FOODS

CHAPTER 10

COLD PLASMA PROCESSING METHODS: IMPACT OF DECONTAMINATION ON FOOD QUALITY

NIKHIL K. MAHNOT, SAYANTAN CHAKRABORTY, KULDEEP GUPTA, and SANGEETA SAIKIA

ABSTRACT

Cold plasma (CP) has emerged as a novel processing technology for the preservation of food and food products. Food spoilage and foodborne illness have been a great concern for both industry and consumers. CP has proved to be a promising and a feasible non-thermal intervention for food processing. This is due to its decontamination efficacy against food borne pathogenic and spoilage microorganisms, minimal impact on food constituents without generating any toxic by-products, cost efficiency and sustainability. CP has also been shown to degrade microbial toxins and pesticides. These multidimensional effects have been ascribed to the generation of various reactive gas species and ultraviolet (UV) mediated chemical and microbial mechanisms. CP's application has been widely looked into various areas of food industry viz. fruits, vegetables, cereals, meat, poultry, and packaging with promising results. There are multiple approaches for plasma generations with the fundamental approach being excitation of individual gas or gas mixes to the plasma state which can be tuned for different applications. CP's characteristic effects have assisted in making food and food products safer for consumers and have opened avenues for further development.

Advances in Food Process Engineering: Novel Processing, Preservation, and Decontamination of Foods.
Megh R. Goyal, N. Veena, & Ritesh B. Watharkar (Eds.)
© 2023 Apple Academic Press, Inc. Co-published with CRC Press (Taylor & Francis)

10.1 INTRODUCTION

The World Health Organization (WHO) has suggested that the key to a healthy life relies upon access to a sufficient amount of nutritionally rich and safe foods. Globally, foodborne illnesses has been associated with a total of 31 contaminating agents comprising of bacteria, virus, parasites, toxins, and chemicals; and one in every 10 person falls sick on consuming such contaminated foods [65]. The World Bank has estimated an occurrence of combined losses of $110 billion in production and medical treatment due to foodborne illness each year [21]. Thus, food safety is seemingly important.

In general terms, there are three specific goals for any food processing company; firstly, to reduce microbial risks, next product quality maintenance and thirdly shelf-life enhancement. The food sector employs diverse technologies, the selection of which is based on properties of specific food commodities or agriculture crops to be converted into finished products. The most widely trusted technology is thermal-based processing, which greatly mitigates microbial risks and confer considerable shelf-life to products. But the catch is, a high heat treatment greatly reduces quality (sensory and nutritional) characteristics. However, in the past two decades, the food industry is trending towards minimally processed foods, i.e., foods that are processed in a way to retain fresh-like characteristics. This is commonly achieved by employment of mild thermal treatments or utilizing non-thermal interventions. Unlike thermal treatment, non-thermal technologies involve the processing of foods at ambient or at milder temperature conditions. Some of these technologies include the use of ultrasound (US), high pressure processing, pulsed electric field, pulsed UV-technology, irradiation, ozone processing, etc., to name a few. Such processes preserve much of the natural like characteristics, provides microbial safety while maintaining acceptable shelf-life. These technologies have been widely studied and have demonstrated favorable applications. However, there are certain limitations concerning to achieving sterility as well as employment of complex machinery and maintenance costs.

The way in which a food is processed is of importance nowadays due to rising environmental safety, sustainability concerns alongside consumer safety. The food sector is an energy intensive sector and regulatory review in various countries has led the sector to look into sustainable processing technologies with long term safety of the ecosystem. As a consequence, the food sector is under constant pressure to adapt and innovate accordingly. Big manufacturing companies nowadays utilize the environment friendly

Cold Plasma Processing Methods 257

tag as it creates a positive brand image, thereby greater sales. Over the last few years, cold plasma (CP) processing has emerged as a new non-thermal technology with multifaceted application in maintaining safety and quality food and food products along with environment safety.

The current chapter discusses the knowhow of CP technology, mechanisms, efficacy in decontaminating microorganisms, and mitigation of contaminants to achieve food safety along with sustainability.

10.2 PLASMA SCIENCE: PHYSICS AND CHEMISTRY

The concept of plasma was first described by Irving Langmuir and his co-workers in the 1920s. However, the existence of plasma physics has been traced back to the 1600s during the time plasmas were known as *gas discharges* [48]. In general terms plasma is often stated as the fourth state of matter. We all have an understanding of the three states of matter: solid, liquid, and gas, each following a hierarchical increase in energy levels, followed by a change of state into the next on achieving a particular energy level. The increase in energy level cause loosening of interaction among molecules till the molecules become completely apart causing a change of state.

Plasma is repeatedly stated as the ionized form of a given gaseous state. When sufficient energy is applied to a gas system, the intermolecular and intra-atomic structures break completely, leading to the release of electrons and generation of ions, and thus plasma is generated. Consequently, any energy source which can achieve such an effect can be readily utilized as a source for plasma generation. Typically, plasma comprises of a co-existing mixture of excited atomic, molecular, ionic and radical species, reactive species, charged ions, electrons, molecules in neutral or excited states along with the release of quanta of electromagnetic radiation [42]. Common examples of plasma as one may observe in daily life include natural lightning or man-made plasma including neon lighting, plasma TV, LEDs (light-emitting diode) and arc generated during welding applications. Now, as we have established about what is plasma, let's understand the physics and chemistry involved.

10.2.1 PLASMA PHYSICS

Plasma can also be referred to as a quasi-neutral gas of charged particles which shows *collective* behavior [4, 16]. Although the particles in plasma are charged, but overall, the charge densities cancel out at equilibrium, thus the quasi-neutral nature. There is a collective interaction between multiple charge particles occurring over long lengths and time scales. Putting this into perspective, let's consider two charge particles which are very close to each other. These two particles would interact with each other individually through their Coulomb electric field. However, when these particles move apart from each other beyond a mean particle separation distance ($n^{-1/3}$, where n is the charged particle density), these charged particles further interact simultaneously with other adjacent charged particles. Thus, indicating collective interaction.

Distinctively, plasma can be classified into thermal and non-thermal plasma (NTP). Ionization is the first step for plasma generation. The mechanism of ionizing very much relies upon the efficient collisions between electrons and other neutral species. For collisions, some of the electrons have to achieve high kinetic energies to exceed the ionization potential of the gas. This can be achieved by heating the gas at sufficiently high temperatures (in the order of 2×10^4 K), such generated plasma is considered as "thermal plasma."

In thermal plasma, the constituent neutral species (T_n), ions (T_i), and electrons (T_e), are in thermodynamic temperature equilibrium ($T_i \approx T_e \approx T_n \le 2\times10^4$ K) thus the overall plasma temperature is very high. NTP can have varying degree of non-equilibrium among the species, based on which they can be subdivided into quasi-equilibrium plasma (temperature range: 100–150°C) and non-equilibrium plasma (<60°C). In quasi-equilibrium plasma a local thermodynamic equilibrium exists between the constituent species. On the other hand, in non-equilibrium plasma, the collisions between the hot electrons and other particles are not frequent enough. Consequently, thermalization between the electrons and neutral species does not occur efficiently and overall, the system remains near room temperature. In simple terms, the electron remains at a much higher temperature than the constituent ions or neutral species ($T_i \approx T_n << T_e$, $T_e \le 10^4$ K) [41]. Ascribable to such characteristics, non-equilibrium plasma is typically referred to as CP or NTP in general. Physicists often relate CP operations to occur at temperatures above hundreds or thousands of degrees above ambient temperature. As a result, the term "cold plasma" has now been assigned to generation of plasma at room temperature and at one atmosphere pressure unlike other NTPs. Such CPs can be typically generated using electrical discharges in gases. As such plasma generation occurs without extreme conditions, the importance of its application in the food industry becomes pertinent.

Cold Plasma Processing Methods   259

10.2.2 PLASMA CHEMISTRY

The plasma generation is driven by collision interaction between energized electrons with other atoms and molecules of a gas subjected to a sufficient high electric field (HEF). These collisions are important for the plasma chemistry. When an electron collides with other atoms or molecules, a wide range of effects are observed viz. ionization, dissociation, association, and excitation. For understanding, let's consider a situation where an electron e, which can collide with a molecule AB or an atom M in a gas. The following primary processes are found to take into effect:

- **Impact Ionization:** $e + M \rightarrow M^+ + 2e$
- **Dissociative Ionization:** $e + AB \rightarrow A^+ + B + 2e$
- **Dissociative Electron Attachment:** $e + AB \rightarrow A^- + B$
- **Electron Attachment:** $e + M \rightarrow M^-$
- **Electron Impact Molecular Disassociation:** $e + AB \rightarrow A + B$ (either A or B could be excited electronically).
- **Electron Impact Excitation:** $e + M \rightarrow M^* + e$ (excitation to higher atomic, molecular or vibration levels).

These are primarily the possible ways in which an electron can collide/interact with other molecules or atoms. Again, another interesting effect is penning ionization which has been observed with plasma discharges involving helium, neon, and krypton [31]. Penning ionization $[A + B^* \rightarrow A^+ + B + e]$ requires the interaction of a metastable species (B^*) with a target neutral molecule (A), having ionization energy lower than the internal energy of the metastable species. Photon and ultraviolet (UV) emission are also other additional processes during plasma generation. These processes occur when exited species release energy and come to the respective ground states. However, it has been suggested that the photon flux during atmospheric CP generation is not sufficient to achieve photocatalysis [23]. Additionally, the impact of UV on DNA damage and surface sterilization is well proven; this marks some decontamination ability of plasma. The probability of the above interaction to materialize is said to be dependent upon the electrons mean energy.

Electron mean energy is dependent upon the gas composition, temperature, and pressure in the discharge and it has been observed that input energy also plays a role. Thus, one can vary these functions to achieve specific process outcomes for any application in particular. The mean electron energy in NTP typically ranges between 2 eV and 5 eV. The plasma chemistry is basically driven by such generated species having lower activation energies consequently, upsurging reaction rates [41]. This leads to secondary effects;

secondary collision involving large or heavy molecules generated on electron collisions. Common secondarily generated molecules include ozone, hydrogen peroxide (H_2O_2), nitric oxide, nitrites, etc. As it is possible to achieve such processes without involving extreme temperatures, CP technique provides a distinctive environment to initiate and assist chemical reactions. Thus, forecasting itself as an interesting technique to be applied to food and agriculture.

10.3 METHODS OF COLD PLASMA (CP) GENERATION

Basic plasma system relies upon a feed gas which can be contained or be free-flowing. Frequently used feed gases are oxygen (O_2), nitrogen (N_2), air, helium (He), neon (Ne), argon (Ar) or their mixtures. Some common methods in practice to induce ionization include corona discharges, dielectric barrier discharges (DBDs), microwave discharges (MD) and plasma jets. Selection of these methods is based on the final application. However, these configurations simply follow one principle that the feed gas needs to be ionized into plasma at one atmosphere pressure or in certain cases below atmospheric pressure.

10.3.1 CORONA DISCHARGES

Corona discharges are stream of charged particles viz. ions and electrons accelerated by an electric field in a confined space. The reason of such streams is due to the asymmetrical nature of the electrodes, where one electrode may be a tip or pit point or a thin wire and the other a plate type. Close to such tips, electric fields of high intensities are formed, and the active corona is generated. Such plasma is not self-sustainable as most of the discharge volume is below the HEF. The plasma effects are mostly due to secondary processes such as photoionization or transport of charged carrier from the HEF region to rest of the confined space. Such discharges are commonly applied for treating fruits and vegetables with ozone in industrial setup [41].

10.3.2 DIELECTRIC BARRIER DISCHARGE (DBD)

DBD method involves excitation of feed gas into plasma by applying an oscillating current between two electrodes separated by a durable dielectric material which may be a glass or plastic. This dielectric medium avoids

Cold Plasma Processing Methods 261

charge transport and uniform non-equilibrium plasma is generated between the areas of the electrode unlike corona discharges. Thereby, one can change the electrode area and shape for higher volume applications which in turn is limited by the voltage supply. Thus, a DBD proves to be a flexible and a scalable plasma generation method. DBD has seen its application in improving air quality, ozone generation, water treatment along with microbial inactivation in juices [37, 38, 59].

10.3.3 MICROWAVE DISCHARGE (MD)

Microwave discharges (MDs) are generated by the use of electromagnetic waves rather than electric current to ionize gases in a chamber. Microwave plasmas are mostly utilized for low pressure applications mostly for biomaterials [55]. The electromagnetic frequency typically exceeds hundreds of MHz, which is produced using a cyclotron. The major merit of MD method is its ability to create intense discharges in chemically active gases. Microwaves have been utilized more actively in plasma jet methods for medical application [20] but quite limited in the food sector.

10.3.4 PLASMA JET

Plasma jet is not much of a source for plasma generation; rather it's merely a modification of the above mentioned configurations. Unlike the other system this is an open type system, i.e., the feed gas is not confined in a closed chamber. As per the name, the feed gas is blown at a high rate through two co-axial electrodes for ionization. The outer electrode is grounded and the other electrode is connected to the systems like microwave resonator or radio frequency [20, 41]. The propagating ionized gas blows out as waves and forms a stream of reactive gas species. The plasma interaction produces long-lived reactive species which in turn is exposed to the objects to be treated. This approach is important as it acts as a remote means of plasma generation and is favorable for use on living tissues; important for biomedical applications. However, its application in the food sector is limited or not cost-effective. In general, for food application, we come across three distinctive methods of exposing plasma to foods:

- Firstly, the food sample can be placed directly within the plasma field; for example, when a food is kept directly under the electrodes as in case of in a DBD method or in the chamber of MD type discharges;

- Secondly, the placement is at point close to the plasma generation, e.g., jet plasma; and
- Thirdly, a remote treatment, where a food is placed at a significant distance from the short-lived reactive species, here plasma effects are ascribed to secondary processes. Based on the methods, interesting results are obtained pertaining to microbial decontamination.

10.4 BIODECONTAMINATION USING CP

Biodecontamination refers to removal or inactivation or causing fatality of biological entities. Emphasizing on the food sector, the common biological entities that need to be dealt with are bacteria, fungi, and viruses. CP has received much importance as a non-thermal processing technology for decontamination owing to multifaceted properties. A lot of work has been carried out for understanding the decontamination process due to CP. As thermal effects are not a mode of action, one can broadly classify the mechanism under two heads. First, interaction due to reactive gas species and charged particles; and second UV inflicted cellular and genetic damage.

10.4.1 REACTIVE GAS SPECIES

The reactive gas species are considered to be the major entities that cause lethal or sub-lethal damage to microbes. The composition of reactive gas species mix is dependent on the type of gas or gas mixture used, plasma generation method and configurations, treatment time and also on applied energy causing ionization. Common feed gases include natural or synthetic dry or humid air, O_2, N_2, He, and others or their mixtures thereof. Some of the reactive gas species of interest, which have shown decontamination effects are compiled as follows:

1. **Reactive Oxygen Species (ROS):** Atomic oxygen (O), singlet oxygen ($^1O_2$), excited state of oxygen (O_2^*) superoxide anion (O_2^-) [primary collisions] and ozone (O_3) [Secondary effects].
2. **Reactive Nitrogen Species (RNS):** Atomic nitrogen (N), nitrogen positive ion (N_2^+) and excited state of nitrogen (N_2^*) [primary collisions], nitric oxide free radical (NO•), nitric oxide (NO) [secondary effects].
3. When humidity is involved, species like H_2O^+, OH^- anion, OH• radical [primary collisions] and H_2O_2 is also generated due to

secondary effects. Water vapor along with nitrogen gas generates other secondary components viz. nitrite (NO_2), nitrous acid (HNO_2), N_2O_4, and N_2O_5.

It has been widely reviewed that reactive species interact with microbial cell components leading to inactivation. On plasma treatment, microbial cells are bombarded by activated radicals and energetic ion causing surface lesion, a phenomenon described as "etching." Although microbes have their inherent repair mechanism such intense bombardment cannot be counteracted, thus the cells are stressed or injured sub-lethally.

Electropermeabilization, i.e., accumulation of electric charges on cell surface, induces cell rupture causing lethal injury during plasma treatment, a mechanism similar to pulse electric field mode of microbial inactivation [63]. It has also been put forward that ROS tend to cause degradation or damage of by breaking bonds of membrane lipid bilayer and proteins. ROS induces lipid peroxidation, as ROS being more reactive than molecular oxygen. Once the bi-layer is disrupted, the cells fail to transport molecules into and out of the cell for normal maintenance [10]. RNS with oxygen have shown to cause interruption of the anti-oxidant defense mechanism by targeting RpoE (DNA-dependent RNA polymerase), OxyR (oxygen regulated gene), DnaK (chaperone protein DnaK gene), and GroES (heat-shock gene) expression [10, 57] leading to cell stress or inactivation.

Another point of action is DNA damage due to reactive gas species interaction. The most important radical inflicting DNA damage are the OH• radicals that accounts for 90% DNA damage on plasma treatment [63]. OH• radicals also initiate DNA oxidation reactions by reacting with nearby organic compounds leading to DNA damage along with disintegration of cellular membranes and other cell components; consequently, cell death [9]. Additionally, when liquid media are treated with plasma an acidification process is observed owing to diffusion and dissolution of gas species in the medium; scientists suggest it plays a role in decontamination [38]. Ozone and peroxide molecules, the outcomes in plasma processes are well known for their potent antimicrobial nature.

10.4.2 UV PHOTONS

The impact of UV-mediated decontamination effects in plasma processing is quite incongruous among scientists. UV is proven and widely known to have DNA damaging effects. For CP treatment the effects of UV have been suggested to occur synergistically with the varied reactive gas species [7, 38]. Researchers working on microwave plasma and radio frequency plasma have pointed towards UV-mediated effects alone [41]. Mosian et al. [44] showed UV to be the vital reason for microbial inactivation in low pressure plasma applications. In contrast for plasma jet application crucial role of UV could not be assigned to microbial inactivation [66]. Overall, one can conclude that UV may not play a major role but cause microbial inactivation synergistically with other mechanisms.

Scientists have carried out a majority of experiments where bacteria, fungi, and viruses have been subjected to CP treatment with positive effects. CP has successfully demonstrated inactivation of pathogenic bacteria like *Salmonella* spp., *Escherichia* spp., *Listeria* spp., *Bacillus* spp., *Pseudomonas* spp., *Staphylococcus aureus*, *Campylobacter jejuni*, etc., in various food matrix with great efficacy [37, 38, 45, 63]. The fungi are competent invaders of a wide range of foods and have the ability to cause rots and produce mycotoxins, which is a serious concern.

Fungi of concern especially *Aspergillus* spp., *Penicillium* spp., *Cladosporium cladosporiodies*, *Zygosaccharomyces rouxii*, *Alternaria alternata*, *Candida albicans*, etc., have been reviewed to be inactivated using varied CP treatments [15, 43, 55]. These microbes are of concern as they are responsible for spoilage of food products, more importantly cause food poisoning and other life risking conditions. More recently, the application of CP processing has been extended to study virus inactivation on food surfaces. Human norovirus (HuNoV), a potent etiological agent transmitted through consumption of raw foods or foods processed on un-sanitized surfaces. Although much of the work has been carried out on viral surrogates with CP, but it still opens up the possibility towards viral safety in foods [28, 40].

10.5 SCOPE OF COLD PLASMA (CP) IN FOOD SECTOR

CP technology drives its benefits from its ability to achieve food safety at room temperature, without negatively affecting the nutritional and quality attributes. This has been effectively used by food scientists, engineers, and technologists to provide consumer safety. Some of the application aspects in the various food sectors are compiled in this section.

Cold Plasma Processing Methods

10.5.1 FRUITS AND VEGETABLES

Most fruits and vegetables are well known for their major health benefitting constituents and have been recommended to be consumed in adequate amounts and are mostly consumed raw [53, 54]. Thus, safety is of utmost importance when consuming raw foods. CP has been applied as an in-package treatment method to a wide range of fruit and vegetables including the fresh-cut produce category. A pilot-scale study utilizing an ionized H_2O_2 aerosol plasma and demonstrated a reduction of *S. typhimurium* and *L. innocua* on inoculated in apples and tomatoes below detection limits (<0.7 log CFU/pc) [61]. The same group also reported similar reduction results for grapes, tomatoes, apples, cantaloupe, and Romanian lettuce [62].

In another pilot plant study with DBD method involving humid air was efficiently used to reduce *E. coli* and *L. monocytogenes* in strawberries and spinach by 2–2.2 logs and 1.3–1.7 logs, respectively [70]. In one study, a CP treatment (60 kV for 5 min) inhibited the inherent aerobic bacterial counts and enhanced the cutting induced phenolic accumulation in freshly cut dragon fruit [32]. In sliced carrots, about 2 log reductions of total mesophiles, yeast, and molds were observed whilst maintaining the textural properties with good carotenoid retention [39]. CP has been widely studied for treatment of other fruits and vegetables like cabbage, cherry tomatoes, cucumber, baby kale, radicchio leaves, and peas, to name a few [41]. The CP treatment has also been reviewed for its ability to inactivate of norovirus in foods such as blueberries, Tulane virus in Romanian lettuce opening upon further possibilities [13, 28, 40] in viral food safety.

10.5.2 JUICE PROCESSING

Decontamination of fruits and vegetables is a mostly surface process. Decontamination in liquids bit of a challenge. CP mediated decontamination in liquids is mostly due to reactive gas species generation and further dissolution and diffusion processes in the media. CP has proved to be effective in achieving an FDA recommended 5-log standard microbial reduction process [12] in juices pertaining to *S. typhimurium*, *E. coli*, and *L. monocytogenes.* However, the level of microbial reduction is dependent upon the juice matrix under treatment. Direct exposure of orange juice to a DBD type treatment (90 kV–2 min) with M65 gas (O_2: 65%, CO_2: 30%, N_2: 5%) achieved approx.

5 log reduction of *S. typhimurium* and also resulted in around 80% reduction in pectin methylesterase activity [67].

A similar process was followed by researchers to inactivate over 5 log cycle reductions in microbes in tender coconut water with added citric acid; when inoculated with *S. typhimurium*, *E. coli*, and *L. monocytogenes* individually without majorly affecting its physicochemical properties [37, 38]. Yeasts like *Zygosaccharomyces rouxii* have been shown to be reduced by around 6 logs in apple juice using a plasma spray reactor (21.3 kV for 30 min) using normal air as the feed gas [64]. Significant reductions of *E. coli* has also been observed in tomato juices (~1.43 log CFU/mL), sour cherry (~3.34 log CFU/mL), and apple juice (~4.02 log CFU/mL) using plasma jet method (650 W, 30–120 s) [8, 33]. In juices, lowering of contents like polyphenol, anthocyanins, flavonoids, and ascorbic acid have been reported [19, 38, 52]. However, these reductions are lower as compared to conventional thermal treatments. In some cases, the quality of juices improved on plasma treatment, as in the case of beetroot juice, an increase in phenolics was observed and for pomegranate juice, anthocyanin content increased by 21–35% [11, 26]. Attempts are being made to optimize the plasma processes in a way to ensure safety along with nutritive stability.

10.5.3 DAIRY AND DAIRY PRODUCTS

Thermal treatment in the dairy industry is mainly targeted to deliver safety to consumers. However, it leads to undesirable changes in protein, cause non-enzymatic browning, vitamin losses and flavor changes. CP technology has been studied to show promising results in this sector as well. Kim et al. [24] suggested a 10 min plasma treatment (250 W, 15 kHz) for milk using ambient air and DBD method for plasma generation to effectively reduce bacterial counts for *S. typhimurium*, *L. monocytogenes*, and *E. coli* by 2.4 log CFU/mL. Yong et al. [68] applied a similar plasma treatment for the same microbes on cheese slices. They reported 2.8 log, 2.6 log and 3.1 log CFU/g reductions for *E. coli*, *L. monocytogenes*, and *S. typhimurium*, respectively.

Reviewed literature suggests that in dairy products substantial microbial inactivation could be achieved but there are limitations to be addressed. Common being lipid oxidation causing sensory and color changes [5]. Consequently, changing plasma parameters pertaining to identifying specific feed gas or gas composition could help solve the issue. Further research addressing the comparative changes of milk and milk products between CP and thermal treatments is needed.

10.5.4 ANIMAL PRODUCTS: MEAT, POULTRY, FRESH FISH, AND SEA FOOD

We have discussed earlier about the generation of reactive nitrogen species (RNS) during plasma generation and secondary processes leading to formation of nitrite. Moreover, CP treatment has been shown to effectively inactivate *Clostridium botulinum* and its spores [35]. Nitrite is the most common additive for curing in the meat industry and *C. botulinum* a potent hazard, consequently supporting CP treatment as a viable technology for meat processing. Having said that, in-package CP treatment is the common go to for treating meat products. This is beneficial as it not only prevents further contamination of the treated products; also, the long lived reactive gas species, importantly O_3 and H_2O_2 continue to inactivate microbes even after the plasma treatment is stopped.

It has been suggested that common effective feed gas for plasma generation include a mixture of nitrogen/oxygen, helium/oxygen, or ambient air to confer efficient decontamination in meat products. Further, the microbicidal effects on *E.coli*, *L. monocytogenes* and *S. typhimurium* are dependent upon power input; an increase in power increases efficacy for inactivation and distance between electrodes [42]. Plasma treatment has successfully been applied to meat products like ham, bacon, pork loin, pork but, beef jerky, etc., to inactivate both gram-negative and gram-positive bacteria with treatment time varying from 90 s–10 min and affective microbial reduction ranging from 0.35 log CFU/g–3.03 log CFU/g [29, 69]. Recent review on the safety of chicken meat products viz. chicken breast and thighs (skin or skinless), slices, and filets; CP treatments have significantly eliminated pathogens such as *Campylobacter jejuni* among others along with inherent mesophiles and psychrophiles [14].

Raw or cooked eggs or egg components (yolk and egg white) have also been studied extensively in the past decade for safety using CP. Available literature suggests greater than 5 log reduction in *Salmonella* spp. and *L. monocytogenes* using feed gases $He+O_2$, N_2+O_2 and $O_2+CO_2+N_2$ for direct and indirect mode of plasma processing [14]. It is important to note, the addition of moisture in a gas mixture or increasing O_2 concentration confers enhanced inactivation in in-package CP treatments. More recently, global demand for fish and seafood consumption is on the rise as exotic meal consumption is becoming a part of food habit. So, concern regarding their safety is on the rise.

Fish and the seafood industry are also finding CP processing as a promising non-thermal intervention. Fillets of Atlantic herring and mackerel, semi-dried pacific saury, dried filefish, and squids have been subjected to CP treatment with ambient air as the feed gas. The plasma processes significantly reduced *Photobacterium phosphoreum, P. citrinum, Pseudomonas* spp., *Staphylococcus* spp., *C. cladosporioides* and other marine bacteria, psychrotrophic, coliforms, lactic acid bacteria (LAB), yeast, and molds [27].

In shrimps, initial load of both *Staphylococcus* spp. and *Salmonella* spp. along with other background microflora were significantly reduced on N_2/O_2 plasma treatment and extend the shelf-life by 14 days [60]. In slices of Asian sea bass, *S. aureus, E. coli, P. aeruginosa, Vibrio parahaemolyticus, S. aureus*, and *L. monocytogenes* were reduced significantly using Ar/O_2 plasma treatment [46]. Although CP has shown promising in the meat industry for food safety but quality changes, lipid degradation, protein denaturation and changes in color are a limitation. Further optimization of the technology is still necessary.

10.5.5 CEREALS, LEGUMES, OILSEEDS, AND NUTS

Cereals, legumes, oilseeds, and nuts are low moisture foods. The safety of these products is predominantly limited by the presence of mycotoxin producing filamentous fungi. Mycotoxins have been associated with serious health issues with infants listed at a higher risk than adults. Mycotoxin mitigation through CP is achieved either by inactivating the producing microbe or by degrading mycotoxin. Both this course of action is mediated through plasma chemistry (reactive gas species and UV-light).

Los et al. [36] demonstrated inactivation of *B. atrophaeus* and *P. verrucosum* in barley and wheat grains. The authors reported a 2.4 and 2.1 log CFU/g in barley and for wheat 1.5 and 2.5 log CFU/g for *B. atrophaeus* and *P. verrucosum*, respectively. Successfully decontamination of the chickpea seeds, beans, soybean, oats, barley, rye, corn, and lentils contaminated with *A. parasiticus* and *Penicillium* sp. to < 1% of initial load depending on plasma treatment times (30 s–30 min) has been reported [56]. In maize grains, fluidized air plasma effectively reduced *A. flavus* and *A. parasiticus* by approx. 5 log CFU/g [7]. An ambient air corona discharge plasma jet exposure (operating at 20 kV, 58 kHz with air, 3 min) on rapeseed surface caused 1.8 and 2.0 log CFU/g decrease in yeast and mold counts, respectively [51].

Aspergillus spp. is the most common fungi which contaminate nuts. In Hazelnuts, fluidized air plasma inactivated *A. flavus* and *A. parasiticus* by 4.5 log and 4.19 log CFU/g, respectively [6, 7]. Complete inactivation of

Cold Plasma Processing Methods

A. brasiliensis was noted with a low pressure DBD plasma treatment with Ar/O$_2$ (10:1 v/v) on pistachios in 15 s [49]. In date palm, Ar gas jet plasma treatment completely inhibited *A. niger* spores [47]. A similar, Ar-plasma jet exposure caused complete destruction of *A. flavus* in fresh and dried walnuts within 11 and 15 min, respectively [1]. Literature review suggests that CP efficiently degrades several mycotoxins, viz. aflatoxins, trichothecenes, vomitoxin, zearalenone, sterigmatocystin, and fumonisins, and also inactivates mycotoxin producing fungi [15, 58]. Although in the initial stages of application, CP has shown promise to develop mycotoxin safe foods, such as: *B. atrophaeus*, and *P. verrucosum*.

10.5.6 HERBS AND SPICES

Safety of herbs and spices is quite a challenge among food technologist. This is due to susceptibility towards fungal contamination, bacterial attack and volatile losses on conventional heat treatment. In the above sections we have very well learnt that CP effectively kills fungi including bacteria and that too at room temperature. Thus, shifting towards CP intervention for herbs and spices seems logical. A high density microwave-combine plasma treatment with helium as the ionizable gas was effective in reducing *B. cereus* spores by 2.7 log spores/cm^2 in vacuum dried red pepper flakes [25]. In another work natural microbiota was significantly reduced in paprika powder and pepper seeds by 3 log CFU/g and in oregano by 1.6 log CFU/g using a remote plasma treatment [17].

In infected dried peppermint powder, substantial *E. coli* O157:H7 reduction was observed on low pressure radio frequency O$_2$-plasma treatment [22]. Similarly, low pressure radio frequency oxygen plasma was applied to saffron that significantly eradicated *Aspergillus* spp., *Rhizopus* spp., and *Penicillium* spp. in 15 min [18]. A treatment involving pulsed light with CP, non-thermally inactivated spores of *A. flavus* (~1.3 log spores/g), *B. pumilus* (~2.3 log spores/g), and *E. coli* O157:H7 (>3.8 log CFU/g) in red pepper flakes [30]. Available research shows promising aspects of CP in the safety of herbs and spices.

10.6 COLD PLASMA (CP) AND SUSTAINABILITY

The world population is set to reach 10 billion by 2050, i.e., approx. 3 billon more are to be provided with food. Which the current climate change and decline in available agricultural land, the food-Agri sector needs to opt for sustainable food production and processing strategies. A reduction in incidence of foodborne illnesses, product recalls and food safety will increase sustainability significantly. As discussed earlier, traditional technologies effectively reduce microbes and make food safe but also affect quality and are also energy-intensive. Food producers and processors also use chemical methods to overcome microbial contamination. However, it is not very safe in terms of immune reactivity, chances of retaining chemical residue and also effluent laden with chemical needs to be treated, which otherwise is a hazard. Thereby, alternatives of chemicals are a much focused area nowadays.

CP has found its applications at various stages so food production and processing, including enhancement of seed germination, water decontamination, decontamination of food, equipment, and packaging materials [2, 3, 34, 50]. CP operations achieve these effects by consuming less energy, operating at ambient conditions and generation of short-lived primary and secondary reactive species, thus leaving behind no major chemical residues; an added advantage. This makes CP treatment green and sustainable application for food, agricultural, and industrial sectors.

10.7 SUMMARY

CP is gaining attention owing to its multifaceted effects on different sectors of the food industry. Although this technique is in its nascent stages, it has proved its efficacy in delivering safe food by eradicating pathogenic microbes: bacteria, fungi (yeast and molds), and degrading mycotoxins. In all generated plasma, a variety of chemical and physical interaction takes place, which still needs to be understood well enough for specific application. The intricate plasma chemistry; reactive gas species, secondary species, and UV light generation effectively interacts with biological materials and confer much of its benefits. Not to forget its ability to bring changes at ambient conditions. However, the technology has certain limitations pertaining to the quality of products. Attributes like flexibility, scalability, and sustainability, the technique has opened unlimited opportunities in the food sector. Consequently, much know-how still remains to be generated. Further, with the emergence of viral pandemics, such as SARS-COV-2 and its ability to survive on varied surfaces; challenges on viral decontamination of food surfaces needs to be taken up.

KEYWORDS

- **biodecontamination**
- **cold plasma**
- **food safety**
- **microwave discharge**
- **non-thermal technology**
- **reactive gas species**
- **ultraviolet**

REFERENCES

1. Amini, M., & Ghoranneviss, M., (2016). Effects of cold plasma treatment on antioxidants activity, phenolic contents and shelf life of fresh and dried walnut (*Juglans regia* L.) cultivars during storage. *LWT, 73*, 178–184.
2. Bourke, P., Ziuzina, D., Boehm, D., Cullen, P. J., & Keener, K., (2018). The potential of cold plasma for safe and sustainable food production. *Trends in Biotechnology, 36*(6), 615–626.
3. Chemat, F., & Vorobiev, E., (2019). *Green Food Processing Techniques: Preservation, Transformation, and Extraction* (1st edn., p. 586). London–United Kingdom: Elsevier.
4. Chen, F. F., (2006). *Plasma Physics and Controlled Fusion* (2nd edn., p. 421). New York: Springer.
5. Coutinho, N. M., Silveira, M. R., Rocha, R. S., Moraes, J., Ferreira, M. V. S., Pimentel, T. C., Freitas, M. Q., et al., (2018). Cold plasma processing of milk and dairy products. *Trends in Food Science & Technology, 74*, 56–68.
6. Dasan, B. G., & Boyaci, I. H., (2017). Effect of cold atmospheric plasma on inactivation of *Escherichia coli* and physicochemical properties of apple, orange, tomato juices, and sour cherry nectar. *Food and Bioprocess Technology, 11*, 334–343.
7. Dasan, B. G., Boyaci, I. H., & Mutlu, M., (2017). Nonthermal plasma treatment of *Aspergillus spp.* spores on hazelnuts in an atmospheric pressure fluidized bed plasma system: Impact of process parameters and surveillance of the residual viability of spores. *Journal of Food Engineering, 196*, 139–149.
8. Dasan, B. G., Mutlu, M., & Boyaci, I. H., (2016). Decontamination of *Aspergillus flavus* and *Aspergillus parasiticus* spores on hazelnuts via atmospheric pressure fluidized bed plasma reactor. *International Journal of Food Microbiology, 216*, 50–59.
9. Dobrynin, D., Fridman, G., Friedman, G., & Fridman, A., (2009). Physical and biological mechanisms of plasma interaction with living tissue. *New Journal of Physics, 11*(11), 115020.
10. Dolezalova, E., & Lukes, P., (2014). Membrane damage and active but nonculturable state in liquid cultures of *Escherichia coli* treated with an atmospheric pressure plasma jet. *Bioelectrochemistry, 103*, 7–14.

11. Dzimitrowicz, A., Jamroz, P., Cyganowski, P., Bielawska-Pohl, A., Klimczak, A., & Pohl, P., (2021). Application of cold atmospheric pressure plasmas for high-throughput production of safe-to-consume beetroot juice with improved nutritional quality. *Food Chemistry, 336*, 127635.

12. FDA, (2001). Hazard analysis and critical control point (HACCP); procedure for the safe and sanitary processing and importing of juice: Final rule. *Federal Register, 66*(13), 6137–6202.

13. Filipić, A., Gutierrez-Aguirre, I., Primc, P., Mozetič, M., & Dobnik, D., (2020). Cold plasma, a new hope in the field of virus inactivation. *Trends in Biotechnology, 38*(11), 1278–1291.

14. Gavahian, M., Chu, Y. H., & Jo, C., (2019). Prospective applications of cold plasma for processing poultry products: Benefits, effects on quality attributes, and limitations. *Comprehensive Reviews in Food Science and Food Safety, 18*(4), 1292–1309.

15. Gavahian, M., & Cullen, P. J., (2019). Cold plasma as an emerging technique for mycotoxin–free food: Efficacy, mechanisms, and trends. *Food Reviews International, 36*(2), 193–214.

16. Gibbon, P., (2016). Introduction to plasma physics. In: Holzer, B., (ed.), *Proceedings of the 2014 CAS–CERN Accelerator School: Plasma Wake Acceleration* (Vol. 1, pp. 51–66). Geneva–Switzerland: CERN.

17. Hertwig, C., Reineke, K., Ehlbeck, J., Erdoğdu, B., Rauh, C., & Schlüter, O., (2015). Impact of remote plasma treatment on natural microbial load and quality parameters of selected herbs and spices. *Journal of Food Engineering, 167*(Part A), 12–17.

18. Hosseini, S. I., Farrokhi, N., Shokri, K., Khani, M. R., & Shokri, B., (2018). Cold low pressure O_2 plasma treatment of *Crocus sativus*: An efficient way to eliminate toxicogenic fungi with minor effect on molecular and cellular properties of saffron. *Food Chemistry, 257*, 310–315.

19. Hou, Y., Wang, R., Gan, Z., Shao, T., Zhang, X., He, M., & Sun, A., (2019). Effect of cold plasma on blueberry juice quality. *Food Chemistry, 290*, 79–86.

20. Isbary, G., Shimizu, T., Li, Y. F., Stolz, W., Thomas, H. M., Morfill, G. E., & Zimmermann, J. L., (2013). Cold atmospheric plasma devices for medical issues. *Expert Review of Medical Devices, 10*(3), 367–377.

21. Jaffee, S., Henson, S., Unnevehr, L., Grace, D., & Cassou, E., (2019). The safe food imperative: Accelerating progress in low- and middle-income countries. *Agriculture and Food Series* (p. 168). Washington, DC: World Bank Group.

22. Kashfi, A. S., Ramezan, Y., & Khani, M. R., (2020). Simultaneous study of the antioxidant activity, microbial decontamination, and color of dried peppermint (*Mentha piperita* L.) using low pressure cold plasma. *LWT, 123*, 109121.

23. Kim, H. H., Teramoto, Y., Negishi, N., & Ogata, A., (2015). A multidisciplinary approach to understand the interactions of nonthermal plasma and catalyst: A review. *Catalysis Today, 256*(Part 1), 13–22.

24. Kim, H. J., Yong, H. I., Park, S., Kim, K., Choe, W., & Jo, C., (2015). Microbial safety and quality attributes of milk following treatment with atmospheric pressure encapsulated dielectric barrier discharge plasma. *Food Control, 47*, 451–456.

25. Kim, J. E., Choi, H. S., Lee, D. U., & Min, S. C., (2017). Effects of processing parameters on the inactivation of *Bacillus cereus* spores on red pepper (*Capsicum annum* L.) flakes by microwave-combined cold plasma treatment. *International Journal of Food Microbiology, 263*, 61–66.

Cold Plasma Processing Methods   273

26. Kovačević, D. B., Putnik, P., Dragović-Uzelac, V., Pedisić, S., Jambrak, A. R., & Herceg, Z., (2016). Effects of cold atmospheric gas-phase plasma on anthocyanins and color in pomegranate juice. *Food Chemistry, 190*, 317–323.

27. Kulawik, P., & Tiwari, B. K., (2018). Recent advancements in the application of non-thermal plasma technology for the seafood industry. *Critical Reviews in Food Science and Nutrition, 59*(19), 3199–3210.

28. Lacombe, A., Niemira, B. A., Gurtler, J. B., Sites, J., Boyd, G., Kingsley, D. H., Li, X., & Chen, H., (2017). Nonthermal inactivation of norovirus surrogates on blueberries using atmospheric cold plasma. *Food Microbiology, 63*, 1–5.

29. Lee, J., Lee, C. W., Yong, H. I., Lee, H. J., Jo, C., & Jung, S., (2017). Use of atmospheric pressure cold plasma for meat industry. *Korean Journal for Food Science of Animal Resources, 37*(4), 477–485.

30. Lee, S. Y., Park, H. H., & Min, S. C., (2020). Pulsed light plasma treatment for the inactivation of *Aspergillus flavus* spores, *Bacillus pumilus* spores, and *Escherichia coli* O157:H7 in red pepper flakes. *Food Control, 118*, 107401.

31. Li, Q., Zhu, W. C., Zhu, X. M., & Pu, Y. K., (2010). Effects of penning ionization on the discharge patterns of atmospheric pressure plasma jets. *Journal of Physics D: Applied Physics, 43*, 382001.

32. Li, X., Li, M., Ji, N., Jin, P., Zhanga, J., Zhenga, Y., Zhang, X., & Li, F., (2019). Cold plasma treatment induces phenolic accumulation and enhances antioxidant activity in fresh-cut pitaya (*Hylocereus undatus*) fruit. *LWT, 115*, 108447.

33. Liao, X., Li, J., Muhammad, A. I., Suo, Y., Chen, S., Ye, X., Liu, D., & Ding, T., (2018). Application of a dielectric barrier discharge atmospheric cold plasma (Dbd-Acp) for *Escherichia coli* inactivation in apple juice. *Journal of Food Science, 83*(2), 401–408.

34. Lindsay, A., Byrns, B., King, W., Andhvarapou, A., Fields, J., Knappe, D., Fonteno, W., & Shannon, S., (2014). Fertilization of radishes, tomatoes, and marigolds using a large-volume atmospheric glow discharge. *Plasma Chemistry and Plasma Processing, 34*(6), 1271–1290.

35. López, M., Calvo, T., Prieto, M., Múgica-Vidal, R., Muro-Fraguas, I., Alba-Elías, F., & Alvarez-Ordóñez, A., (2019). A review on non-thermal atmospheric plasma for food preservation: Mode of action, determinants of effectiveness, and applications. *Frontiers in Microbiology, 10*, 622.

36. Los, A., Ziuzina, D., Akkermans, S., Boehm, D., Cullen, P. J., Impe, J. V., & Bourke, P., (2018). Improving microbiological safety and quality characteristics of wheat and barley by high voltage atmospheric cold plasma closed processing. *Food Research International, 106*, 509–521.

37. Mahnot, N. K., Mahanta, C. L., Farkas, B. E., Keener, K. M., & Misra, N. N., (2019). Atmospheric cold plasma inactivation of *Escherichia coli* and *Listeria monocytogenes* in tender coconut water: Inoculation and accelerated shelf-life studies. *Food Control, 106*, 106678.

38. Mahnot, N. K., Mahanta, C. L., Keener, K. M., & Misra, N. N., (2019). Strategy to achieve a 5-log *salmonella* inactivation in tender coconut water using high voltage atmospheric cold plasma (HVACP). *Food Chemistry, 284*, 303–311.

39. Mahnot, N. K., Siyu, L. P., Wan, Z., Keener, K. M., & Misra, N. N., (2020). In-package cold plasma decontamination of fresh-cut carrots: Microbial and quality aspects. *Journal of Physics D: Applied Physics, 53*, 154002.

40. Min, S. C., Roh, S. H., Niemira, B. A., Sites, J. E., Boyd, G., & Lacombe, A., (2016). Dielectric barrier discharge atmospheric cold plasma inhibits *Escherichia coli* O157:H7, *Salmonella, Listeria monocytogenes*, and *Tulane* virus in romaine lettuce. *International Journal of Food Microbiology, 237*, 114–120.

41. Misra, N. N., & Jo, C., (2017). Applications of cold plasma technology for microbiological safety in the meat industry. *Trends in Food Science & Technology, 64*, 74–86.

42. Misra, N. N., Schlüter, O., & Cullen, P. J., (2016). *Cold Plasma in Food and Agriculture: Fundamentals and Applications* (1st edn., p. 380). London–United Kingdom: Academic Press.

43. Misra, N. N., Yadav, B., Roopesh, M. S., & Jo, C., (2019). Cold plasma for effective fungal and mycotoxin control in foods: Mechanisms, inactivation effects, and applications. *Comprehensive Reviews in Food Science and Food Safety, 18*(1), 106–120.

44. Moisan, M., Barbeau, J., Crevier, M. C., Pelletier, J., Philip, N., & Saoudi, B., (2002). Plasma sterilization: Methods mechanisms. *Pure and Applied Chemistry, 74*(3), 349–358.

45. Niemira, B. A., (2012). Cold plasma decontamination of foods. *Annual Review of Food Science and Technology, 3*, 125–142.

46. Olatunde, O. O., Benjakul, S., & Vongkamjan, K., (2019). High voltage cold atmospheric plasma: Anti-bacterial properties and its effect on quality of Asian sea bass slices. *Innovative Food Science & Emerging Technologies, 52*, 305–312.

47. Ouf, S. A., Basher, A. H., & Mohamed, A. A., (2015). Inhibitory effect of double atmospheric pressure argon cold plasma on spores and mycotoxin production of *Aspergillus Niger* contaminating date palm fruits. *Journal of the Science of Food and Agriculture, 95*(15), 3204–3210.

48. Piel, A., (2010). *Plasma Physics: An Introduction to Laboratory, Space, and Fusion Plasmas* (1st edn., p. 398). Verlag–Berlin: Springer Science & Business Media.

49. Pignata, C., D'Angelo, D., Basso, D., Cavallero, M. C., Beneventi, S., Tartaro, D., & Gilli, G., (2014). Low-temperature, low-pressure gas plasma application on *Aspergillus brasiliensis, Escherichia coli* and pistachios. *Journal of Applied Microbiology, 116*(5), 1137–1148.

50. Porto, C. L., Ziuzina, D., Los, A., Boehm, D., Palumbo, F., Favia, P., Tiwari, B., Bourke, P., & Cullen, P. J., (2018). Plasma activated water and airborne ultrasound treatments for enhanced germination and growth of soybean. *Innovative Food Science & Emerging Technologies, 49*, 13–19.

51. Puligundla, P., Kim, J. W., & Mok, C., (2017). Effect of corona discharge plasma jet treatment on decontamination and sprouting of rapeseed (*Brassica napus* L.) seeds. *Food Control, 71*, 376–382.

52. Rodríguez, O., Gomes, W. F., Rodrigues, S., & Fernandesa, F. A. N., (2017). Effect of indirect cold plasma treatment on cashew apple juice (*Anacardium occidentale* L.). *LWT, 84*, 457–463.

53. Saikia, S., Mahnot, N. K., & Mahanta, C. L., (2015). Effect of spray drying of four fruit juices on physicochemical, phytochemical, and antioxidant properties. *Journal of Food Processing and Preservation, 39*(6), 1656–1664.

54. Saikia, S., Mahnot, N. K., & Mahanta, C. L., (2016). Phytochemical content and antioxidant activities of 13 fruits of Assam, India. *Food Bioscience, 13*, 15–20.

55. Scholtz, V., Pazlarová, J., Soušková, H., Khuna, J., & Julák, J., (2015). Nonthermal plasma: A tool for decontamination and disinfection. *Biotechnology Advances, 33*(6), 1108–1119.

56. Selcuk, M., Oksuz, L., & Basaran, P., (2008). Decontamination of grains and legumes infected with *Aspergillus spp.* and *Penicillium spp.* by cold plasma treatment. *Bioresource Technology, 99*(11), 5104–5109.

Cold Plasma Processing Methods

57. Shaw, P., Kumar, N., Kwak, H. S., Park, J. H., Uhm, H. S., Bogaerts, A., Choi, E. H., & Attri, P., (2018). Bacterial inactivation by plasma treated water enhanced by reactive nitrogen species. *Scientific Reports, 8*, 11268.

58. Shi, H., Ileleji, K., Stroshine, R. L., Keener, K., & Jensen, J. L., (2017). Reduction of aflatoxin in corn by high voltage atmospheric cold plasma. *Food and Bioprocess Technology, 10*(6), 1042–1052.

59. Shimizu, K., Kristof, J., & Blajan, M. G., (2019). Applications of dielectric barrier discharge microplasma; Chapter 4. In: Nikiforov, A., & Chen, Z., (eds.), *Atmospheric Pressure Plasma – from Diagnostics to Applications* (pp. 71–98). London–United Kingdom: IntechOpen.

60. Silva, D. A. D. S., Campelo, M. C. D. S., Rebouças, L. D. O. S., Vitoriano, J. D. O., & Junior, C. A., (2019). Use of cold atmospheric plasma to preserve the quality of white shrimp (*Litopenaeus vannamei*). *Journal of Food Protection, 82*(7), 1217–1223.

61. Song, Y., Annous, B. A., & Fan, X., (2020). Cold plasma-activated hydrogen peroxide aerosol on populations of *Salmonella typhimurium* and *Listeria innocua* and quality changes of apple, tomato, and cantaloupe during storage: A pilot-scale study. *Food Control, 117*, 107358.

62. Song, Y., & Fan, X., (2020). Cold plasma enhances the efficacy of aerosolized hydrogen peroxide in reducing populations of *Salmonella typhimurium* and *Listeria innocua* on grape tomatoes, apples, cantaloupe, and romaine lettuce. *Food Microbiology, 87*, 103391.

63. Thirumdas, R., Sarangapani, C., & Annapure, U. S., (2014). Cold plasma: A novel non-thermal technology for food processing. *Food Biophysics, 10*, 1–11.

64. Wang, Y., Wang, T., Yuan, Y., Fan, Y., Guo, K., & Yue, T., (2018). Inactivation of yeast in apple juice using gas-phase surface discharge plasma treatment with a spray reactor. *LWT–Food Science and Technology, 97*, 530–536.

65. WHO, (2015). *WHO Estimates of the Global Burden of Foodborne Diseases: Foodborne Disease Burden Epidemiology Reference Group 2007–2015* (1st edn., p. 265). Switzerland: World Health Organization.

66. Xu, G. M., Zhang, G. J., Shi, X. M., Ma, Y., Wang, N., & Li, Y., (2009). Bacteria inactivation using DBD plasma jet in atmospheric pressure argon. *Plasma Science and Technology, 11*(1), 83–88.

67. Xu, L., Garner, A. L., Tao, B., & Keener, K. M., (2017). Microbial inactivation and quality changes in orange juice treated by high voltage atmospheric cold plasma. *Food and Bioprocess Technology, 10*, 1178–1791.

68. Yong, H. I., Kim, H. J., Park, S., Alahakoon, A. U., Kim, K., Choeb, W., & Jo, C., (2015). Evaluation of pathogen inactivation on sliced cheese induced by encapsulated atmospheric pressure dielectric barrier discharge plasma. *Food Microbiology, 46*, 46–50.

69. Yong, H. I., Lee, H., Park, S., Park, J., Choe, W., Jung, S., & Jo, C., (2017). Flexible thin-layer plasma inactivation of bacteria and mold survival in beef jerky packaging and its effects on the meat's physicochemical properties. *Meat Science, 123*, 151–156.

70. Ziuzina, D., Misra, N. N., Han, L., Cullen, P. J., Moiseev, T., Mosnier, J. P., Keener, K., et al., (2020). Investigation of a large gap cold plasma reactor for continuous in-package decontamination of fresh strawberries and spinach. *Innovative Food Science & Emerging Technologies, 59*, 10229.

CHAPTER 11

MICROBIAL DECONTAMINATION OF FOODS WITH COLD PLASMA: A NON-INVASIVE APPROACH

IRFAN KHAN, NAZIA TABASSUM, and ABDUL HAQUE

ABSTRACT

Today, people are using nonthermal technologies in the food processing sector for microbial decontamination and to extend the storability. The cold plasma (CP) technology is one of the most effective nonthermal technologies that are being in practice. CP is a modern nonthermal food preservation technology that has attracted significant popularity over the past decade. It is a promising approach to achieve the best of food safety standards and extended storability without hampering food quality. This robust technology inactivates or kills vegetative and spores forming bacteria with the help of energetic reactive chemical species by using noble gases such as helium, neon, argon, and sometimes with nitrogen. It is less toxic to the food items and a cost-effective operation as well. In this chapter, an effort has been made to summarize the benefits and advantages of CP technology and its limitations in the food processing sector. Although the applications of CP in food is in its juvenile stage and it has to move thousands of miles from sophisticated laboratory to the traditional platform. The CP technology would prove better than conventional technology to achieve food safety standards as suggested by the statutory organizations soon.

Advances in Food Process Engineering: Novel Processing, Preservation, and Decontamination of Foods.
Megh R. Goyal, N. Veena, & Ritesh B. Watharkar (Eds.)
© 2023 Apple Academic Press, Inc. Co-published with CRC Press (Taylor & Francis)

11.1 INTRODUCTION

At present, consumers have become more conscious about their health and they demand foods with natural attributes with extended shelf life. That's why, in the past few decades, an increased emphasis has been on the minimal processing technologies and nonthermal processing technologies. In this chapter, an effort has been made to summarize the basics of cold plasma (CP), their generation methods, and the preservation of plant-based foods, livestock products, etc. The heat treatments are the oldest and most traditional methods used for the preservation of food. But they do have certain undesirable effects on the food quality, nutrients loss, and reduction in the organoleptic properties as well [26]. On the other hand, cold plasma technology (CPT) is one of the most effective emerging food preservation technologies that inactivate the microorganisms without altering the natural essence of food, i.e., maintain quality and nutrients, sensory attributes. Although, some researchers claimed that CPT have minimum effects on the physicochemical, textural, and sensory quality of the food which has been discussed in the later section of this chapter. While very few studies have been done on the interaction and behavior of CP species and food components at the molecular level. This will open the doors of future opportunities in the field of CPT. It is the need of today and the future to invent and utilize the cost-effective preservation technologies to reduce the burden of post-harvest losses in the form of revenue.

The potential of CPT in food preservation has been determined experimentally, and this is the reason why CPT would perhaps turn out to be the most reliable and cost-effective solution of food preservation in the coming future [18]. The trend is now shifting towards nonthermal food technologies from thermal technologies. Several other nonthermal technologies are already in practice today such as pulsed electric field, nuclear magnetic resonance, high-intensity ultrasound (US), etc., and atmospheric CP have recently been added to this list. The nonthermal, economical, and environment-friendly nature of CP make it more viable and advantageous over existing preservation technologies. The CPT could open new avenues for the food industries as it is the most dominating sector that employs a larger fraction of the population worldwide. Quality and safety are two wheels with which the food industries decide their journey. Therefore, industries need to adopt new technologies in order to give customers a higher quality of food. This chapter addresses the impact of CPT on the organoleptic quality and food safety of foods.

11.2 PLASMA AS A FOURTH STATE OF MATTER

Plasma is generally recognized as an ionized form of gas consisting of an equal number of positive ions and negative ions or electrons [7, 8, 14, 21, 29, 42]. It has already been discussed that the ionized form of gas is known as plasma, i.e., when a sufficient amount of energy is given to the gas especially the ideal gases, the gas molecules become charged by breaking of bonds, as a result, anions, and cations are formed (Figure 11.1). There are numerous examples of plasma that exists naturally viz., sun, is the typical example of the plasma system. The uppermost layer of our atmosphere is ionized by the sun and is called plasma that helps in long-range radio communication [19].

Approximately 99% of the matter in the universe exists in a plasma state, which is produced by various mechanisms such as interstellar gases are ionized by ultraviolet (UV) radiations coming from stars while the neutral atoms of stars are ionized due to high temperature [29].

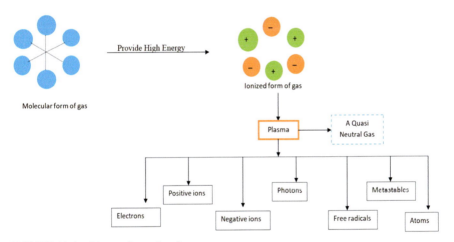

FIGURE 11.1 Plasma formation from gas.

11.3 COLD PLASMA (CP): CLASSIFICATION AND GENERATION

We have discussed the general understanding of plasma as the fourth state of matter in the previous section. On a broad scale, plasma can be broadly categorized as equilibrium/thermal or hot plasma and nonthermal or low temperature or CP. Both electrons and heavy particles have high temperatures in thermal plasma and are present in thermodynamic temperature equilibrium. NTP is further divided into two categories, i.e., non-equilibrium plasma (T < 60°C) and quasi-equilibrium plasma (T → 100–150°C). In the case of non-equilibrium plasma, the cooling of neutral and ionic species dominates over electron energy transfer which keeps the gas at low temperature [52]. This is why non-equilibrium plasma is called NTP or CP (Figure 11.2).

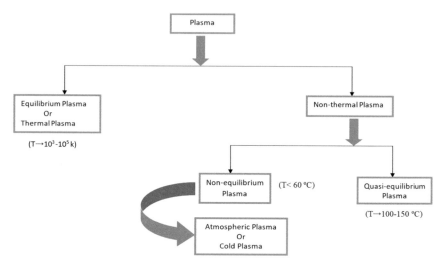

FIGURE 11.2 Different types of plasma [52].

CP is characterized by a high temperature of electrons and low temperature of heavy particles which is closer to the room or atmospheric temperature. The term "cold plasma" is introduced to distinguish between atmospheric plasma and other NTP. So, the CP is also known as atmospheric CP. While in the case of quasi-equilibrium plasma a thermodynamic equilibrium exists locally among the species. The CP is generated at low energy input at atmospheric pressure. The CP produced at atmospheric temperature and pressure makes it more flexible to use for foods and other biological materials.

The generation of CP requires energy from an external source such as electric discharge, magnetic energy, etc. For the generation of thermal plasma, a huge amount of energy is required while comparatively very low energy is needed for the generation of atmospheric CP as mild conditions are required for the operation. Mainly four methods are employed for the CP generation viz., gliding arc discharge, corona discharge, dielectric barrier discharge (DBD), and atmospheric plasma jet discharge method (Figure 11.3). Among the four methods, the corona discharge method is used for the generation of atmospheric CP and is best suited for the sterilization of biological systems such as food. In the corona discharge method, the CP is produced by a non-uniform electric field at a pressure equivalent to atmospheric conditions.

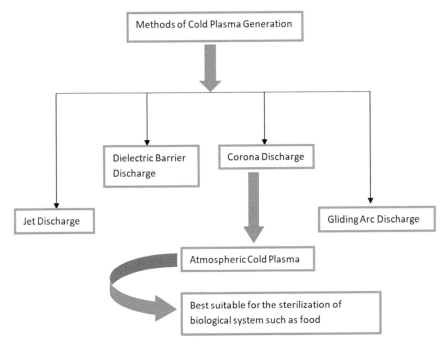

FIGURE 11.3   Methods of cold plasma generation.

11.4   APPLICATIONS OF CPT IN THE FOOD SECTOR

CP has found numerous applications in varying fields of the food sector. It has demonstrated high efficiency in reducing the number of bacteria in agricultural produce such as plant-based products, livestock products, etc. It has also found application in the food packaging industry, wherein sterilization of packaging materials or decontamination of in-package food has been exploited.

11.4.1   PROCESSING OF FRUITS AND VEGETABLES

The fruits and vegetables may be the carriers to various disease-causing agents, such as bacteria, viruses, yeast, molds, etc. The CPT can be employed to decontaminate the fruits and vegetables, reducing their microbial count and provide for the emerging fresh-cut produce industry. It has been proved to be a better alternative than the conventional methods used in the food processing. Few of the earlier works were performed on apples, cantaloupe, and lettuce to reduce the population of *E. coli* O157:H7, *Salmonella typhimurium,* and *Listeria monocytogenes,* respectively [13, 51].

Scientists performing CP experiments have concluded that plasma antimicrobial efficiency can be influenced by the form and volume of plasma produced by reactive species. Different experimental parameters such as temperature, working gas, sample distance, plasma emitter, voltage, electrodes, and treatment time are used to determine the species [76]. Plasma processing performed at higher flow rates or power levels proved to be more effective but also negatively impacted the tissue quality in corn salad leaves and degraded the color in spinach [33]. This limited the process to a "high-intensity-short-time" [4]. Similarly, plasma-activated water (PAW) has been revealed to have exceptional anti-bacterial activity and soon could be established as a substitute sanitizer for the fresh produce industry. PAW was applied to strawberry fruits inoculated with *S. aureus* for 15 minutes and kept for 0 or 4 days at 20°C. Significant reduction (1.6–2.3 log) was observed which was interestingly further reduced after 4 days of storage. More importantly, the increased PAW treatment time had no significant effect on the quality of the treated strawberries [40].

The perishable commodities, viz., fruits, and vegetables are on the greater risk of microbial spoilage owing to their greater fraction of moisture content among the proximate constituents. It has been found that around 30% of *Salmonella* inoculums affixing itself onto green pepper slices in the exposure of less than 30 s [64]. Their removal by washing became even more complicated as these pathogens tend to swiftly and irrevocably attach themselves to the surface of fruits and vegetables and stay there for long durations. Recently, CP has been significantly found to reduce the population of these spoilage microorganisms and could be crucial in enhancing the storability of perishables. The fruits and vegetables are perfect models for plasma treatment as they have a high moisture content that proves to be a favorable environment for plasma activity. Therefore, moist microorganisms are more prone to destruction by plasma treatment as compared to dry organisms [16]. Highly reactive free radicals are formed upon ionization of water molecules in the cells and kill the microorganisms by damaging the deoxyribonucleic acid (DNA) in the nucleus of the microbial cells [67]. These free radicals cause oxidation of DNA, chromosomes, nucleus, and cell membranes resulting in their destruction [71].

Microbial Decontamination of Foods with Cold Plasma 283

Degradation in the quality of fresh-cut produce also takes place due to enzymatic browning which could also be prevented by the application of CP. For example, enzyme trypsin can be deactivated by alteration in structure leading to its deactivation [16]. DBD plasma treatment was observed to reduce browning or darkened area as well as the polyphenol-oxidase activity in fresh-cut apples [70] and fresh-cut kiwi [57].

One of the major applications of fresh produce is the juice industry and the CP has demonstrated positive outcomes during the processing of heat-sensitive products. More than 5 log reduction of *E. coli, Candida albicans,* and *S. aureus* were observed in orange juice [63] and of *Citrobacter freundii* in apple juice [69] after plasma processing. The processed juices, when assessed for qualitative changes, showed no significant alterations, thus suggesting the huge potential of plasma in decontaminating fresh produce and extending its shelf life, while also preserving the nutritional and organoleptic properties. CP therapies had an effect on fruit and vegetable-related pathogenic microbes (Table 11.1).

11.4.2 PROCESSING OF CEREALS

Immediately after harvest, cereal grains are susceptible to microbial contamination from microbes, such as *Aspergillus, Fusarium, Alternaria, Bacillaceae, Lactobacillaceae, Pseudomonadaceae,* etc. They produce undesirable color and odors and reduce the nutritive values of the grain. Reduced cooking and baking quality and production of mycotoxins that make the products inedible for consumption are also some of the damaging effects of these pathogens. These pathogenic microbes in low moisture foods often show resistance to heat that may be lethal to cells in high moisture environments [6]. Various experiments have shown the effectiveness of CP against these pathogens in low-moisture foods.

The infestation of pests has always been one of the major concerns for the grain processors. For long chemical pesticides and insecticides were used until their detrimental effect on the consumer's health came to light. More recently, researchers have looked into the use of CP to help generate ozone that has strong oxidizing properties. Ozone has been reported as an effective fumigant for the killing of insects, microbial inactivation, and mycotoxin destruction while having minimal to no effect on the quality of grain. Ozone leaves no residue and has been used as a substitute of chlorine, potassium bromate, and benzoyl peroxide for the wheat flour treatment [44, 58].

TABLE 11.1 Effect of Atmospheric Cold Plasma Treatments on Selected Pathogenic Bacteria and Quality of Fruits and Vegetables

Product	Treatment	Maximal Log Reduction			Impact on Quality	References
		E. coli 0157:H7	*Salmonella typhimurium*	*Listeria monocytogenes*		
Almonds	Air; 47 kHz; 549 W	1.34 log in 20 sec	–	–	No considerable changes in color, aroma, and surface features	[50]
Black peppercorns	Air + Ar; 2 kHz; 30 kV	–	5.0 log in 80 sec	–	Little change in color	[68]
Cherry tomatoes	Ar; 50 Hz; 70 kV	3.1 log in 1 min	6.3 log in 10 Sec	6.7 log in 2 min	Not determined	[77]
Cucumber	Ar + O$_2$; 10 kV; 8 W	4.7 log in 1 min	–	–	No considerable impact was found in physicochemical properties	[5]
Orange juice	Air; 60 kHz; 20 kV	5.0 log in 8 sec	–	–	No significant effect on pH value, vitamin C content, and turbidity	[63]
Strawberries	Air; 50 Hz; 70 kV	3.5 log in 5 min	3.8 log in 5 min	4.2 log in 5 min	Not determined	[77]
Tomato	Ar + O$_2$; 10 kV; 8 W	3.3 log in 1 min	–	–	Not determined	[5]

Microbial Decontamination of Foods with Cold Plasma   285

The "dielectric barrier discharge" is the most efficient method to produce ozone [1, 48]. A recent study conducted on wheat flour (soft and hard) gave improved results for the strength of dough and mixing time after CP treatment. This was due to the fact that sulfhydryl groups of protein underwent oxidization and successive disulfide bond formation between cysteine moieties [44]. CP was now seen to change flour functionality, but more regulation was needed to enhance flour functionality. It would have been necessary for this improvement.

Another research showed accelerated lipid oxidation and reduction in the total free fatty acids and phospholipids of wheat flour treated with CP, though the low levels of CP proved powerless in reducing the microbial count [3]. Lipids form a minor constituent of cereals and therefore it is unlikely for plasma processing to cause significant oxidation. The contrary was the case for lipid rich cereals viz., chia, and oats seeds or when bran was also included [20]. The experiments conducted on white and brown rice showed undesirable oxidative developments which further increased on increasing the treatment time [38]. The oxidation effect was attributed to long treatment times and high power levels. Therefore, it could be concluded that it is essential to optimize the processing parameters of CP for cereal processing especially, lipid-rich grains such as whole flour.

11.4.3 FOOD PACKAGING

The accountability of food packaging material can never be neglected in protecting the food from the surrounding environment during storage and transportation. Therefore, it is very necessary to ensure the safety of food packaging material. Apart from reducing microbial load in packaging material or in-package food, CPT also improves various properties of packaging material including adhesion, printability, strength, and polymerization. Originally, the technology of CP was developed for increasing the surface energy of polymer used in packaging material, but later on, this technology became prominent in decontaminating/sterilizing the food and packaging materials required to enclose these foodstuffs (Figure 11.4). Thus, the CP may prove effective in improving the esthetic properties of packaging materials.

CP technique can be successfully adapted for reducing the microbial load from packaging materials. There are two approaches; one is in-package food decontamination/sterilization and the other is the direct sterilization of food packaging material. Direct decontamination of food packaging material can be achieved in two ways, "direct-treatment" and "indirect-treatment." In direct treatment, material to be sterilized is held within the region of plasma discharge whereas, in indirect treatment, the material to be sterilized is held outside the region of plasma discharge so that coating can be done on packaging material, as in surface modification treatment.

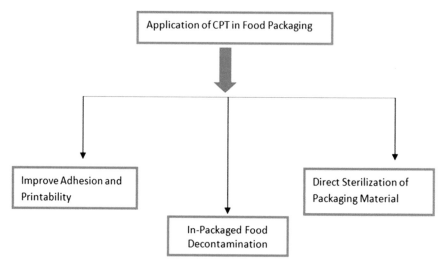

FIGURE 11.4 Applications of cold plasma in food packaging.

11.4.3.1 IMPROVING ESTHETIC CHARACTERISTICS OF PACKAGING MATERIALS

Labeling and good appearance are the basic requirements of any packaging material, and hence it is very essential to enhance such properties of packaging materials that are directly or indirectly related to appearance. Good printability and improved adhesion in food packaging material may be attained by plasma treatment [53]. Concerning surface energy, the crystal structure of polyethylene (PE) terephthalate (PET) film can be determined using CP treatment [25]. Although, PET possesses a wide range of good properties like formability and strength but has low surface energy which needs to be activated for enhancement in dyeing and printing ability and for this purpose, CPT can help a lot. The application of plasma in boosting the adhesiveness of packaging materials is met by bringing the physical and chemical variation on the exterior top of the polymer. Dominant free radicals in plasma deactivate organic contamination, particularly from the surface,

Microbial Decontamination of Foods with Cold Plasma

and hence the top surface of packaging material gets oxidized [34, 62]. This results in chemical composition variation of the surface which ultimately enhances the surface energy of polymer and then the plasma-treated polymer may effectively be used in the packaging of various food commodities.

11.4.3.2 DECONTAMINATION OF IN-PACKAGE FOODS

The decontamination of in-package food is generally carried out by DBD plasma treatment due to simplicity in its design, low running cost, and wide range of operating pressure. The in-package sterilization is specifically more suitable for fresh produce and ready to consume meals [47]. The fruits and vegetables require normal processing temperature so that their nutritional quality is less affected during the treatment processes. In the very early study of DBD plasma treatment (12 kV for 5 minutes) of spinach in a flexible package containing *E. coli* 0157:H7 [33], 1.5–2 log reduction of contaminant was reported after half an hour of treatment whereas 3–5 log reduction was recorded within a whole day. Subsequently, in various studies, the reduction of aerobic microbes by 5 logs was shown in the fresh produce like tomatoes and strawberries after the treatment of CP [45, 46].

The confirmation of the absence of physiological disorder was also made in these studies. The potential of upholding the better quality of plasma-treated in-package strawberries in a modified atmosphere was also reported. Recently, Min et al. showed the 1.1 log CFU/g reduction of *E. Coli* in romaine lettuce treated by DBD CP (at 42.6 kV for 10 minutes) [43]. Another study revealed the reduction of *Listeria monocytogenes* by 2.04 log CFU/g whereas reduction of *E. coli* and *Salmonella typhimurium* was reported by 2.54 log and 2.68 log CFU/g, respectively on pork butt when treated with surface dielectric barrier discharge (SDBD) in-package plasma [26]. Therefore, it can be concluded that decontamination of in-package can be achieved using CPT and it is more effective in conjunction with a modified gas atmosphere.

11.4.3.3 STERILIZATION OF FOOD PACKAGING MATERIAL

Studies were conducted in which cascading structure was used in DBD CP treatment at a frequency of 10–50 kHz and they reported 4–7 log reduction of contaminants (*B. atrophaeus* spores and vegetative cells) in different packaging films and materials like PE and polyethylene terephthalate (PET)

[49]. Thus, this direct approach of CP treatment of food packaging material helped in the sterilization of different packaging material. In another study conducted by Kim et al., 3–4 log reduction of contaminants like *E. coli, L. monocytogenes* and *S. typhimurium* was reported on casing made up of collagen, polypropylene (PP) and PET packaging materials; atmospheric pressure plasma jet was used along with nitrogen and oxygen as carrying gases and treatment was applied for 10 min [31].

A study reported that 3–4 log reduction for contaminants (*E. coli, S. aureus* and *Salmonella typhimurium*) on PP, PE, paper foil, nylon, and parchment paper; DBD plasma was used with oxygen at a frequency of 35 kHz and 10 kV voltage for 10 min [56]. There was no significant change in optical properties and tensile strength of packaging material. However, there is very little literature available on the behavior of biodegradable polymers on treatment with DBD plasma. Polylactic acid and corn starch films were analyzed to study the effect of DBD plasma on these biodegradable packaging materials. The increased surface roughness was reported in polylactic acid films when treated with DBD plasma [54, 55].

The modification in the packaging material using CP is achieved by surface coating. The surface of the polymeric packaging material is modified by the deposition of substance/s having antimicrobial characteristics using CP technique. Such bioactive substances deposited on the surface of packaging material by CP technique readily behave as antimicrobial compounds like lysozyme and vanillin [17, 41]. One study was conducted to analyze the oxygen transmission characteristics of coating material on PET, PP, and high-density polyethylene (HDPE) as well as low-density polyethylene (LDPE). The coating material, SiO_x (oxide of silicon) was coated on aforesaid packaging material using CP treatment. In that study, it was found that diffusivity was small than that of uncoated material [23].

In a study, Nisin as well as Auranta FV which are the compound that is available at commercial level as antimicrobial agents were coated on PE packaging material using cold (nitrogen) plasma treatment. These compounds provided resistance against the microbial activities of molds and yeast. Thus, accumulation of such coating substances on normal packaging materials helps in enhancing the shelf life of food, packed within these packaging materials [11, 28]. These studies provided new insights into the use of CP treatment in ensuring food safety by focusing on packaging and storage levels. This approach could help quite in preserving food for a longer duration. A detailed probe on the effects of such antimicrobial coatings on human health is also required before its implication in food packaging materials.

11.4.4 MEAT AND POULTRY PROCESSING

The most difficult job in meat and meat industry is to guarantee the quality of meat and meat products to consumers. The use of CPT does not check only microbial load but also has a significant effect on the sensory and textural attributes of the product. CPT also helps in maintaining the hygiene of plant and equipment in meat processing industries. It also assists in the curing of meat. Curing is a method adapted principally in color retention of meat; however, it also helps in preservation. This novel technology also contributes to the preservation of fish and seafood which are highly perishable. All these aspects of the application of CP in meat processing are discussed in this section.

The pathogens which are generally observed in meat products are *L. monocytogenes, B. cereus, Salmonella spp., Campylobacter spp., Enterobacter, Clostridium botulinum, E. Coli, Shigella,* etc. [66]. The reduction of the cell population of microbes depends on the cell population at the initial level. It was revealed in a study that using DBD air plasma for 3 minutes on chicken breast resulted in a significant reduction of the population of *C. Jejuni*; in one of the samples, it was reduced to 2.45 log from 4 log CFU [15]. A reduction in aerobic bacteria was found on bacon when treated with the helium (He) plasma with the oxygen-helium mixture, operating at 125 W; the reduction observed was from 7.08 to 2.5 log CFU/g in just 90 seconds [30]. Another study revealed that 2 minutes of CPT (He + O_2 mixture) of smoked salmon fish suppressed the growth of *L. Monocytogenes* [35]. In another analysis, it was reported that the population of *Photobacterium phosphorus*, which is responsible for spoilage of seafood, was significantly reduced in cold-smoked salmon when treated by CP [9].

11.4.4.1 PRESERVING THE SENSORY AND TEXTURAL ATTRIBUTES OF MEAT AND MEAT PRODUCTS

Color enhancement of meat products may also be achieved using CPT. The utilization of nitrite water produced by DBD plasma has been suggested nowadays particularly in the curing of meat. Nitrites are incorporated as a curing agent to enhance the bright pink color of meat products. Nitrites are also helpful in inhibiting the *Clostridium botulinum* [27]. Nitrite in the batter of meat can be oxidized with oxygen sequestering, and interaction with myoglobin will form nitric oxide myoglobin which imparts pink color in the cured meat [24]. Plasma nitrites have been involved in achieving the red color of meat products as indicated by bathochromic shift (red color) in the spectrum of myoglobin when it was given nitrogen plasma treatment [2]. It was also reported in another study that when plasma-treated water was used for curing the emulsion sausage no mutagenicity was found [32]. Hardness/firmness, gumminess, chewiness, cohesiveness, and adhesiveness are the attributes that are important in defining the texture of meat and meat products. It was found that these textural attributes were not much affected by plasma treatment while studying beef loin and pork butt [26].

11.4.4.2 MAINTAINING THE SANITATION AND HYGIENE OF MEAT PROCESSING PLANT AND EQUIPMENT

Apart from the above-discussed CP applications of meat processing, the hygiene of meat processing equipment and tools are equally important. By using DBD style configuration plasma operating at the atmospheric configuration, the decontamination of cutting knife (rotary type), which is used in the meat industry, was explained for the first time by Leipold et al. [39]. Another approach was given by Wang et al. to modify the surface of equipment which are in direct contact with food during processing. By utilizing the vinyl ether RF plasma environment, a structure like PE glycol was developed on stainless steel, which showed the remarkable decrease of adherence of *L. Monocytogenes* biofilm on such modified stainless steel equipment [72].

11.4.5 MILK AND DAIRY PRODUCTS

For milk and milk products, pasteurization remains is the best choice for inactivating pathogenic microorganisms and enhancing the storability. Loss of several nutrients takes place during the high temperature of pasteurization and this disadvantage opens up the door towards other alternatives. CPT is one of those options. Therefore, the CP technique can help in satisfying the consumers by extending shelf life without compromising on flavor (taste and aroma) of milk and milk products along with upholding sensory and textural attributes. A general conclusion cannot be made at this point because this approach is in the emerging stage and limited researches are available on this topic. However, based on several types of research and available literature, the following general remarks can be made. CPT is quite promising in reducing the microbial load and prevention of biofilm formation, just like in other food products discussed earlier. Few studies have been performed on CP treatment of milk and cheese, some of them are discussed below. A 54% reduction in the population of *E. coli* was reported in milk treated with plasma for 3 minutes with insignificant variation in color and pH [22]. In another study, it was revealed that there was a significant reduction of *S. aureus* and *E. coli* in cheese given treatment of CP [37]. However, along with this change, the decrease in L* (lightness) value and reduction in sensory score was also marked.

A study was performed on encapsulated DBD plasma to evaluate its antimicrobial efficiency and a remarkable decrease in the count of *Salmonella typhimurium, L. Monocytogenes* and *E. coli* on cheese was found. Another study of plasma effect on the aforementioned three microorganisms in cheddar cheese was also carried out; the plasma-treated cheddar cheese was reported with decreased pH and L* value, increased thiobarbituric acid content and poor sensory score [73]. In another investigation, the structural change of whey protein was noticed on the plasma treatment of milk for 15 minutes [60]. In another study, the CP treatment output for alkaline phosphatase inactivation was reported [61]. The impact of plasma on milk fat was examined using chromatographic methods and noted the development of several oxidation products along with variability in the content of fatty acids [59].

An investigation was carried out on the chocolate milk drink after treatment with CP and they reported increased particle size and good consistency [12]. CPT is gaining momentum day-by-day in food processing industries because of its nature of disinfection and prevention of biofilm formation along with preserving other quality and nutritional parameters. Available literature advocated that CPT did not significantly impact the organoleptic quality of milk and milk-based products except slight variation in color, which was detectable only through instrumental color measurement, thus this difference was not much detected by human naked eyes. The decrease in pH may be attributed to some of the species generated during plasma treatment like oxides of nitrogen.

There have been only a few studies on the use of CP in the milk sector, so more study is required in this sector. Although CPT has a good efficiency in decreasing microbial population at a lower temperature the biggest challenge is that thermal pasteurization has been well developed and established in the dairy industry; and for the replacement of these thermal processing methods with the emerging technology of CP, the assessment of cost is very necessary along with additional advantages of CPT. Commercialization of CPT in the field of milk sterilization will open up doors for this emerging technology. The effect of CPT on the quality of meat and milk products has been presented below (Table 11.2).

TABLE 11.2 Effect of Atmospheric Cold Plasma Treatments on Selected Pathogenic Bacteria and Quality of Meat and Dairy Products

Product	Treatment	Maximal Log Reduction			Impact on Quality	References
		E. coli **0157:H7**	*Salmonella typhimurium*	*Listeria monocytogenes*		
Bacon	He + O_2; 13.56 MHz; 125 W	3.0 log in 1.5 min	1.0 log in 1.5 min	2.6 log in 1.5 min	No tissue damage; higher lightness value	[30]
Beef	N_2 + O_2; 15 kHz; 2 W	2.6 log in 10 min	2.6 log in 10 min	1.9 log in 10 min	No change in texture and little change in taste and color	[26]
Beef jerky	Ar; 15 kHz; 200 W	2.7 log in 10 min	3.0 log in 10 min	2.4 log in 10 min	Very little changes in color, flavor, and overall acceptability	[75]
Cheddar cheese	Air; 15 kHz; 2 W	3.6 log in 10 min	2.1 log in 10 min	5.8 log in 10 min	Color and appearance not affected	[74]
					Little change in flavor and overall acceptance	
Cheese	Air; 15 kHz; 250 W	2.7 log in 1 min	3.1 log in 45 sec	–	Not determined	[73]
Cooked chicken breast	N_2 + O_2; 50 kHz; 2 kV	–	–	4.7 log in 2 min	Not determined	[36]
Ham	He; 13.56 MHz; 150 W	–	–	1.7 log in 2 min	Not determined	[65]
Pork	N_2 + O_2; 15 kHz; 2 W	2.5 log in 10 min	2.7 log in 10 min	2 log in 10 min	No change in texture and little change in taste and color	[26]
Unfrozen and frozen pork	Air; 58 kHz; 20 kV	1.5 log in 10 min	–	1.0 log in 10 min	Considerable effect on color and appearance	[10]

11.5 SUMMARY

In the present scenario, people have become more aware and health-conscious than before. So, to provide the foods with fresh values, CPT has come into existence. The CPT has been proved as one of the most sophisticated nonthermal non-invasive emerging technologies that will probably replace thermal preservation technologies in the coming future. Although experimental analysis has successfully established the CP, robust technology for the decontamination of foods, more work has to be done on the impact of CP on the quality attributes of food. In spite of robust and versatile technology, the decontamination impact is limited to the surface of the food. No literature has been found on the interaction of CP species and food components at the molecular level that may open some new insights into CPT in the future. Overall, CPT has great potential to provide better preservation results with significantly good quality attributes of food. It will surely occupy a significant market space and generate plenty of employment opportunities in the future.

KEYWORDS

- cold plasma
- deoxyribonucleic acid
- dielectric barrier discharge
- food preservation
- microbial decontamination
- non-invasive
- nonthermal

REFERENCES

1. Amjad, M., Salam, Z., Facta, M., & Ishaque, K., (2012). A simple and effective method to estimate the model parameters of dielectric barrier discharge ozone chamber. *IEEE Transactions on Instrumentation and Measurement, 61*(6), 1676–1683.
2. Attri, P., Kumar, N., Park, J. H., Yadav, D. K., Choi, S., Uhm, H. S., Kim, I. T., et al., (2015). Influence of reactive species on the modification of biomolecules generated from the soft plasma. *Scientific Reports, 5*, 8221.

3. Bahrami, N., Bayliss, D., Chope, G., Penson, S., Perehinec, T., & Fisk, I. D., (2016). Cold plasma: A new technology to modify wheat flour functionality. *Food Chemistry, 202,* 247–253.

4. Baier, M., Foerster, J., Schnabel, U., Knorr, D., Ehlbeck, J., Herppich, W. B., & Schlüter, O., (2013). Direct nonthermal plasma treatment for the sanitation of fresh corn salad leaves: Evaluation of physical and physiological effects and antimicrobial efficacy. *Postharvest Biology and Technology, 84,* 81–87.

5. Baier, M., Gorgen, M., Ehlbeck, J., Knorr, D., Herppich, W. B., & Schlute, O., (2014). Nonthermal atmospheric pressure plasma: Screening for gentle process conditions and anti-bacterial efficiency on perishable fresh produce. *Innovative Food Science and Emerging Technologies, 22,* 147–157.

6. Beuchat, L. R., Komitopoulou, E., Beckers, H., Betts, R. P., Bourdichon, F., Fanning, S., Joosten, H. M., & Ter Kuile, B. H., (2013). Low-water activity foods: Increased concern as vehicles of foodborne pathogens. *Journal of Food Protection, 76*(1), 150–172.

7. Chapman, B., (1980). *Glow Discharge Processes. Sputtering and Plasma Etching* (p. 432). New York: Wiley.

8. Chen, F. F., (1974). *Introduction to Plasma Physics* (p. 329). US: Springer.

9. Chiper, A. S., Chen, W., Mejlholm, O., Dalgaard, P., & Stamate, E., (2011). Atmospheric pressure plasma produced inside a closed package by a dielectric barrier discharge in Ar/CO_2 for bacterial inactivation of biological samples. *Plasma Sources Science and Technology, 20*(2), 025008.

10. Choi, S., Puligundla, P., & Mok, C., (2016). Corona discharge plasma jet for inactivation of *Escherichia coli 0157:H7* and *Listeria monocytogenes* on inoculated pork and its impact on meat quality attributes. *Annals of Microbiology, 66,* 685–694.

11. Clarke, D., Tyuftin, A. A., Cruz-Romero, M. C., Bolton, D., Fanning, S., Pankaj, S. K., Bueno-Ferrer, C., et al., (2017). Surface attachment of active antimicrobial coatings onto conventional plastic-based laminates and performance assessment of these materials on the storage life of vacuum packaged beef sub-primal. *Food Microbiology, 62,* 196–201.

12. Coutinho, N. M., Silveira, M. R., Fernandes, L. M., Moraes, J., Pimentel, T. C., Freitas, M. Q., Silva, M. C., et al., (2019). Processing chocolate milk drink by low-pressure cold plasma technology. *Food Chemistry, 278,* 276–283.

13. Critzer, F. J., Kelly-Wittenberg, K., South, S. L., & Golden, D. A., (2007). Atmospheric plasma inactivation of foodborne pathogens on fresh produce surfaces. *Journal of Food Protection, 70*(10), 2290–2296.

14. D'Agostino, R., (1990). *Plasma Deposition, Treatments, and Etching of Polymers* (1st edn., p. 544). San Diego–CA: Academic Press.

15. Dirks, B. P., Dobrynin, D., Fridman, G., Mukhin, Y., Fridman, A., & Quinlan, J. J., (2012). Treatment of raw poultry with nonthermal dielectric barrier discharge plasma to reduce *Campylobacter jejuni* and *Salmonella enterica*. *Journal of Food Protection, 75*(1), 22–28.

16. Dobrynin, D., Fridman, G., Friedman, G., & Fridman, A., (2009). Physical and biological mechanisms of direct plasma interaction with living tissue. *New Journal of Physics, 11*(11), 115020.

17. Fernandez-Gutierrez, S., Pedrow, P. D., Pitts, M. J., & Powers, J., (2010). Cold atmospheric pressure plasmas are applied to the active packaging of apples. *IEEE Transactions on Plasma Science, 38*(4), 957–965.

Microbial Decontamination of Foods with Cold Plasma

18. Fernandez, A., Shearer, N., Wilson, D. R., & Thompson, A., (2012). Effect of microbial loading on the efficiency of cold atmospheric gas plasma inactivation of *Salmonella enterica* serovar *Typhimurium. International Journal of Food Microbiology, 152*(3), 175–180.

19. Frank-Kamenetskii, D., (1972). *Plasma: The Fourth State of Matter* (p. 160). US: Springer.

20. Gavahian, M., Chu, Y. H., Khaneghah, A. M., Barba, F. J., & Misra, N. N., (2018). A critical analysis of the cold plasma-induced lipid oxidation in foods. *Trends in Food Science & Technology, 77*, 32–41.

21. Grill, A., (1994). *Cold Plasma in Materials Fabrication: From Fundamentals to Applications* (p. 272). New York: IEEE Press.

22. Gurol, C., Ekinci, F. Y., Aslan, N., & Karachi, M., (2012). Low-temperature plasma for decontamination of *E. coli* in milk. *International Journal of Food Microbiology, 157*(1), 1–5.

23. Hedenqvist, M. S., & Johansson, K. S., (2003). Barrier properties of SiO_x-coated polymers: Multi-layer modeling and effects of mechanical folding. *Surface and Coatings Technology, 172*(1), 7–12.

24. Honikel, K. O., (2008). The use and control of nitrate and nitrite for the processing of meat products. *Meat Science, 78*(1, 2), 68–76.

25. Jacobs, T., De Geyter, N., Morent, R., Van, V. S., Dubruel, P., & Leys, C., (2011). Plasma modification of PET foils with different crystallinity. *Surface and Coatings Technology, 205*(2), S511–S515.

26. Jayasena, D. D., Kim, H. J., Yong, H. I., Park, S., Kim, K., Choe, W., & Jo, C., (2015). Flexible thin-layer dielectric barrier discharge plasma treatment of pork butt and beef loin: Effects on pathogen inactivation and meat quality attributes. *Food Microbiology, 46*, 51–57.

27. Jung, S., Kim, H. J., Park, S., Yong, H. I., Choe, J. H., Jeon, H. J., Choe, W., & Jo, C., (2015). Color developing capacity of plasma-treated water as a source of nitrite for meat curing. *Korean Journal for Food Science of Animal Resources, 35*(5), 703–706.

28. Karam, L., Casetta, M., Chihib, N., Bentiss, F., Maschke, U., & Jama, C., (2016). Optimization of cold nitrogen plasma surface modification process for setting up antimicrobial low-density polyethylene films. *Journal of the Taiwan Institute of Chemical Engineering, 64*, 299–305.

29. Keidar, M., Yan, D., & Sherman, J. H., (2019). Plasma is the fourth state of matter; Chapter 1. In: Keidar, M., Yan, D., & Sherman, J. H., (eds.), *Cold Plasma Cancer Therapy* (p. 1–3). California, USA: Morgan & Claypool Publishers.

30. Kim, B., Yun, H., Jung, S., Jung, Y., Jung, H., Choe, W., & Jo, C., (2011). Effect of atmospheric pressure plasma on inactivation of pathogens inoculated onto bacon using two different gas compositions. *Food Microbiology, 28*(1), 9–13.

31. Kim, H. J., Jayasena, D. D., Yong, H. I., Alahakoon, A. U., Park, S., Park, J., Choe, W., & Jo, C., (2015). Effect of atmospheric pressure plasma jet on the foodborne pathogens attached to commercial food containers. *Journal of Food Science and Technology, 52*, 8410–8415.

32. Kim, H. J., Sung, N. Y., Yong, H. I., Kim, H., Lim, Y., Ko, K. H., Yun, C. H., & Jo, C., (2016). Mutagenicity and immune toxicity of emulsion-type sausage cured with plasma-treated water. *Korean Journal for Food Science of Animal Resources, 36*(4), 494–498.

33. Klockow, P. A., & Keener, K. M., (2009). Safety and quality assessment of packaged spinach treated with a novel ozone-generation system. *LWT-Food Science and Technology, 42*(6), 1047–1053.

34. Kostov, K. G., Nishime, T. M. C., Castro, A. H. R., Toth, A., & Hein, L. R. O., (2014). Surface modification of polymeric materials by cold atmospheric plasma jet. *Applied Surface Science, 314*, 367–375.

35. Lee, H. B., Noh, Y. E., Yang, H. J., & Min, S. C., (2011). Inhibition of foodborne pathogens on polystyrene, sausage casings and smoked salmon using nonthermal plasma treatments. *Korean Journal of Food Science and Technology, 43*(4), 513–517.

36. Lee, H. J., Jung, H., Choe, W., Ham, J. S., Lee, J. H., & Jo, C., (2011). Inactivation of *Listeria monocytogenes* on agar and processed meat surfaces by atmospheric pressure plasma jets. *Food Microbiology, 28*(8), 1468–1471.

37. Lee, H. J., Jung, P. S., Choe, W., Ham, J. S., & Jo, C., (2012). Evaluation of a dielectric barrier discharge plasma system for inactivating pathogens on cheese slices. *Journal of Animal Science and Technology, 54*(3), 191–198.

38. Lee, K. H., Woo, K. S., Yong, H. I., Jo, C., Lee, S. K., Lee, B. W., Oh, S. K., et al., (2018). Assessment of microbial safety and quality changes of brown and white cooked rice treated with atmospheric pressure plasma. *Food Science and Biotechnology, 27*(3), 661–667.

39. Leipold, F., Kusano, Y., Hansen, F., & Jacobsen, T., (2010). Decontamination of a rotating cutting tool during operation by means of atmospheric pressure plasmas. *Food Control, 21*(8), 1194–1198.

40. Ma, R., Wang, G., Tian, Y., Wang, K., Zhang, J., & Fang, J., (2015). Nonthermal plasma-activated water inactivation of food-borne pathogen on fresh produce. *Journal of Hazardous Materials, 300*, 643–651.

41. Mastromatteo, M., Lecce, L., De Vietro, N., Favia, P., & Del Nobile, M. A., (2011). Plasma Deposition processes from acrylic/methane on natural fibers to control the kinetic release of lysozyme from PVOH monolayer film. *Journal of Food Engineering, 104*(3), 373–379.

42. Milella, A., (2008). Plasma processing of polymers. In: Mark, H. F., (ed.), *Encyclopedia of Polymer Science and Technology* (4th edn.). Hoboken, New Jersey: John Wiley & Sons, Inc. https://doi.org/10.1002/0471440264.pst560.

43. Min, S. C., Roh, S. H., Niemira, B. A., Boyd, G., Sites, J. E., Uknalis, J., & Fan, X., (2017). In-package inhibition of *E. coli* O157:H7 on bulk romaine lettuce using cold plasma. *Food Microbiology, 65*, 1–6.

44. Misra, N. N., Kaur, S., Tiwari, B. K., Kaur, A., Singh, N., & Cullen, P. J., (2015). Atmospheric pressure cold plasma (ACP) treatment of wheat flour. *Food Hydrocolloids, 44*, 115–121.

45. Misra, N. N., Keener, K. M., Bourke, P., Mosnier, J. P., & Cullen, P. J., (2014). In-package atmospheric pressure cold plasma treatment of cherry tomatoes. *Journal of Bioscience and Bioengineering, 118*(2), 177–182.

46. Misra, N. N., Moiseev, T., Patil, S., Pankaj, S. K., Bourke, P., Mosnier, J. P., Keener, K. M., & Cullen, P. J., (2014). Cold plasma in modified atmospheres for post-harvest treatment of strawberries. *Food and Bioprocess Technology, 7*, 3045–3054.

47. Misra, N. N., Pankaj, S. K., Frias, J. M., Keener, K. M., & Cullen, P. J., (2015). The effects of nonthermal plasma on chemical quality of strawberries. *Postharvest Biology and Technology, 110*, 197–202.

48. Moiseev, T., Misra, N. N., Patil, S., Cullen, P. J., Bourke, P., Keener, K. M., & Mosnier, J. P., (2014). Post-discharge gas composition of a large-gap DBD in humid air by UV-Vis absorption spectroscopy. *Plasma Sources Science and Technology, 23*(6), 065033.

Microbial Decontamination of Foods with Cold Plasma 297

49. Muranyi, P., Wunderlich, J., & Heise, M., (2007). Sterilization efficiency of a cascaded dielectric barrier discharge. *Journal of Applied Microbiology, 103*(5), 1535–1544.

50. Niemira, B. A., (2012). Cold plasma reduction of salmonella and *Escherichia coli* O157:H7 on almonds using ambient pressure gases. *Journal of Food Science, 77*(3), M171–M175.

51. Niemira, B. A., & Sites, J., (2008). Cold plasma inactivates salmonella Stanley *and Escherichia coli* O157: H7 inoculated on golden delicious apples. *Journal of Food Protection, 71*(7), 1357–1365.

52. Nisha, R. B., & Narayan, R., (2019). Review on cold plasma technology: The future of food preservation. *International Journal of Chemical Studies, 7*(3), 4427–4433.

53. Pankaj, S. K., Bueno-Ferrer, C., Misra, N. N., Milosavljevi, V., O'Donell, C. P., Bourke, P., Keener, K. M., & Cullen, P. J., (2014). Application of cold plasma technology in food packaging. *Trends in Food Science & Technology, 35*(1), 5–17.

54. Pankaj, S. K., Bueno-Ferrer, C., Misra, N. N., O'Neill, L., Jiménez, A., Bourke, P., & Cullen, P. J., (2014). Characterization of polylactic acid films for food packaging as affected by dielectric barrier discharge atmospheric plasma. *Innovative Food Science & Emerging Technology, 21*, 107–113.

55. Pankaj, S. K., Bueno-Ferrer, C., Misra, N. N., O'Neill, L., Tiwari, B. K., Bourke, P., & Cullen, P. J., (2015). Dielectric barrier discharge atmospheric air plasma treatment of high amylose corn starch films. *LWT—Food Science and Technology, 63*(2), 1076–1082.

56. Puligundla, P., Lee, T., & Mok, C., (2016). Inactivation effect of dielectric barrier discharge plasma against foodborne pathogens on the surfaces of different packaging materials. *Innovative Food Science & Emerging Technology, 36*, 221–227.

57. Ramazzina, I., Berardinelli, A., Rizzi, F., Tappi, S., Ragni, L., Sacchetti, G., & Rocculi, P., (2015). Effect of cold plasma treatment on physico-chemical parameters and antioxidant activity of minimally processed kiwifruit. *Postharvest Biology and Technology, 107*, 55–65.

58. Sandhu, H. P. S., Manthey, F. A., & Simsek, S., (2012). Ozone gas affects the physical and chemical properties of wheat (*Triticum aestivum* L.) starch. *Carbohydrate Polymers, 87*(2), 1261–1268.

59. Sarangapani, C., Keogh, D. R., Dunne, J., Bourke, P., & Cullen, P. J., (2017). Characterization of cold plasma treated beef and dairy lipids using spectroscopic and chromatographic methods. *Food Chemistry, 235*, 324–333.

60. Segat, A., Misra, N. N., Cullen, P. J., & Innocente, N., (2015). Atmospheric pressure cold plasma (ACP) treatment of whey protein isolate model solution. *Innovative Food Science & Emerging Technologies, 29*, 247–254.

61. Segat, A., Misra, N. N., Cullen, P. J., & Innocente, N., (2016). Effect of atmospheric pressure cold plasma (ACP) on activity and structure of alkaline phosphatase. *Food and Bioproducts Processing, 98*, 181–188.

62. Shaw, D., West, A., Bredin, J., & Wagenaars, E., (2016). Mechanisms behind the surface modification of polypropylene film using an atmospheric-pressure plasma jet. *Plasma Sources Science and Technology, 25*, 065018–065024.

63. Shi, X. M., Zhang, G. J., Wu, X. L., Li, Y. X., Ma, Y., & Shao, X. J., (2011). Effect of low-temperature plasma on microorganism inactivation and quality of freshly squeezed orange juice. *IEEE Transactions on Plasma Science, 39*(7), 1591–1597.

64. Solomon, E. B., & Sharma, M., (2009). Microbial attachment and limitations of decontamination methodologies; Chapter 2. In: Sapers, G. M., Solomon, E. B., & Matthews,

K. R., (eds.), *The Produce Contamination Problem: Causes and Solutions* (pp. 21–45). Burlington–USA: Academic Press.

65. Song, H. P., Kim, B., Choe, J. M., Jung, S., Moon, S. Y., Choe, W., & Jo, C., (2009). Evaluation of atmospheric pressure plasma to improve the safety of sliced cheese and ham inoculated by 3-strain cocktail *Listeria monocytogenes. Food Microbiology, 26*(4), 432–436.

66. Stoica, M., Stoean, S., & Alexe, P., (2014). Overview of biological hazards associated with the consumption of the meat products. *Journal of Agroalimentary Processes and Technology, 20*(2), 192–197.

67. Sudheesh, C., & Sunooj, K. V., (2020). Cold plasma processing of fresh-cut fruits and vegetables; Chapter 14. In: Siddique, M. W., (ed.), *Fresh-Cut Fruits and Vegetables: Technologies and Mechanisms for Safety Control* (pp. 339–356). London–United Kingdom: Academic Press.

68. Sun, S., Anderson, N. M., & Keller, S., (2014). Atmospheric pressure plasma treatment of black peppercorns inoculated with salmonella and held under controlled storage. *Journal of Food Science, 79*, 2441–2446.

69. Surowsky, B., Froehling, A., Gottschalk, N., Schlueter, O., & Knorr, D., (2014). Impact of cold plasma on *Citrobacter freundii* in apple juice: Inactivation kinetics and mechanisms. *International Journal of Food Microbiology, 174*, 63–71.

70. Tappi, S., Berardinelli, A., Ragni, L., Dalla, R. M., Guarnieri, A., & Rocculi, P., (2014). Atmospheric gas plasma treatment of fresh-cut apples. *Innovative Food Science & Emerging Technologies, 21*, 114–122.

71. Thirumdas, R., Sarangapani, C., & Annapure, U. S., (2015). Cold plasma: A novel nonthermal technology for food processing. *Food Biophysics, 10*(1), 1–11.

72. Wang, Y., Somers, E., Manolache, S., Denes, F., & Wong, A., (2003). Cold plasma synthesis of poly(ethylene glycol)-like layers on stainless-steel surfaces to reduce attachment and biofilm formation by *Listeria monocytogenes. Journal of Food Science, 68*, 2772–2779.

73. Yong, H. I., Kim, H. J., Park, S., Alahakoon, A. U., Kim, K., Choe, W., & Jo, C., (2015). Evaluation of pathogen inactivation on sliced cheese induced by encapsulated atmospheric pressure dielectric barrier discharge plasma. *Food Microbiology, 46*, 46–50.

74. Yong, H. I., Kim, H. J., Park, S., Alahakoon, A. U., Kim, K., Choe, W., et al., (2015). Pathogen inactivation and quality changes in sliced cheddar cheese treated using flexible thin-layer dielectric barrier discharge plasma. *Food Research, 69*, 57–63.

75. Yong, H. I., Lee, H., Park, S., Park, J., Choe, W., Jung, S., et al., (2017). Flexible thin-layer plasma inactivation of bacteria and mold survival in beef jerky packaging and it's effect on the meat's physicochemical properties. *Meat Science, 123*, 151–156.

76. Ziuzina, D., & Misra, N. N., (2016). Cold plasma for food safety. In: *Cold Plasma in Food and Agriculture* (pp. 223–252). Academic Press.

77. Ziuzina, D., Patil, S., Cullen, P. J., Keener, K. M., & Bourke, P., (2014). Atmospheric cold plasma inactivation of *Escherichia coli, Salmonella entrica* serovar *Typhimurium and Listeria monocytogenes* inoculated on fresh produce. *Food Microbiology, 42*, 109–116.

CHAPTER 12

FOOD IRRADIATION: CONCEPTS, APPLICATIONS, AND FUTURE PROSPECTS

SUBHAJIT RAY

ABSTRACT

In this chapter, irradiation of food commodities focuses on the history of irradiation in the food processing industries, basic concept, principles, sources of irradiation, e.g., radioactive metals viz. Cobalt (Co), Cesium (Cs) for emission of gamma rays, X-rays, accelerated electron beams, etc., mechanisms involving destruction of DNA of pathogens, insects, and parasite disinfestation, inhibition of sprouting, delay in ripening of fruits, etc., methods of irradiation, e.g., radappertization, radicidation, radurization by applying appropriate dose, application of irradiation on food commodities, e.g., fruits (viz. banana, lichi, mango, etc.), onions, potatoes, fish, meat, poultry, frozen seafood, spices, etc., for the purpose of extension of shelf stability, influence of radiation processing on food quality, preservation stability of radiation processed foods, economical, and safety aspects and future prospects of food irradiation technology.

12.1 INTRODUCTION

In those sections of the globe where the conveyance of food material is troublesome and where preservation of food materials under refrigeration is insufficient or exceptionally costly, the application of irradiation may foster generous dispensation of some food materials than would otherwise be feasible.

Advances in Food Process Engineering: Novel Processing, Preservation, and Decontamination of Foods.
Megh R. Goyal, N. Veena, & Ritesh B. Watharkar (Eds.)
© 2023 Apple Academic Press, Inc. Co-published with CRC Press (Taylor & Francis)

In this method, a further diverse and perhaps nutritionally exceptional food may become acquirable to the inhabitants. The utility of food irradiation in treating foods has been demonstrated on a technological basis. Radiation is defined as a phenomenon of energy emission as moving high energy sub-atomic particles. Preservation of food through irradiation technique is considered to be one of the important method of extension of storage stability of food. The type of radiation in food preservation is electromagnetic in nature. Food irradiation is a physical means of treatment comparable to the processing of food by heating or freezing it. The process involves exposing whole or packed food materials, to electromagnetic waves in a specially fabricated enclosed room to protect the user for a finite period of time [30].

The most common and approved sources of gamma rays for food processing in commercial scale is radioisotopes, e.g., Cobalt-60 and Cesium-137 [4]. Gamma rays and X-rays are electromagnetic radiations of very short wavelength (2,000 A°). There are also machine sources which can produce electrons or X-rays for food processing. It is important to note that exposing food to either Cobalt-60 or Cesium-137 radiation sources or electron beams (10 MeV maximum energy) or X-rays (5 MeV maximum energy) does not induce radioactivity [12]. Irradiation processing eliminates the use of harmful chemicals for the treatment of food products and can be preserved without the loss of freshness as well as wholesomeness. The measurement of irradiation dose applied to different kinds of food products is represented in Table 12.1.

TABLE 12.1 Units of Measurements of Irradiation Dose and Equivalent Conversion Unit

1 curie	1 Becquerel (Bq)
1 rad	10^2 erg/g
1 Krad	10^3 rad/10^5 erg/g
1 Mrad	10^6 rad/10^3 Krad/10^8 erg/g
1 Gy	10^2 rad/1 J/kg
1 kGy	10^5 rad/10^3 J/kg
1 kGy	0.1 Mrad

This chapter focused on the various aspects, including concepts, basic principles, potential applications, and future prospects of food irradiation process and established as an important preservation technology to combat the food loss problems as well as in expanding trade in food and agriculture commodities.

12.2 HISTORY OF FOOD IRRADIATION IN FOOD PROCESSING

Food processing and preservation has gained significant importance in terms of extension of shelf stability, storage life and minimization of wastage. Ionizing radiation results in better quality of food products in terms of extended shelf-life and wholesomeness as compared with conventional food preservation technologies, e.g., canning, drying, freezing, fermentation, addition of synthetic food additives, controlled atmosphere storage (CAS), modified atmospheric packaging (MAP), etc. Irradiation is carried out by means of UV radiation and radio emission (Figure 12.1). UV ray induced irradiation with 200–280 nm wavelength and low infiltration power causes diminishment of microbiological pollution in air and germicidal effect, i.e., destruction of DNA molecules. The penetrating power of alpha ray and beta ray are poor and a depth of 4 to 5 cm, respectively. However, ionizing radiation with special emphasis on gamma-ray radiation has excellent penetration power up to 30–40 cm of depth. A comparative study on the penetration ability of these rays can be represented in Figure 12.1. Ionizing radiation can strike negatrons out of the pounded material, disrupting the structure of the molecule and thereby ion formation or free electrons.

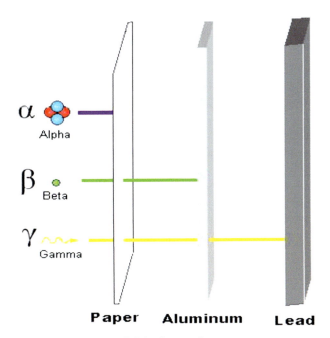

FIGURE 12.1 Penetrative power of alpha, beta, and gamma rays.

A recent report [19] shows that worldwide 60 irradiated food products were approved commercially in more than 50 countries. Export of irradiated food products, e.g., sweet potato, shellfish, lettuce, spinach, papaya, etc., has gained significant importance in USA [15, 21, 22].

European countries, including Belgium, France, Netherland, etc., has been successfully employed commercial processing of irradiated foods viz. poultry, herbs, and spices, dehydrated blood, fish, shellfish, meat, and vegetable [9, 18]. The potential of irradiation processing of food, e.g., spices, dried vegetables, fruits (e.g., mango, banana), fermented sausage, herbs, sweet tamarind, and others in Asian countries like India, Bangladesh, Pakistan, Malaysia, Indonesia, and Thailand was successfully commercialized [9, 18, 35].

12.3 BASIC CONCEPTS OF IRRADIATION TECHNOLOGY

Radiation is energy that can be transferred from one body to another across empty space. Radiant energy behaves as continuous waves when it moves swiftly through space. However, this same energy also acts like a particle, when it bursts into or out of matter. In the action of bursting into or out of matter, energy is given off or absorbed in definite packets or unit amounts. Each unit amount is a quantum of energy. Food irradiation with the implementation of ionizing radiation can destroy the pathogenic microorganisms, minimize the risk of microbial population, eliminate the insects and parasites, inhibit the sprout formation of onions, potatoes, etc., delay in ripening process thereby achievement of extension of shelf stability of perishable food products. Ionizing radiation has the ability to destroy the DNA molecules of the microbial cells and as a result, the microbes become inactivated [8].

Two basic processes occur when ionizing radiation act upon any sort of matter. The reason behind the creation of ions, molecular excitement or molecular fragmentation is none but the involvement of primary process. Involvement of interaction of the primary process derived products is due to the secondary process and can facilitate the generation of compounds from those originally occurs. The biological effects of radiation on a living organism are associated mainly with the impairment of metabolic process responsible for cell division [3].

High doses of radiation can be lethal. At low doses, living organisms can recover from radiation injury. The effect of irradiation processing on living creature depends upon time to appear. A unique advantage of ionizing radiation lies in its strong penetrating power. Materials to be treated can be

Food Irradiation: Concepts, Applications

exposed to radiation in its final packaging, thus avoiding any potential risk of recontamination. A very specific consequence of the penetrating power of the ionizing radiation is its suitability for quarantine treatment of fresh fruits where chemical insecticides or the medium as steam converging the heat cannot reach, radiation easily penetrates.

12.4 PRINCIPLES OF FOOD IRRADIATION FOR SPOILAGE PREVENTION

During the last four decades, the possibility of using ionizing radiation for processing or preservation of food has aroused great interest in many countries. Research has been conducted with respect and distribution both to the technology of the process and to the suitability of the irradiated food for human consumption. As a result, it has been demonstrated that radiation processed food materials are techno-commercially feasible for production and safe for consumption. It has been established that technology of food preservation by irradiation has a significant role for the storage stability, transportation, and marketability of food products. Now in considering food irradiation as a preservation process, we should understand about the spoilage of food. Spoilage of food commodities results from three classes of changes viz, physical, chemical, and biological.

12.4.1 PHYSICAL CHANGES

This type of spoilage mainly is the result of discoloration, chopping or breakdown during production, delivering, and gripping operations. Drying of a food or the reverse, the uptake of moisture can alter appearance and other sensory properties. Moisture absorption can easily increase the microbial spoilage.

12.4.2 CHEMICAL CHANGES

The deterioration due to chemical process conventionally occurs from counteraction of food ingredients with each other or from counteraction of the food with the surroundings. Therefore, sugar, and protein present in a foodstuff can interact to produce undesirable flavor and color changes also known as non-enzymatic browning or Maillard reaction. Lipid molecule can

304   Advances in Food Process Engineering

react with atmospheric oxygen and results in the development of off-flavor usually known as rancidity. Enzymes present in a food can change flavor.

12.4.3 BIOLOGICAL CHANGES

Biological spoilage includes (i) microbiological, e.g., Bacteria, yeasts, molds, (ii) insects and parasites, and (iii) senescence. Biological spoilage can manifest itself by sensory changes, e.g., appearance alteration, odor, and flavor changes and other observable changes. In certain cases, the spoilage may be a condition of hazard to the health of the consumer through the presence of bacterial toxin or mycotoxins or of an infectious organism such as bacterium or parasite. For living foods such as raw fruit and vegetables, spoilage may be the consequence of over ripening or sprouting.

12.5 SOURCES OF IRRADIATION

The irradiation processing of food materials through ionizing radiation is generally considered to gamma-rays from radioactive sources like Cs-137 or Co-60 and Roentgen rays consisting of energy that is transmitted in the form of radiation up to 5 million electron volts (MeV) and 10 MeV, respectively. These irradiations are selected due to the fact that:

- They generate the necessary effects related to food materials;
- They do not influence the emission due to radionuclide in packaged foodstuffs.

Other types of radio emission in some perspective do not favor the needs of radiation induced food processing. Exception is the variation in penetration, non-particulate radiation and negatrons are equivalent in radiation processed food and can be used through interchange.

12.5.1 GAMMA RAYS

Radionuclide sources especially Co-60 and Cs-137 [4] are used as gamma-ray irradiator for the processing of food to provide extended shelf stability (Figure 12.2). Cs-137 is a fission product. The occurrence of Cs-137 is finite around the globe and contrarily Co-60 can be formed quickly by neutron induced nuclear fission of Co-59 [14] in an atomic pile. Co-60 prills are

Food Irradiation: Concepts, Applications

enclosed in rod made by stainless steel. This geometrical arrangement reduces self-assimilation and evolution of heat. Co-60 preserved in a reservoir rack under water not exceeding the depth of 6 m. Therefore, the water layer assimilates irradiation. In case of radioactive metals induced discharge of radiating rays produces the stable isotope. The result will be the minimization in the number of radionuclides over a duration of time. Machine sources usually provide higher dose rate compared to gamma-ray sources. The effect of gamma-rays on a bacterial cell can be represented in Figure 12.2.

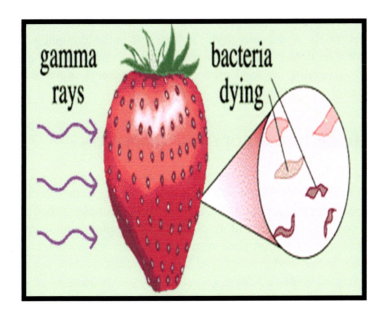

FIGURE 12.2 Effect of gamma rays on bacterial cells.

12.5.2 X-RAYS

Roentgen rays are assimilated and interaction with matter takes place. Here the quantum of electromagnetic radiation creates energy which is utilized to discharge electrons via atomic emission. The discharged high energy accelerated electrons causes ionization of molecules. Radioactive source is not required for X-ray emission on the target materials for the enhancement of storage stability. For X-ray emission electron accelerators or machine sources of accelerated energies up to 5 MeV is considered. Now due to higher penetration ability, X-rays can be used for relatively thick products.

12.5.3 ACCELERATED ELECTRON BEAMS

Accelerated electron beams are usually generated from machine sources. Various types of high energy electron accelerators of energies up to 10 MeV are usually considered for the direct emergence of electron beams. There are two categories of electron accelerators, e.g., direct current accelerator and the other is microwave or radio frequency or linear accelerator. In these types of accelerators, electrons are expedited rapidly as compared to light in emptied tubes. Electrons ejected from suitable sources are forced from the negatively charged terminals of the tube and results in affinity by positively charged terminals. A magnetic device able to scan at the terminal point of the accelerator tube diverts the monoenergetic electron beam onto the radiated materials. Generally, the food product of less thickness is treated for this purpose [14].

12.6 CHARACTERISTICS OF FOOD IRRADIATION

Irradiation technology of food preservation is considered to be a physical treatment technique of food materials, and it is similar to the heat treatment methods, e.g., drying, canning, etc., and freezing of foods for food preservation. The only unique feature of irradiation is the employment of a particular type of energy. Irradiation dose results practically no such increment of temperature in the food product. The highest irradiation dose preferably to be implemented is about 5 Mrad (50 killoGray). This quantity of energy is similar to about 0.012 kcal (50 J). Therefore, if all the irradiation process degrades to thermal energy, the enhancement of temperature of food product will be about 12°C. Due to this fact, irradiation has been termed as cold process or cold pasteurization process [32].

12.7 MECHANISMS OF FOOD IRRADIATION

Microbial population as well as insects and parasites when irradiated then the destruction or damage takes place and it will exclusively depend upon irradiation dose. Irradiation induced damage to the cytoplasmic membrane of cellular structure and DNA molecules are very much significant. The radiation resistance of different microbial species is different and due to this fact, the irradiation dose varies. The vegetative cell shows less radiation resistance than spores. Therefore, ionizing radiation with the active participation of

Food Irradiation: Concepts, Applications

gamma rays, X-rays, or accelerated electron beams has a multipurpose role in the processing and preservation of foods which result the qualitatively upgraded food products in terms of pathogen free, disinfested, sproutless, and maturity.

12.7.1 DESTRUCTION OF PATHOGENIC MICROORGANISMS

Irradiation of foods significantly destroys pathogenic microorganisms in different kinds of food products [5, 6, 10, 23, 33]. Basically, medium level radiation dose on fleshy foods like meat and chicken fulfills two major objectives. Irradiation offers a good and effective way of eradication of pathogenic bacteria, e.g., *Salmonella* spp., *Vibrio* spp., *Shigella* spp., and *Campylobacter* spp., very conventional food poisoning bacterial species in poultry, seafoods, etc. Irradiation of spices with a dose of 5–10 kGy is effective to reduce salmonellosis caused by *Salmonella spp.*

12.7.2 DISINFESTATION OF INSECTS AND PARASITES

A very significant contribution of irradiation of food is the control of storage pests in cereals, pulses, flour-based products, dried fruits, and vegetables [5, 6, 10, 11, 17, 25, 27, 28, 33], dehydrated fleshy foods like fish and aroma producing spices. Insect inactivation in preserved cereal-based products requires an irradiation dose of 0.5 kGy. Various parasites can be efficiently removed by means of irradiation in food products [5, 6, 10, 20, 25, 29]. Meat parasites, e.g., parasitic nematode causing trichinellosis, pork tapeworm causing neurocysticercosis and intracellular parasitic protozoa causing toxoplasmosis can be effectively eliminated by low dose irradiation of 0.3–0.4 kGy.

12.7.3 INHIBITION OF SPROUTING

Germination is a potential factor of financial loss in a variety of vegetables, e.g., potato, onion, garlic, etc. It has been proved that radiation induced food processing is a suitable way of storing these food commodities for a prolonged period of time with no such germination losses [2, 5, 7, 28, 29, 31]. Irradiation technology is an effective and safe alternative to using chemical germination suppressor and sprout inhibitors, e.g., $C_4H_4N_2O_2$, $C_4H_9NO_2$, $C_{10}H_{12}ClNO_2$, etc., to fulfill the purpose. Irradiation can prevent *Phthorimaea operculella* outbreak ubiquitous in predestined potato production zones.

12.7.4 DELAY IN RIPENING

Radiation treatment of climacteric fruits in the fully developed fresh state to a quantity of 0.25–0.40 kGy can retard maturation or senescence up to 6–8 days [5, 6, 10, 20, 25]. Irradiation becomes beneficial for equatorial fruits, e.g., *Musa xparadisiaca, Mangifera indica, Carica papaya*, litchi, etc., at the time of their delivery and storage. Radiation treated fruits can be subjected to ordinary maturation after the delaying time. Radiation treatment of apricot with a quantity of 3 kGy expedites the development of red colored flavonoids and tetraterpenoids, thereby upgrading their hue and appearance.

12.8 RESEARCH METHODS IN FOOD IRRADIATION

12.8.1 IRRADIATION DOSE AND DOSIMETRY

Generally, there is a minimum dose requirement of irradiation process for the purpose of food preservation. In case of every molecular constituents of food materials needs radiation treatment, will based upon the impetus of therapy. In several instances, radiation processing of the surface will be good enough, on another way the total food material essentially requires the minimum quantity [13]. Dose requirements for the fulfillment of different objectives are given in Table 12.2. The overall dose requirement for every food material must be contemplated as specific for the food. Dose requirement for reduction of microorganisms is represented in Table 12.3.

12.8.1.1 RADAPPERTIZATION

Generally, it is applicable to the canning industry deals with radiation sterilization or where commercial sterility is achieved. Recommended dose levels of irradiation is 30–40 kGy. Dose level of 10 kGy utilized in processed food industry is very much important to the decontamination followed by destruction of pathogens in healthcare and hygiene products for personal use [26]. The implementation of radiation dose chiefly relies upon the category of food products and results occurred in target [5, 16]

Food Irradiation: Concepts, Applications 309

as represented in Table 12.4. It has been also observed from the outcome of various reports [13, 36] and establish that the consumption of radiation processed foodstuff is completely harmless and secured for the customers [13, 24]. The destruction of *Clostridium botulinum* strains by irradiation technique in some food products is represented in Table 12.5.

TABLE 12.2 Dose Requirement of Different Implementations of Radiation Treated Foods

Implementation	Required Quantity (kGy)
Extermination of parasitic infection of meat	0.1–3
Ionizing radiation subjected to potential reduction of pathogenic microorganisms of frozen muscle foods and other food materials and fodders	3–10
Microbial decontamination in food and constituents, e.g., spices, starch, enzyme prepared, etc.	3–20
Mild ionizing radiation dose of putrescible food commodities, e.g., horticultural, and agricultural produce, muscle foods, fish	0.5–10
Pest fumigation of grain-based food products, wheat flours, raw, and dehydrated fruits, etc.	0.2–0.8
Radappertization of fish, meat, and poultry products	25–60
Retardation of germination of *Solanum tuberosum* and *Allium cepa*	0.03–0.12

TABLE 12.3 Quantity Requirement of Ionizing Radiation for Sufficient Decrease of Different Microbes

Microbes	Quantity (kGy)
Aerobic rod shaped bacterial species	3
Aflatoxin producing *A. flavus*	3
Clostridium spp.	25
Escherichia spp.	2
Pseudomonas spp.	<1
S. cerevisiae	10
Salmonella spp.	4
Spores of *B. cereus*	25
Staphylococcus spp.	5–10
Streptococcus spp.	5
Ultramicroscopic infectious disease-producing agent	>30

TABLE 12.4 Effect of Irradiation on Food Products

Food Products	Effects
Banana, papaya, and Lichi	Delay in ripening
Bulbs and tubers of *Allium sativum*, *Solanum tuberosum* and *Allium cepa*	Inhibition of sprout
Cereals and their products, dry vegetables	Parasite and insect disinfestations
Meat (mutton, bacon, chicken, and fish), liquid milk, fruit juices, etc.	Microbial decontamination and enzyme inactivation
Meat, vegetables	Extension of shelf stability

TABLE 12.5 Irradiation Dose on Food Items for *Clostridium botulinum* Destruction

Food Items	Dose (kGy)
Bacon	23
Beef	47
Chicken	45
Cod fish cake	32
Corned beef	25
Ham	37
Pork	51
Pork sausage	24–27
Shrimp	37

12.8.1.2 RADICIDATION

It is also termed as radiation pasteurization. A typical dose level of 2.5–10 kGy is considered to be effective to reduce the viable specific non-spore-forming pathogens, other than viruses. Refrigerated and frozen chicken carcass treated with irradiation dose of 2.5 kGy was highly effective in destroying *salmonella* spp. Whole cocoa bean treated with irradiation dose of 5 kGy results 99.9% destruction of bacterial flora. It has been observed that irradiation dose of 4 kGy can reduce the spores of mycotoxin producer *Penicillium citrinum* and aflatoxin producer *Aspergillus flavus* of about 5 logs/gm and 7 logs/gm, respectively. Moreover, radicidation have been approved to destroy pathogens in fresh poultry, cod fish, spices, and condiments.

Food Irradiation: Concepts, Applications 311

12.8.1.3 RADURIZATION

Preservation period of sea fish may be prolonged 2–6 fold by radurization with dose levels from 1–4 kGy. Shellfish and fish will also produce the same result. An application of 2.5 kGy irradiation dose on hamburger killed all *Pseudomonas* spp., gram-negative rod-shaped *Enterobacteriaceae* family and gram-positive rod-shaped *Brochothrix thermosphacta*. Sprout inhibition of potato, onion, etc., and insect disinfestation of cereal-based products and pulses can be achieved by this technique of irradiation.

12.9 POTENTIAL OF FOOD IRRADIATION

The application potential of food irradiation process is due to the perfect and efficient inhibition of microbial DNA synthesis thereby cell division is diminished. The decrease in microbial load is achieved with no such deleterious effects on the food materials by the application of appropriate dose [13]. Hence microbes, pests, egg cell and plant tissue can be intercepted to reproduce thereby securing stability of food as described in Table 12.6.

TABLE 12.6 Preservative Effects of Ionizing Radiation

Effects	Results
Delay of maturing and senescence of horticultural and agricultural produce	Extension of storage stability of horticultural and agricultural produce
Destruction and sterilizing pests	Pest decontamination of foodstuffs
Interception of development and multiplication of pests disseminated by foodstuffs	Interception of parasitic ailment of food products
Reduction or elimination of microorganisms	Reduced infection of foodstuffs, enhanced preservation period of foods, interception of food intoxication
Retardation of germination	Extension of storage period of tubers, minimization of malting wastage

Therefore, irradiation results in food product improvement in terms of functional or sensory quality characteristics. The most important cause for the application of food irradiation is to destroy living organisms that cause putrefaction and deterioration or are a threat to the health of the consumer. The amount of energy caused by ionizing radiation used to bring about the control or elimination of these organisms varies according to the resistance

312   Advances in Food Process Engineering

of the particular species and according to the numbers or load of organisms present. The usefulness of irradiation in overcoming microbial obstacle has been distinctly demonstrated by considering different types of food material, and this can be of considerable human health importance, specifically in the case of pathogens such as *Salmonella* spp.

12.10   INFLUENCE OF IONIZING RADIATION ON QUALITATIVE CHARACTERISTICS OF FOODS

The unpleasant variations that take place in specific radiation treated food-stuffs is due to either straightway by ionizing radiation or concomitantly as a result of post-irradiation changes. Water undergoes radiation degradation or radiolytic decomposition into hydrogen peroxide, hydrogen and hydroxyl radicals, hydrogen and hydroxyl ion, etc. [1].

$$H_2O \rightarrow e^-_{aq}, HO^{\cdot}, H^{\cdot}, HO_2^{\cdot}, H_3O^+, OH^-, H_2O_2, H_2$$

By irradiation under aerobic environments, bad flavors and odors are considerably reduced as a result of the deficiency of oxygen to produce hydrogen peroxides (H_2O_2). The most efficient way to reduce bad flavors is to undergo radiation treatment at a temperature below freezing point. The influence of temperature below freezing point is to minimize or stop radiolytic decomposition and its subsequent reaction components. It has been noticed that ionizing radiation of elevated doses results in the evolution of radiation odors in specific foodstuffs, e.g., meat. With regard to B vitamins levels of Co-60 irradiation between 2 kGy and 6 kGy affected partial destruction of the following B vitamins, e.g., thiamine, niacin, pyridoxine, and vitamin B_{12} in oyster mushroom. The increase level of vitamin B_2, B_5 and B_9 by the effect of ionizing radiation is due to liberation of obligated vitamins. The production of C_2H_4 in apples is hindered by ionizing radiation. Therefore, the material produced is unable to develop as quickly as sample or control which is not radiated. The production of C_2H_4 in immature lemons was investigated upon ionizing radiation, which results in a rapid maturation than in control.

12.10.1   EFFECT ON NUTRITIONAL CONSTITUENTS

Irradiation processing of foods has a significant impact on the nutritional constituents, e.g., carbohydrate, lipid, proteins, amino acids, and vitamins as well as pigments and enzymes. Radiation causes the cleavage of glycosidic

Food Irradiation: Concepts, Applications

bonds in polysaccharides, e.g., α (1, 4) and α (1, 6) glycosidic bond in amylose and amylopectin fragments in starch molecule, β (1, 4) glycosidic bond in cellobiose fragment in cellulose molecule, etc., which results the textural defects in some food materials. Lipids have been found to be radiation-sensitive, and there is a significant destruction of naturally occurring antioxidants by the influence of ionizing radiation. Lipids containing more unsaturated fatty acids are prone to autooxidation and linoleic acid content changed significantly in comparison to linolenic acid in fish with the intervention of irradiation. Water molecules present in food materials undergo fission by the influence of irradiation, and the result will be the formation of free radicals, e.g., H^+ and OH^-. These free radicals will react with the lipid components and causes damage. There is no such direct impact of food irradiation on denaturation of protein but protein molecule undergoes several changes, including breakdown of peptide linkage, polymerization, coagulation, and precipitation.

Irradiation causes the cleavage of aromatic ring-containing amino acids, e.g., phenylalanine, tyrosine, etc., and can produce some kind of typical flavor due to disruption of sulfur containing amino acids, e.g., cysteine, methionine, etc. Irradiation processing has a potential impact on the sensitivity of different fat-soluble and water-soluble vitamins and can be compared to the destruction caused by heat treatment. It has been reported [20] that food vitamins, e.g., vitamin A (retinol, fat-soluble), vitamin B_1 (thiamine, water-soluble), vitamin C (ascorbic acid, water-soluble) and vitamin E (tocopherol, fat-soluble) are highly sensitive, beta-carotene (vitamin A precursor, oil-soluble), vitamin K (menadione, oil-soluble) are moderately sensitive and water-soluble vitamins like folic acid, pantothenic acid, water-soluble vitamin B_2, B_3, B_6, vitamin B_{10}, vitamin B_{12} (cyanocobalamin), vitamin B complex and ergocalciferol are little sensitive during irradiation processing. A significant change in natural pigment, e.g., chlorophyll, carotene, xanthophyll, anthocyanin, anthoxanthin, etc., is observed in fruits and vegetables when they are exposed to a particular irradiation dose. Radiation preservation of raw meat results in the color bloom due to the formation of bright red colorant oxymyoglobin. Enzyme inactivation takes place by the irradiation treatment caused by X-ray of moderate dose. Gamma-ray irradiation can inactivate the sulfhydryl group-containing enzyme urease [34].

12.10.2 *EFFECT ON SENSORY OR ORGANOLEPTIC PROPERTIES*

The actual irradiation dose on food products depends upon an equilibrium between the requirement and tolerance level without any undesirable changes.

High doses cause more or less organoleptic changes leading to the development of off-flavors and/or textural variation in foods containing large amount of water. Therefore, these unacceptable changes can be considered as serious constraints in order to maintain food quality. Whether foodstuffs derived from animal resources containing a significant amount of protein, the maximum permissible quantity of irradiation conventionally is fixed by the development of the irradiation flavor in permeability.

It has been realized that the radiation induced changes in the texture of some food products can be a limiting factor, e.g., in fresh fruits and vegetables. Polysaccharides in an accelerated spoilage with, e.g., naturally occurring plant polysaccharide can be decomposed by ionizing radiation. The phenomenon is frequently accomplished by the liberation of free Ca and it causes soft texture of the product. Therefore, the viewed enhancement in penetrability can explain the information that the inherent immunity of the plant tissues caused by ionizing radiation is frequently diminished due to rapid deterioration when microorganisms are provided a scope to attack the products after ionizing radiation. However, this primary softness of texture caused by ionizing radiation on several foodstuffs is frequently overcompensated due to the incident that during prolonged post-irradiation preservation, this softness of texture occurs at a considerable slower rate compared to native interstitial tissues. The physicochemical outcome of ionizing radiation can be considered as improvement of quality of irradiated foods. Textural softening caused by irradiation and enhancement in penetrability is preferable during reduction of the cooking time of for instance dried vegetables is required or increased yield of juice can be obtained from irradiated fruits.

12.11 PRESERVATION PERIOD OF RADIATION TREATED FOODSTUFFS

Foodstuffs treated to radappertization process of food irradiation can be anticipated to be stable during storage of canned food. Radappertization does not destroy enzymatic activity. Employing 45 kGy enzymes inactivated in chicken, bacon, and fresh pork, the products to be acceptable after two years. Foods preserved at 70°C were admissible further than other foods preserved at 37.7°C. The influence of ionizing radiation on broiled beef slices, hamburger or frankfurter can be stored for 12 years at refrigerated storage temperature. The above mentioned foodstuffs were packaged with flavor protectives and with the treatment of irradiation dose of 10.8 kGy. The usual deteriorating microflora of marine products is remarkably delicate

Food Irradiation: Concepts, Applications

towards irradiation that 99% of the overall microflora of these products is conventionally killed by the dose level of 2.5 kGy.

12.12 ECONOMICAL AND SAFETY ASPECTS OF FOOD IRRADIATION

However, the process of ionizing radiation provides marked benefits it cannot restore the allowable standards of manufacturing of food products in the context of good manufacturing practices (GMP). The procedure either be employed to screen significant microbial infection or decomposition of foodstuffs, nor can it be utilized to decontaminate any pre-formed toxic components. The procedure as such is comparatively concentrated on capital as compared to labor, but as good as cost-effective to alternative capital concentric manufacturing of food products. The effects of large acute whole-body radiation exposure differ from person to person. Single exposure of skin alone to 6,000–8,000 mSv of X-ray, beta or gamma rays can result in skin erythema. The human reproductive organs are radio sensitive and single exposure in the range of 5,000–6,000 mSv can produce permanent sterility. Acute and prolonged over exposure of the body may result in cancer, life-shortening and genetic disorders.

International Commission on radiation protection (ICRP) as an independent non-governmental international regulatory authority has been recommended the radiation protection standard. As per United Nations risk assessment (UNRA) studies, the health hazards produced by the chemical industry and thermal power plant are much more than those by the nuclear industry in terms of environmental pollution. The safety of irradiated food has been examined carefully both at the national and international levels. A number of test systems have been used to evaluate the wholesomeness and assess the different aspects of safety of irradiated foods for human consumption. Multigeneration feeding studies conforming to the internationally accepted protocols have been undertaken with wheat, shrimp, mackerel, and whole diet exposed to various radiation doses. These extensive studies involving the feeding and examination of thousands of test animals have failed to reveal any evidence of toxicity including mutagenicity and carcinogenicity. International atomic energy agency (IAEA)/world health organization (WHO) joint expert committee has recommended that the foodstuffs irradiated up to an average irradiation dose of 1 Mrad can be accepted as safe from the health point of view and do not present any toxicological hazards. Therefore, considering the worldwide acceptance of various kinds of irradiated food, India also focused

316   Advances in Food Process Engineering

the research and development activity towards the radiation preservation of economically important perishable food commodities.

12.13 FUTURE PROSPECTS OF FOOD IRRADIATION TECHNOLOGY

The final breakthrough of radiation treated food materials relies upon their effectiveness and recognition by the consumers. A fresh food product or method occasionally encounters primary obstruction. Hence it is to be anticipated that foodstuffs treated by ionization radiation may satisfy the primary customer hurdle. However, ionizing radiation can contribute microbiologically safe and healthy food products. Due to deficiency of proper learning, awareness, and significant inadequacy of open option of radiation treated foodstuffs its advantages still have not attained widely among population. It is admissible that by way of proper learning and initiation of product consumer can be illuminated on the matter. A productive strand has been undertaken towards the initiation of radiation treated foodstuffs where customer consortiums have been associated in nations. Nations that ponder commercialization of radiation induced food processing should attempt for complete customer involvement through fruitful common facts and awareness by the administration and commercial production sectors.

12.14 SUMMARY

Food irradiation is a method of preservation which has taken a parallel giant stride during the past six decades in the West. Nowadays, India, Bangladesh, and so many countries in the world have approved food irradiation technology for the extension of shelf life of food commodities. A major impetus to the campaign to use nuclear energy in food preservation, which was launched simultaneously in several countries, was provided by the realization that an increased agricultural output had to be adequately supplemented with suitable post-harvest techniques. Radiation proves one of the most suitable ways to prevent food losses due to adverse climatic conditions and the unsatisfactory storage and handling facilities. Food preservation technologies depend upon basically three mechanisms viz. inhibition, inactivation, and avoiding recontamination of microbes before and after processing, where irradiation of food belongs to the category of inactivation of microbes. Radiation-induced food processing is the regulated implementation of energy of radiation

Food Irradiation: Concepts, Applications

followed by ionization, e.g., gamma beam from irradiating source of Co-60 or Cs-137, Roentgen beam and stimulated negatrons to farm products, foodstuffs, and their constituents for enhancing cleanliness and protection and elongating preservation and transportation period. Ionizing radiations are consisting of wavelength of 2,000 A° or less, e.g., α-particle, β-particle, γ-rays, x-rays, cosmic rays, etc. Ionizing energy or radiation provides energy to kill cellular construction, which includes DNA in bacterial cells, parasites, pests, and molds without appreciably raising the temperature and usually known as cold sterilization. Application of irradiation dose ranges between 0.03 kGy and 50 kGy on food material depends upon the characteristic effects, e.g., inhibition of sprouting, disinfestation of insects and parasites and reduction or elimination of microbes. Ionization radiation can contribute wholesomeness of foodstuffs and without any well-being hazard. Several countries have taken steps towards the commercialization of food irradiation technology for the preservation of onions, potatoes, frozen seafood, spices, etc. Therefore, keeping in view of the above facts, irradiation technology has gained significant importance over the other conventional and non-conventional technology towards the enhancement of shelf stability of food materials.

KEYWORDS

- **controlled atmosphere storage**
- **food commercialization**
- **food irradiation**
- **food preservation**
- **good manufacturing practices**
- **microbial inactivation**
- **modified atmospheric packaging**

REFERENCES

1. Arena, V., (1971). *Ionizing Radiation and Life: An Introduction to Radiation Biology and Biological Radiotracer Methods* (p. 543). St. Louis: Mosby.
2. Aziz, N. H., Souzan, R. M., & Azza, A. S., (2006). Effect of gamma-irradiation on the occurrence of pathogenic microorganisms and nutritive value of four principal cereal grains. *Applied Radiation and Isotopes, 64*(12), 1555–1562.

3. Bongirwar, D. R., (1993). Improved dehydration processes and role of atomic energy in food preservation techniques. In: *Proceedings in National conference on Recent Advances in Biochemical Engineering and Process Biotechnology* (pp. 59–68).

4. CAC (Codex Alimentarius Commission), (2013). *Codex General Standard for Irradiated Foods* (p. 2). CODEX STAN (106-1983. Rev. 1-2003).

5. CENA (Nuclear Energy Center in Agriculture). Available in: http://www.cena.usp.br (accessed on 11 June 2022).

6. Crawford, L. M., & Ruff, E. H., (1996). A review of the safety of cold pasteurization through irradiation. *Food Control, 7*(2), 87–97.

7. Curzio, O. A., Croci, C. A., & Ceci, L. N., (1986). The effects of radiation and extended storage on the chemical quality of garlic bulbs. *Food Chemistry, 21*(2), 153–159.

8. Diehl, J. F., (1995). *Safety of Irradiated Foods* (2nd edn., p. 464). Boca Raton: CRC Press.

9. Eustice, R., (2011). *Food Irradiation: A Global Perspective and Future Prospects.* ANS Nuclear Café. https://www.ans.org/news/article-733/food-irradiation-a-global-perspective-future-prospects/ (accessed on 11 June 2022).

10. Evangelista, J., (2005). *Alimentos: Um Estudo Abrangente* (p. 450). São Paulo: Atheneu.

11. Fan, X., & Mattheis, J. P., (2001). 1-methylcyclopropene and storage temperature influence responses of "gala" apple fruit to gamma irradiation. *Postharvest Biology and Technology, 23*(2), 143–151.

12. Farkas, J., (2003). Food irradiation, Chapter 26. In: Mazumdar, A., & Hatano, Y., (eds.), *Charged Particle and Photon Interactions with Matter: Chemical, Physicochemical, and Biological Consequences with Applications* (pp. 785–812). Boca Raton: CRC Press.

13. Farkas, J., (2006). Irradiation for better foods. *Trends in Food Science & Technology, 17*(4), 148–152.

14. Fellows, P. J., (2016). *Food Processing Technology: Principles and Practice* (4th edn., p. 1152). Duxford–United Kingdom: Woodhead Publishing.

15. Follett, P. A., & Weinert, E. D., (2012). Phytosanitary irradiation of fresh tropical commodities in Hawaii: Generic treatments, commercial adoption, and current issues. *Radiation Physics and Chemistry, 81*(8), 1064–1067.

16. Gava, A. J., (1977). *Princípios de Tecnologia de Alimentos* (p. 284). São Paulo: NBL Editora.

17. Hallman, G. J., (1999). Ionizing radiation quarantine treatments against tephritid fruit flies. *Postharvest Biology and Technology, 16*(2), 93–106.

18. International Atomic Energy Agency (IAEA), (2001). *Consumer Acceptance and Market Development of Irradiated Food in Asia and the Pacific* (pp. 17–19). IAEA-TECDOC-1219. Vienna–Austria: IAEA Publication Series.

19. Institute of Food Science and Technology (IFST), (2015). *Food Irradiation.* https://www.ifst.org/resources/information-statements/food-irradiation (accessed on 11 June 2022).

20. Kilcast, D., (1994). Effect of irradiation on vitamins. *Food Chemistry, 49*(2), 157–164.

21. Kume, T., & Todoroki, S., (2013). Food irradiation in Asia, the European Union, and the United States: A status update. *Radioisotopes, 62*(5), 291–299.

22. Kume, T., Furuta, M., Todoriki, S., Uenoyama, N., & Kobayashi, Y., (2009). Status of food irradiation in the world. *Radiation Physics and Chemistry, 78*(3), 222–226.

23. Loaharanu, P., (1996). Irradiation as a cold pasteurization process of food. *Veterinary Parasitology, 64*(1, 2), 71–82.

24. Marin-Huachaca, N. S., Lamy-Freund, M. T., Mancini-Filho, J., Delincée, H., & Villacencio, A. L. C. H., (2002). Detection of irradiated fresh fruits treated by E-beam or gamma rays. *Radiation Physics and Chemistry, 63*(3–6), 419–422.

Food Irradiation: Concepts, Applications

25. Moy, J. H., & Wong, L., (2002). The efficacy and progress in using radiation as a quarantine treatment of tropical fruits: A case study in Hawaii. *Radiation Physics and Chemistry, 63*(3–6), 397–401.

26. Oetterer, M., Dionosio, A. P., & Takassugui, R. G., (2009). Ionizing radiation effects on food vitamins: A review. *Brazilian Archives of Biology and Technology, 52*(5), 1267–1278.

27. Patil, B. S., Vanamala, J., & Hallman, G., (2004). Irradiation and storage influence on bioactive components and quality of early and late season "Rio red" grapefruit (*Citrus paradisi* Macf.). *Postharvest Biology and Technology, 34*(1), 53–64.

28. Pellegrini, C. N., Croci, C. A., & Orioli, G. A., (2000). Morphological changes induced by different doses of gamma irradiation in garlic sprouts. *Radiation Physics and Chemistry, 57*(3–6), 315–318.

29. Pezzutti, A., Marucci, P. L., Sica, M. G., Matzkin, M. R., & Croci, C. A., (2005). Gamma-ray sanitization of Argentinean dehydrated garlic (*Allium sativum* L.) and onion (*Allium cepa* L.) products. *Food Research International, 38*(7), 797–802.

30. Refai, M. K., Aziz, N. H., El-Far, F., & Hassan, A. A., (1996). Detection of ochratoxin produced by *A. ochraceus* in feedstuffs and its control by γ radiation. *Applied Radiation and Isotopes, 47*(7), 617–621.

31. Rios, M. D. G., & Penteado, M. D. V. C., (2003). Determination of alpha-tocopherol (Vitamin E) in irradiated garlic by high performance liquid chromatography (HPLC). *Quimica Nova, 26*(1), 10–12.

32. Roberts, T., (1998). Cold pasteurization of foods by irradiation. *Virginia Cooperative Extension,* 1–6.

33. Sommers, C., Fan, X., Niemira, B., & Rajkowski, K., (2004). Irradiation of ready-to-eat foods at USDA'S eastern regional research center: 2003 update. *Radiation Physics and Chemistry, 71*(1, 2), 511–514.

34. Tanaka, S., Hotano, H., & Ganno, S., (1959). Actions of radiations on enzymes: Part I-biochemical effects of gamma radiation on urease. *The Journal of Biochemistry, 46*(4), 485–493.

35. Taubenfeld, R., (2013). Irradiated food coming to a supermarket near you. *Chain Reaction, 118*, 10, 11.

36. Wang, J., & Chao, Y., (2003). Effect of gamma irradiation on quality of dried potato. *Radiation Physics and Chemistry, 66*(4), 293–297.

INDEX

α

α-alumina, 178
α-lactalbumin, 190, 191
α-particle, 317

β

β-carotene, 28, 31, 48, 96, 120, 121
β-cyclodextrins, 112
β-lactoglobulin, 190
β-particle, 317

γ

γ-linolenic acid, 31
γ-rays, 317

κ

κ-carrageenan, 105, 107, 110, 116, 12

A

Aspergillus, 264, 268, 269, 283
  flavus, 268, 269, 309, 310
Absorption, 16, 58, 74, 237, 238, 240, 241,
  244, 245, 249, 303
Accelerated electron beams, 306
Acetic acid fermentation, 219
Acetone-butanol-ethanol (ABE), 220, 225
Acidic electrolyzed water treatment, 10
Acidification process, 263
Acoustic sensors, 86
Active
  composites, 101, 102
  ingredient, 94, 96–99, 111, 119, 129
Acylglycerols, 113
Adenosine triphosphate (ATP), 219, 234
Adhesion, 3, 15, 208, 285, 286
Adulteration, 13, 35
Advance
  analytical techniques, 86
  automation interventions, 87

drying techniques, 87
food dehydration techniques, 75
  heat-pump coupled drying, 80
  high electric field (HEF) drying, 84
  impinging stream drying, 85
  infrared (IR) drying, 83
  microwave-assisted drying, 77
  refractance window (RW) drying, 75
  spray-freeze-drying (SFD), 81
  superheated steam drying, 79
  ultrasound (US) assisted drying, 82
Aerobic atmosphere, 219
Aflatoxins, 34, 269
Agarose, 122
Agglomeration, 65, 102, 104
Agricultural
  commodities, 300
  crops, 72
  products, 69–71, 80, 85
Agro-byproduct, 61
  utilization, 61
Agro-food industries, 114
Agro-industries, 171
Agro-produce byproducts, 61
Alcoholic
  beverage dealcoholization, 152
  fermentation, 219
Alfotoxin B$_1$, 34
Alginates, 97, 107, 116
Alkaline
  cleaning agent, 186
  phosphatase inactivation, 291
Alternaria, 264, 283
  alternata, 264
Aluminum, 6, 118, 178, 190
  oxide (Al2O3), 118, 178, 190, 204
Amphipathic compounds, 110
Amylopectin, 59, 115, 313
  molecules, 59
Amylose, 59, 115, 313
Anaerobic treatment, 207

Index

Anion-exchange membrane (AEM), 153
Annihilation, 101
Anthocyanin, 60, 112, 143, 144, 165–167, 266, 313
Anti-allergic effects, 96
Anti-bacterial activity, 12, 282
Anti-carcinogenic, 167
Anti-inflammatory, 96
Antimicrobial
  coatings, 288
  peptides, 111
Anti-nutritional
  components, 65, 240
  factors, 221
Antioxidant, 60
  activity, 13, 31, 96, 112, 161, 165–167, 226
  compounds, 95, 113, 167
  defense mechanism, 263
  properties, 207
Anti-solvent, 26, 28
Anti-yeast activity, 10
Apple juice extraction, 8
Arachis hypogaea, 35
Arbitrary units (AU), 228, 233
Argon (Ar), 260, 268, 269, 277
Aroma
  compounds, 31–33, 154, 169, 193, 198, 203
  recovery (beverages), 203
  rejection, 203
  ring-containing amino acids, 313
Artificial
  flavors, 31
  intelligence neural network, 217
  method, 72
Ascorbic acid, 38, 60, 112, 165, 266, 313
Assessment (fermentation kinetics), 230
  effects of temperature-pH, 230
Asymmetrical nature, 260
Atmospheric
  configuration, 290
  pressure, 198, 201, 221, 260, 280, 288
Atomic
  emission, 305
  force microscopy (AFM), 118
Augusto model, 248, 249
Aureobasilium pullulans, 17
Automation interventions, 87

Auto-oxidation, 313
Auxiliary nutritive substance, 115
Azeotropic, 154

B

Bacillaceae, 283
Bacillus spp., 264
  subtilis, 10
Backflushing, 186
Bacteria, 29, 38, 95, 97, 103, 151, 156, 180, 187, 190, 199, 201, 202, 206, 219, 229, 256, 262, 264, 267–270, 277, 281, 289, 307
Bacteriocin, 111
  activity, 228, 233
  production, 228, 231–233
Bagasse, 206
Baranyi model, 226
Barrel
  configuration, 52, 65
  surface, 53, 55
Beer, 180, 201, 202
  processing, 201
  production, 201
Beeswax, 117
Bentonite, 158, 161, 193
Benzoyl peroxide, 283
Bergamot, 169
Beta-carotene, 313
Beverage, 7, 10, 96, 107, 111, 144, 166, 171, 178, 190, 208, 209, 240
  industries, 209
  processing, 177
Bifidobacterium, 97
Bioaccessibility, 93
Bioactive
  compound, 8, 16, 28, 29, 38, 93–95, 97, 100–102, 114, 122, 130, 143, 144, 151, 166, 167, 171
  molecules, 38
  substances, 288
Bioavailability, 12, 93, 95, 120–122, 130, 221
Bio streams, 178
Bio-capsules, 113
Bioconversion, 217, 220
Biodecontamination, 262, 271
Biodegradability, 207
  polymers, 288

Index   323

Biogas production, 206
Biological
    active
        compounds, 107, 166
        proteins, 143
    matrix, 73
    oxygen demand (BOD), 205
    reactions, 70, 73
    spoilage, 304
Biomass
    concentration, 224, 225, 227, 228, 233
    derived sugar, 169
Biomaterials, 261
Biomedical
    applications, 261
    field, 3
Biopolymeric, 105, 156
    suspension, 107
Bioprocessing industries, 233
Bioreactor model, 221
Biotechnical processes, 171
Biotransformation, 129, 177
    nanoparticles, 129
Bi-polar membrane (BPM), 153
Boiling point, 26, 73, 198
Breakfast cereals, 63
Brewing industry wastes, 205
Broccoli, 96
Brochothrix thermosphacta, 311
Brownian motion, 118
Bulk density, 58

C

Campylobacter, 289, 307
    jejuni, 264, 267, 289
Clostridium botulinum, 267, 289, 309, 310
Calcium chloride, 105, 116
Campden Chorleywood Food Research
    Association (CCFRA), 48
Candelilla wax, 117
Candida albicans, 264, 283
Capillary action, 241, 242
Carbohydrates, 31, 39, 115
Carbon dioxide (CO_2), 25–29, 31–40, 81,
    112, 152, 219, 265, 267
Carboxymethylcellulose (CMC), 116
Carcinogenicity, 315
Cardamom, 31, 32

Cardiovascular diseases, 60, 96
Carica papaya, 308
Carnauba wax, 117
Carotene, 313
Carotenoid, 39, 95, 96, 120, 143, 144
    retention, 265
Carrageenan, 116
Cation exchange membrane (CEM), 153
Cavitation, 5, 8, 10, 16–18
Cell
    biomass production, 232
    dry mass (CDM), 227, 228, 232–234
    membranes, 70, 263
Cellulose
    acetate (CA), 145, 146, 166, 189, 190,
        204, 243
    esters, 149, 151
    nanofiber (CNF), 13
Centrifugation, 40, 119, 144, 156, 158, 161,
    165, 171, 199
Ceramic membrane, 178, 179, 187, 192,
    194, 202, 204
Cereal, 31, 33, 35, 36, 46–48, 60, 63, 96,
    122, 219, 238, 255, 268, 285, 307, 310
    foods, 64
Cesium (Cs), 299, 300, 304, 317
Chamomile, 31, 32
Chaperone protein DnaK gene, 263
Cheddar cheese, 4, 12, 114, 291
Cheese-making industry, 189
Chemical
    agents, 73, 144, 161, 178, 180, 191, 193, 208
    arrangements, 73
    changes, 59, 303
    cleaning
        methods, 186
        protocols, 186
    composition, 47, 287
    contamination, 25
    industry, 315
    insecticides, 303
    interaction, 185
    ionization mass spectrometry, 34
    oxygen demand (COD), 205–207
Chickpea, 238, 268
Chitosan, 12, 105, 107, 110, 116, 120, 122,
    158
Chlorophyll, 313

324 Index

Chocolate production, 240
Chromatographic methods, 291
Chromium, 6, 151
Chromosomes, 282
Cinnamon, 32, 119
  maltrodextrin, 119
  oil, 32
Citric acid, 38, 222, 266
Citrobacter freundii, 283
Cladosporium cladosporiodies, 264
Classical extraction techniques, 40
Close-boiling mixtures, 154
Coacervation techniques, 116
Coagulation, 144, 158, 189, 190, 207, 313
Coaxial airflow process, 106
Coffee decaffeination, 31
Cohesiveness, 15, 290
Cold
  extrusion system, 51
  plasma (CP), 257, 258, 271, 277, 278,
    280, 281, 286, 293
    corona discharges, 260
    dielectric barrier discharge (DBD), 260
    generation methods, 260
    microwave discharge (MD), 261
    plasma jet, 261
    technology (CPT), 278, 281, 285–287,
      289–291, 293
  smoked salmon, 289
  stabilization, 200
  sterilization
    technique, 149, 151
    technology, 169
Colloidal particles, 93, 191–194, 199, 202
Color
  degradation, 129, 167, 194
  reduction, 161, 192
Colorants, 158
Coloring pigments, 37
Combustion reactions, 79
Commercial, 170, 316, 317
  CPT, 291
  distilled oils, 32
  production sectors, 316
Compact installation, 165
Complex, 56
  coacervation, 109, 110
  phytochemicals, 4

Compound annual growth rate (CAGR), 46,
  63, 65, 66
Compression zone, 48
Computational fluid dynamics (CFD), 87
Concentration polarization (CP), 148,
  184–187, 207, 208, 255, 257–260, 262,
  264–270, 277–283, 285–291, 293
Condiments, 310
Confectionaries, 63
Confocal laser scanning microscopy, 119
Controlled
  atmosphere storage (CAS), 301, 317
  delivery, 100
  release, 107, 110, 130
Conventional
  clarification
    processes, 158
    techniques, 158
  clarifiers, 191
  cutting methods, 10
  distillation, 155
  drying, 71, 75, 79, 86, 87
    technologies, 75, 86, 87
  evaporative concentration, 165
  extraction method, 32
  food poisoning bacterial species, 307
  freeze-dried process, 81
  heat treatment, 269
  methodologies, 86, 208, 209
  methods, 4, 9, 16, 17, 26, 29, 30, 32, 33,
    36, 74, 111, 157, 169, 170, 178, 185,
    204, 282
  processing, 13, 26
  raw material, 64
  separation technologies, 170
  steam distillation, 32
  techniques, 26, 182, 191–193
  technologies, 69, 144, 171, 178, 277, 317
  thermal
    concentration technologies, 166
    processing, 167, 193
    treatments, 266
Copernicia prunifera, 117
Core materials, 96
Corn bran, 29
Corona
  discharge, 260, 261
    method, 280, 281
  wind, 84

Index

Cosmetic industry, 29, 93, 94
Cosmic rays, 317
Cost-effective
microencapsulation, 103
preservation technologies, 278
technique, 191
Coulomb electric field, 258
Critical
flux, 186
pressure, 39, 112
Cross
contamination, 75
flow
filtration, 152
velocity (CFV), 155, 167, 184, 187, 202
Cryogenic fluid, 81
Crystallization, 3, 4, 12, 38, 191, 192
Cunnighamella echinulata, 31
Curcuma longa, 38, 114
Curcumin, 120
Customer consortiums, 316
Cyclodextrin, 111, 112, 120
Cysteine, 285, 313
Cytoplasmic membrane, 306

D

Dairy, 3, 4, 13, 33, 102, 121, 122, 167, 170, 171, 177, 178, 185, 187–191, 205, 208, 209, 266, 291
effluents, 205
industries, 205
processing, 177
products, 30
raised wastewater, 205
De-acidification, 143, 177
Dead-end configurations, 152
Dealcoholization, 170, 177, 180, 200, 201, 203
beer, 202
product, 203
Decontamination process, 262
Degree of,
puffing, 54, 58
solubility, 64
Dehydration, 9, 69–78, 82–87, 152, 177
kinetics, 78
methods, 69, 71

process, 71–73, 76, 78, 84
techniques, 71, 72, 75, 82
technologies, 69–72, 75, 84, 86, 87
Demineralization, 143, 151, 153, 170, 180, 181, 191, 208, 209
whey, 180, 191
Denaturation (protein), 35, 47, 59, 64, 240, 313
Deoxyribonucleic acid (DNA), 70, 259, 263, 264, 282, 293, 299, 301, 302, 306, 311, 317
De-pectinized juices, 194
Depressurization, 29, 112
Desalination, 143, 152, 169, 171
seawater, 152
Design (cutting knife), 56
Desorption, 74, 244
Detection (food pollutants), 35
Deterioration process, 70
Diacetoxyscirpenol, 34
Diafiltration (DF), 167, 189, 205
Dialysis (DI), 147, 149, 153, 203
Diatomaceous earth, 178, 199, 202
materials, 199, 202
Dichloromethane, 36
Die
pressure, 57
temperature, 57
Dielectric
barrier discharge (DBD), 260, 261, 265, 266, 269, 280, 283, 285, 287–291, 293
behavior, 73
food system, 84
Dietary fiber, 59
Diffusion coefficient, 118
Digitalization, 61
Dilution rate, 222–224
Direct
contact membrane distillation (DCMD), 207
decontamination, 285
ultrasonication, 5, 6
Discoloration, 303
Discrete molecular interactions, 111
Dissociation, 259
electron attachment, 259
Dissolvability, 115
Docosahexaenoic acid (DHA), 15, 18
Doehlert approach, 226
Double emulsion, 12, 107, 109, 201

326 Index

Downward concave shape (DCS), 237,
  241–243, 245, 248–250
Drug industry, 94
Dry
  extrusion, 64
  systems, 71, 87
  weight (DW), 202
Durable dielectric material, 260
Dynamic
  light scattering (DLS), 118, 119
  steady-state equilibrium, 47

E

Escherichia spp., 264, 309
  coli, 12, 38, 265–269, 282, 283, 287–289,
    291
Easy-to-cook food items, 63, 65
Eco-friendly technology, 45
Effective
  diffusivity, 243
  preservation strategy, 70
Efficiency, 8, 16, 26, 39, 69, 71, 74–79, 82,
  84–87, 98, 102, 113, 118, 119, 143, 148,
  152, 157, 158, 165, 167, 169, 171, 182,
  190, 205, 207, 208, 222, 232, 255, 281,
  282, 291
Eicosapentaenoic acid (EPA), 15
Electric
  discharges, 258
  equipment, 72
  potential gradient, 181
  production, 206
Electrodes, 84, 260, 261, 267, 282
Electrodialysis (ED), 149, 153, 169, 172,
  177, 178, 181, 185, 187, 191, 192, 200,
  208, 209
  ultrafiltration (EDUF), 182
Electrohydrodynamic (EHD), 84, 87
Electromagnetic, 77, 78, 257, 261, 300, 305
  energy, 78
  field, 78
  frequency, 261
  radiation, 77, 257, 300, 305
Electron
  attachment, 259
  impact
    excitation, 259
    molecular disassociation, 259
  mean energy, 259

Electropermeabilization, 263
Electrospinning, 105
Electrospun, 105
Electrostatic
  charge, 191
  extrusion, 105
  field, 105, 191
  repulsion, 191
Elementary
  exponential equation, 226
  mathematics, 226
  mode analysis, 225
Empirical model, 237, 239, 243, 244, 250
Emulsification, 3, 16, 106, 115, 120, 129
Emulsions delivery carriers, 108
Encapsulated, 94–96, 101, 115, 119, 120,
  123, 129
  agent, 36
  aromas, 117
  bioactive compounds, 130
  compounds, 94
  probiotics, 110
  systems, 129
  technology, 106, 113, 130
Encapsulates
  different forms, 97
    coated matrix type encapsulates, 99
    matrix type encapsulates, 98
    reservoir type encapsulates, 97
Endoenzymes, 102
Endosperm, 237, 245
End-product inhibition, 226
Energy
  balance, 49
  consumption, 4, 9, 10, 16, 65, 69, 71, 80,
    83, 143, 152, 158, 161, 165, 170, 171,
    177, 178, 184, 192–194, 198, 200–202,
    208
  efficient process, 7
Enhanced water diffusivity, 83
Enterobacter, 10, 289
  aerogenes, 10
Enterobacteriaceae, 311
Environmental
  conditions, 97, 217, 221, 223, 228, 232
  impact, 25, 171, 202, 206
  parameters, 218
  preservation, 26

Index 327

safety, 129, 256
sustainability, 204
Enzymatic, 204, 304
browning, 111, 166, 266, 283, 303
concentration, 223
hydrolysis, 206
inhibitors, 59
pretreatment, 199
processing, 204
protein hydrolysis, 16
reactions, 15, 71
softening, 16
treatment, 158, 161, 193, 194, 202
Equilibrium moisture content (EMC), 74, 241–245, 248, 250
Essential oils, 13, 32, 95, 96, 101, 110, 111, 113, 116, 121
Esthetic properties, 285
Etching, 263
Ethanol rejection, 203
Ethno-artistic manner, 218
Ethylene, 112
vinyl alcohol (EVOH), 112
European Food Safety Authority (EFSA), 114
Evaporation
moisture, 47
techniques, 161
Excitation, 255, 259, 260
Exoenzymes, 102
Exothermic reactions, 84
Exotic meal consumption, 267
Expansion ratio, 58
External energy source, 57
Extract, 3, 4, 8, 12, 13, 17, 26–38, 40, 61, 112, 144, 152, 167, 169, 180, 207, 238, 240
bioactive compounds, 38
deoxynivalenol, 34
heat-labile compounds, 38
lipids, 36
turmeric rhizomes, 33
Extruder, 15, 46–58, 63, 65, 66
Extrusion, 15, 45–47, 59, 62–65, 67, 98, 99, 105, 106
3D
food printing, 62
printers, 62

cooking, 46–48, 52, 55, 58–61, 64–66
starch, 59
food printer, 62
methods, 105
process, 45, 47, 49, 52, 57, 60, 65
technology, 45–47, 49, 60–65
extrusion-cooking, 47
raw materials for extrusion-cooking, 47
temperature, 51, 58

F

Fat
alcohols, 117
randomization, 39
rich foods, 36
Feed, 55
stream, 155
zone, 48, 53
Fermentation, 8, 12, 13, 15, 28, 61, 113, 149, 154, 199–202, 207, 217–235, 238–240, 301
apple pomace, 13
dairy products, 121
kinetics, 225, 234
parameters, 228
process, 113, 201, 217, 218, 222, 223, 225, 227, 229, 230, 233, 234, 240
products, 13
sausage, 302
system, 224, 225
Fiber, 60, 146
Fick second law of diffusion, 243, 246
Filtration, 3, 8, 38, 40, 119, 145, 149, 152, 171, 180, 183, 184, 192–194, 199, 202–204
process, 8, 203
Financial performance, 170
Fining agents, 178, 180, 193, 194, 199
First-generation technologies, 75
Flaked cereal, 63
Flash
distillation, 169
evaporation, 54
Flavonoid, 38, 226, 266, 308
Flavor, 119
compounds, 32, 169
encapsulation, 119
Flocculation, 144, 158, 180, 192, 193

328 Index

Fluidized bed
  coating, 103, 104
  drying, 71
  granulation, 115
  spray drying, 115
Fluorinated polyolefins, 149
Flux balance analysis, 225
Food
  agri sector, 270
  applications, 40, 78, 84, 86, 117
  borne
    illness, 255, 256, 270
    mycotoxin, 226
  commercialization, 317
  commodities, 70, 256, 287, 299, 303, 307, 309, 316
  customization, 62
  Drug Administration (FDA), 114, 265
  drying technique, 71
  engineering, 61, 86
  grade
    compounds, 95
    encapsulants, 101
    polymers, 117
  industry, 3, 4, 7, 8, 10–13, 15–17, 25–28, 31, 33, 35, 38–40, 45, 46, 54, 61, 74, 93–96, 102, 105, 110, 111, 114, 115, 120, 129, 143, 152–154, 167, 171, 178–182, 184, 185, 187, 201, 204, 206–209, 255, 256, 258, 270, 278, 308
  irradiation, 299, 300, 302, 303, 311, 313, 314, 316, 317
    process, 300, 311
  packaging, 285, 287
    material, 285, 286, 288
  pasteurization, 28, 35
  pigments, 102, 105, 111
  pollutants, 35
  preservation, 3, 4, 8, 16, 69, 70, 219, 277, 278, 293, 300, 301, 303, 306, 308, 316, 317
  processing, 4, 10, 12, 15–18, 25, 27, 28, 40, 45, 60, 72, 83, 84, 87, 93, 143, 144, 171, 177–179, 182, 204, 207, 208, 217, 255, 256, 277, 282, 291, 299–301, 304, 307, 316
    applications, 84, 144
    industry, 25
    sector, 87, 144, 277

product processing, 4, 177
  quality, 13, 17, 87, 277, 278, 299, 314
  safety, 28, 35, 256, 257, 264, 265, 268, 270, 271, 277, 278, 288
  stuffs decomposition, 315
  waste, 10, 29, 204, 208, 209
Fortified waters, 107
Forward
  axial velocity, 55
  osmosis (FO), 149, 154, 161, 166, 172, 177, 178, 180, 182, 189, 198, 208, 209
Fouling, 147–149, 152, 154–156, 161, 177, 178, 182, 184–186, 190–192, 199, 200, 202, 205, 207–209
  control, 186
Fractionation, 28, 30, 118, 177, 180, 187, 189, 190, 209
Fragile food items, 11
Free
  fatty acids, 12, 36, 285
  flowing powder, 95
Freeze
  drying, 9, 71, 75, 76, 81, 82, 99, 106, 110
  point, 73, 200, 312
Frozen seafood, 299, 317
Fruit
  decontamination, 265
  juice, 8, 111, 158, 166, 167, 185, 193, 198, 199, 208, 209
Fumonisins, 269
Functional
  modification, 17
  properties, 4, 13, 15, 16, 65, 66, 95, 169, 190, 217, 218
Fusarium, 283

G

Galactomannans, 115
Gamma rays, 304
  irradiation, 313
  irradiator, 304
  radiation, 301
  sources, 305
Gas
  chromatography (GC), 32, 36, 40
  diffusion (GD), 147, 152, 154
  discharges, 257
  separation, 149, 152

Index 329

Gastric acidity, 100
Gastrointestinal tract, 122, 129, 130
Gelatin, 97, 102, 104, 107, 110, 116, 119, 122, 158, 161, 193, 199
Gelatinization, 47, 58, 59, 64, 240
Gelation agent, 107
Gellan, 105, 107, 116, 120, 122
Generally recognized as safe (GRAS), 105, 114, 130
Genetic
  damage, 262
  disorders, 315
Genipin, 109
Geometrical arrangement, 305
Germicidal effect, 301
Germination, 240, 307
  process, 240
Ginger, 31, 32
Glass transition temperature, 104
Global breakfast cereal market, 63
Glutaraldehyde, 116, 120
Glycerides, 97, 107, 117
Glycolysis pathway, 219
Gompertz
  functions, 226
  model, 226, 232
Good manufacturing practice (GMP), 315, 317
Grains, 65, 70, 72, 75, 85, 237–242, 244, 245, 248–250, 268, 283, 285
Gripping operations, 303
Ground beef, 29
Growth limiting factors-inhibitor, 217
Gum
  acacia, 102, 110, 115, 116, 122
  tragacanth, 115
Gustatory cells, 111

H

Heart-related problems, 96
Heat
  air drying, 74
  exchange rate, 55
  induced techniques, 189
  instability, 116
  mass transfer, 70, 71, 76, 82, 85, 87, 104, 221
  pump
    assisted drying, 80
    condenser, 80
    coupled drying technique, 80
    dried apple, 81
  screw conveyor, 55
  sensitive
    commodities, 80
    components, 77, 84, 86
    compounds, 154, 193, 194
    food compounds, 177
    nutrients, 75, 79, 84
  shrinking, 116
  transfer coefficient, 51
  treatment methods, 306
Helium (He), 259, 260, 262, 267, 269, 277, 289
Hemodialysis, 153
Herbs spices, 269
Heterogeneous, 93, 145, 241, 245
Hetero-lactic fermentation, 219
High
  cost, 56
  density polyethylene (HDPE), 288
  electric field (HEF), 69, 84, 87, 259, 260
  energy
    sub-atomic particles, 300
    utilization process, 17
  frequency ultrasonication, 17
  intensity ultrasonic waves, 7
  performance liquid chromatography (HPLC), 32, 39, 119
  pressure homogenization, 120
  protein low-fat dietary food, 28
  speed bottling, 7
  temperature
    short-time (HTST), 47, 52, 60, 187
    treatments, 187
Hollow-fiber module (HFM), 146, 147
Homogeneous, 145
  liquid sample, 81
  process, 241
Homo-lactic fermentation, 219
Hot
  air drying, 71
    technologies, 75
  extrusion, 52, 65
  fluid cylindrical chambers, 148
  plasma, 279
Human norovirus (HuNoV), 264

330   Index

Hydration, 239, 240, 249, 251
   kinetics, 237, 239, 241–246, 248–250
   model, 238
Hydraulic pressure, 149, 154, 157, 166, 181
   difference, 149, 152
   gradient, 149
   processing, 154
Hydric potential theory, 249
Hydrocolloids, 116
Hydrodynamic
   chromatography, 118
   conditions, 186
Hydrogen peroxide (H_2O_2) 10, 260, 262,
   265, 267, 312
Hydrophilic, 102, 109, 151, 154, 155, 170, 186
   compounds, 109
Hydrophobic, 17, 155
   compounds, 109
   interactions, 155
   membrane, 170, 182, 198, 203
   microporous membranes, 182
   substances, 111
Hydrosoluble, 110
Hydroxyl radicals, 312
Hygiene products, 308
Hypertension, 96
Hypertonic solution, 182

I

Immobilized yeast cell technique, 113
Immunoglobulin (IG), 190
Impinging stream drying, 69, 85, 87
Improved textural property, 10
Inactivation of,
   heat-resistant bacteria, 38
   mesophilic bacteria, 10
   microorganisms, 37
Indirect ultrasonication, 5
Industrial application, 86, 171, 221
Infrared (IR), 71, 75, 76, 82–84, 86, 87
   dried food materials, 84
   radiation, 75, 76, 83, 84
Ingredient incorporation, 221
Inorganic
   components, 152
   materials, 117
In-package
   sterilization, 287
   treatment method, 265

Instant pasta, 64
Instrumental color measurement, 291
Intense bombardment, 263
Internal pore blocking, 185
International
   atomic energy agency (IAEA), 315
   Commission on Radiation Protection
     (ICRP), 315
Inter-related complex process, 218
Intra-atomic structures, 257
Intracellular parasitic protozoa, 307
Ionic polarization, 78
Ionization, 34, 258–262, 282, 305, 316, 317
   radiation, 301–304, 306, 309, 311–316
Ionotropic gelation, 116
Irradiation
   processing, 300, 312, 313
     foods, 312
   technology, 299, 306, 307, 316, 317
Isothermal process, 182, 198

J

Juice processing, 265
   industry, 3

K

Kappaphycus alwarezii, 8
Key-controlled variables, 57
Kinetic, 221
Kiwifruit, 169
Kluyeromyces marxianus, 15
Krypton, 259

L

Listeria spp., 264
   monocytogenes, 265–268, 282, 287–291
Labeling, 286
Lab-scale experiments, 171
Lactic acid
   bacteria (LAB), 114, 219, 234, 240, 268
   synthesis, 225
Lactobacillaceae, 283
Lactobacillus, 13, 97, 233
   amylovorus DCE 471, 231, 233
   delbrueckii, 15
   strains, 13
Lactococcus lactis CECT 539, 231

Index

331

Lactoferrin, 190
Large-scale-industrial applications, 105
Latent heat of vaporization, 73
Lavender, 31, 32
  flavor, 32
Lecithin, 38, 40
Legumes, 72, 122, 238, 242, 268
Lemongrass, 96
Length to diameter (L-D), 56
Light-emitting diode (LEDs), 257
Linoleic acid, 313
Lipid, 117
  degradation, 268
  oxidation, 70, 266, 285
  transesterification, 39
Lipophilic compounds, 12
Liposomes, 110, 111
Liquid emulsion membranes, 201
Livestock products, 278, 281
Logistic model, 226
Long-range radio communication, 279
Low-density polyethylene (LDPE), 288
Luedeking-Piret model, 232
Lycopene, 28, 31, 38, 96, 120

M

Macroemulsions, 108, 109
Macro-mixing, 8
Macromolecular compounds, 158
Macroscopic structure, 249
Macroturbulence, 8
Magnetic device, 306
Maintenance cost, 71, 129, 170
Malted grains, 240
Maltodextrin, 97, 115, 119, 120, 122
Mangifera indica, 308
Manosonication (MS), 5, 10
Manothermosonication (MTS), 5, 10
Manpower utilization, 4
Marinated, 16
  pickled products, 16
Marine products, 77, 314
Marketability (food products), 303
Mascarpone, 190
Mass
  balance, 49, 221, 246
  transfer
    coefficient, 247, 248
    mechanism, 82, 149

Mathematical
  model, 49, 221, 225, 230, 233, 237–239,
    241, 249, 250
  modeling, 217, 234
    approaches, 217
  relations, 221
Meager mechanical requirements, 155
Meat
  analog, 64
  industry, 105, 267, 268, 289, 290
  poultry processing, 289
  processing, 267, 289, 290
  tenderization, 3, 12, 15, 17
Mechanical
  dehydrator, 110
  effects, 4, 5
  energy, 57, 58
  stability, 179
  stresses, 110, 146
Medical application, 261
Mega-molecular compositions, 111
Melassogenic ions, 192
Melt
  rheology, 56
  viscosity, 57
Membrane
  applications, 209
  bioreactor (MBR), 177, 205–207
  cost, 170
  distillation (MD), 149, 154, 155, 155, 161,
    165, 169, 171, 177, 178, 182, 185, 193,
    194, 198, 199, 203, 207–209, 260, 261
  lipid bilayer, 263
  performance, 186, 200
  porosity, 185
  process (MP), 143, 144, 156, 172, 177,
    178, 207, 208
    technology (MPT), 144, 149, 150,
      153–155, 157, 158, 161, 167,
      169–171
  separation processes, 179
  technologies, 143, 151, 169, 171, 177,
    178, 184, 185, 187, 189–194, 198, 201,
    202, 204, 206, 207, 209
  treatment, 161
Menthol, 33
Mesophiles, 265, 267
  bacteria, 167

Index

Mesquite gum, 115
Metabolic
  activity (microbes), 220
  network structure information, 225
  process, 219, 302
  rate, 225
Metabolites, 96, 217, 222–224, 226, 227, 234
Metastable species, 259
Meter zone, 48, 53, 55
Methionine, 313
Microbial
  activity, 194, 217, 240
  biomass production, 222
  cells, 96, 97, 263, 282, 302
  contamination, 52, 270, 283
  decontamination, 47, 262, 277, 293
  deterioration, 72
  growth, 11, 187, 217, 218, 220–222, 226, 229, 230, 233
    kinetics, 217
  inactivation, 10, 17, 18, 261, 263, 264, 266, 283, 317
  inoculate, 223
  load reduction, 149
  population, 224, 226, 227, 291, 302, 306
  reduction, 28, 265, 267
  rejection, 167
  spoilage, 12, 161, 282, 303
Microbiological, 178, 187, 199–202, 208, 209, 301, 304
  pollution, 301
  stabilization, 201, 202
  sterilization, 178, 208, 209
  techniques, 200
Micro-channels, 82
Microemulsions, 108, 109, 120
Microencapsulates, 101, 103
Microencapsulation, 99, 100, 104, 109, 115, 116, 130
  technology, 99, 100
Microfiltration (MF), 146, 147, 149, 151, 152, 155, 158, 161, 167, 169, 171, 172, 177–180, 185, 187, 189–194, 199, 201, 202, 208, 209
Microfluidization, 120
Micronutrient fortification, 60
Micropores, 145, 154
Microsolutes, 153

Microwave, 77, 78, 261
  assisted
    dehydration, 78
    drying technology, 78, 79
  combine plasma treatment, 269
  discharges (MDs), 260, 261, 271
  drying, 78
  plasmas, 261
  process, 78
Milk
  dairy products, 290
  products, 266, 290, 291
Millets, 238, 240
Million electron volts (MeV), 300, 304–306
Minerals, 100, 102–104, 111, 187, 189, 190, 192, 221, 238
Mini-emulsions, 108
Minutes, 38, 282, 287, 289, 291
Mitigation (contaminants), 257
Mixed strain microbial cultures, 218
Mixer geometries, 107
Modified atmospheric packaging (MAP), 301, 317
Moisture evaporation, 76
Molasses, 206
  decoloration, 206
Molecular
  diffusion, 237
  environment, 122
  fragmentation, 302
  friction, 78
  weight cut off (MWCO), 149, 151, 152, 167, 179, 181, 183, 190, 191, 204
Monod model, 226
Monovalent ions, 151, 181, 191
Mozzarella cheese, 190
Much-retained volatiles, 81
Multi-core
  encapsulates, 98
  type system, 99
Multifaceted properties, 262
Multigeneration feeding studies, 315
Multi-spectral imaging, 86
Musa xparadisiaca, 308
Muscular cell integrity, 15
Mutagenicity, 290, 315
Mycotoxin, 28, 33, 34, 264, 268–270, 283, 304
  destruction, 283
Myoglobin, 289, 290

Index

N

Nanoemulsion, 11, 12, 18, 109, 120
  technique, 121
Nanoencapsulation, 99, 100, 130
Nanofiltration (NF), 118, 149, 151, 153, 155, 161, 167, 169, 171, 172, 177, 178, 180, 181, 185, 189, 191, 193, 194, 198, 201, 205, 206, 208, 209
  technology, 167
Natural
  drying, 72
  flavors, 28, 31, 113
  method, 72
  mineralization, 112
  orange oil nanoemulsion, 10
  resins, 117
N-butane, 40
Negatrons, 301, 304, 317
Neon (Ne), 257, 259, 260, 277
Neurocysticercosis, 307
Niacin, 312
Nicotinamide adenine dinucleotide (NAD$^+$), 219, 220, 234
Nisin production, 231, 232
Nitrite (NO$_2$), 263, 267, 289
Nitrogen (N$_2$), 35, 81, 112, 152, 170, 205, 232, 260, 262, 263, 265, 267, 268, 277, 288, 290, 291
Nitrous acid (HNO$_2$), 263
Non-acetate membrane, 193
Non-alcoholic drinks, 107
Non-destructive
  assessment, 86
  method, 17
  property, 4
Non-equilibrium plasma, 258, 261, 280
Non-exponential empirical model, 239
Non-fat dry milk solids, 122
Non-governmental international regulatory authority, 315
Non-invasive, 293
Non-ionizing radiations, 77
Non-isothermal membrane process, 182
Non-particulate radiation, 304
Non-perishable crops, 72
Non-porous
  membranes, 203
  permselective membrane, 181
  selective membrane, 198

Non-processed commodities, 144
Non-sporeforming pathogens, 310
Non-thermal, 17, 277–279, 293
  intervention, 255, 256, 267
  mechanical method, 4
  methods, 17
  plasma (NTP), 258, 259, 279, 280
  processing technologies, 278
  technology, 169, 257, 271
Non-toxic, 13, 25
Non-uniform
  heating, 79
  product quality, 75
Non-vegetarian foods, 65
Nuclear magnetic resonance, 118, 278
Nutraceuticals, 25, 29, 31, 39, 177, 208
Nutritional
  characteristics, 157, 165, 221
  compounds, 165
  evaluation, 59
  parameters, 64, 291

O

Offseason crops, 70
Oil
  in-water (O/W), 11, 12, 106, 107
    in-oil (O/W/O), 11
  seeds, 30, 60, 72, 268
Olive
  mill wastewater, 207
  oil wastewater effluents, 207
Omega-3 fatty acids, 31
Omoto model, 239, 247
Operating parameters, 86, 158, 161, 184, 185, 192, 194, 200, 202, 204
Optimum
  parameters, 165
  temperature, 37
Oregano, 31, 33, 269
  essential oil, 121
Organic
  acids, 191, 233
  chemicals, 35
  membranes, 145, 146
  molecules, 96, 181
  solutes, 186
  solvents, 25–27, 34, 36, 103, 105, 112, 116, 117
  volatile aroma compounds, 203

Organochlorine, 36
Organoleptic, 4, 9, 71, 161, 178, 187,
    199–201, 221, 278, 283, 291, 314
    characteristics, 71, 199, 200
    properties, 4, 161, 178, 200, 278, 283
    quality, 9, 278, 291
Organophosphorus pesticides, 36
Osmotic
    dehydration, 71, 74
    distillation (OD), 161, 165, 178, 180,
        182, 185, 193, 194, 198, 199, 201–203,
        207–209
    evaporation, 165, 166, 182
    membrane distillation (OMD), 207
    pressure, 154, 166, 194, 203
        difference, 154, 166
Over-drying (product), 85
Oxalic acid, 221
Oxidative
    changes, 96
    developments, 285
    enzymes, 158
Oxygen (O_2), 36, 60, 79, 96, 116, 152, 217,
    229, 260, 262, 263, 265, 267–269, 288,
    289, 304, 312
    derivates, 96
    regulated gene (OxyR), 263
Ozone generation, 261

P

Penicillium, 264, 268, 269
    citrinum, 268, 310
Pseudomonas, 264, 268, 309, 311
    aeruginosa, 12, 268
Packaging, 46, 69, 70, 112, 161, 207, 255,
    270, 281, 285–288, 303
    materials, 270, 281, 285, 286, 288
Pancreatic lipases, 129
Paper production, 206
Paraffin, 117
Parahydroxy benzoic acid, 33
Parasites, 256, 302, 304, 306, 307, 317
Parboiling, 238
Partial vaporization, 181, 198
Particulate solids, 53
Passion fruit, 158, 161, 167, 169
Pasteurization, 17, 35, 38, 143, 167, 180,
    187, 190, 207, 290, 291, 306, 310

Pathogenic
    microbes, 270, 283
    microorganisms, 290, 302, 307, 309
Pectin
    methylesterase activity, 266
    β-lactoglobulin, 116
Pediococcus acidilactici PA003, 231
Peleg
    expression, 239
    model, 244, 245
Penicillin formation, 222
Pepper, 31, 33, 110, 269, 282
Peppermint, 31, 33, 269
    oil, 33
Perishable
    commodities, 282
    crops, 70, 72
    storability, 282
Peroxide molecules, 263
Pervaporation (PV), 147, 149, 153, 154,
    169, 171, 177, 178, 181, 182, 193, 194,
    198, 199, 203, 208, 209
Pesticides, 28, 35, 36, 39, 255, 283
Phase
    modification, 115
    transition energy, 51
Phenolic compounds, 96, 121, 144, 167,
    207, 240
Phenomenological
    characteristics, 247
    model, 237–239, 243, 246, 249, 250
Phenylalanine, 313
Phosphorus, 205
Phospoketolase pathway, 219
Photobacterium
    phosphoreum, 268
    phosphorus, 289
Photoionization, 260
Phthorimaea operculella, 308
Physicochemical
    characteristics, 45, 46, 115, 155
    features, 114
    properties, 17, 26, 66, 187, 239, 266
    stability, 12
Physiological characteristics, 225
Phytic acids, 240
Phytochemicals, 25, 96
Phytonutrients, 25, 221

Index

335

Pimpinella saxifraga essential oil (PSEO), 121
Pineapple, 84, 169
Pistachio, 29
Plant
  exudates, 115
  foods, 278
Plasma, 257, 261, 267, 279, 282, 284, 289, 292
  activated water (PAW), 282
  chemistry, 259, 268, 270
  deactivate organic contamination, 286
  generation, 257–259, 261, 262, 266, 267, 281
  jet, 260
    application, 264
  physics, 257
  treated
    in-package strawberries, 287
    polymer, 287
Plasticizers viscosity, 27
Plate frame modules (PFM), 148
Polar compounds, 27, 39
Polyamide (PA), 145, 151, 155, 170, 189
Polybutadiene, 117
Polydextrose, 115
Polydispersity index (PDI), 11, 118
Polyesters, 149
Polyether
  amide, 151
  sulfone (PES), 151, 153, 179, 190, 192, 204, 205
Polyethylene (PE), 206, 286–288, 290
  terephthalate (PET), 65, 286–288
Polygalacturonan, 240
Polymer, 117
  hydrophobic materials, 155
  matrix, 121, 153
  membranes, 178, 179, 192, 202, 204
  packaging material, 288
  substances, 105, 109
Polymerization, 60, 179, 285, 313
Polyphenol, 37, 38, 95, 100, 111, 112, 121, 157, 161, 165–167, 199, 200, 207, 266, 283
  adsorption, 200
  compounds, 29, 96, 121
  nature, 120
  oxidase, 112
Polypiperazineamide, 155

Polypropylene (PP), 117, 288
  membrane, 203
Polysaccharide, 58, 97, 105, 107, 115, 116, 144, 156, 181, 192, 194, 199, 200, 313, 314
  molecules, 193
Polystyrene, 117
Polysulfone (PS), 145, 146, 151, 153, 155, 179, 190, 192, 204
Polytetrafluoroethylene (PTFE), 179, 192, 207
Polyunsaturated fatty acids-enriched oil, 95
Polyvinylidene fluoride (PVDF), 179, 190, 192, 204
Polyvinylpyrrolidone (PVP), 117
Pore diameter, 145, 185
Pork butt, 287, 290
Positive pressure gradient, 48
Post
  harvest techniques, 316
  irradiation preservation, 314
Power law, 49
Precipitation, 26, 29, 38, 99, 112, 144, 193, 313
Preservation
  operations, 72
  technologies, 278, 293, 301, 316
Pressure
  driven concentration technique, 165
  energy, 51
  plasma applications, 264
Printability, 285, 286
Probiotic
  bacteria, 95, 100, 103, 105, 115
  microorganisms, 97, 121
Process parameters, 57, 66, 85, 217, 218
Product quality, 56, 57, 69, 71, 75, 76, 78–81, 83, 85, 86, 177, 189, 218, 256
Propylene glycol, 119
Protein, 16, 59, 116, 189
  denaturation, 16, 59, 268
  materials, 205
  polysaccharide composites, 109
Pseudomonadaceae, 283
Pulsed electric field treatment, 158
Pumping efficiency, 55
Pure titanium probe, 6
Purification, 28, 143, 152, 181, 204, 207, 208
Pyridoxine, 312
Pythium irregulare, 29

Q

Quasi
- neutral gas, 257
- steady-state assumption, 225

R

Radappertization, 299, 309, 314
Radiant energy, 302
Radiation, 300, 302, 308, 309, 311–313, 316
- injury, 302
- intensity, 84
- processed foods, 299
- properties, 83
- resistance, 306
- sterilization, 308
Radicchio leaves, 265
Radicidation, 299, 310
Radio
- emission, 301, 304
- frequency, 261, 264, 269, 306
Radioactive metals, 299, 305
Radiolytic decomposition, 312
Radionuclide sources, 304
Radurization, 299, 311
Rancidity, 304
Ratkowskys square root model, 230
Reactive
- gas species, 255, 261–265, 267, 268, 270, 271
- nitrogen species (RNS), 262, 263, 267
- oxygen species (ROS), 262, 263
- species, 257, 261–263, 270, 282
Real-time production quantification, 217
Refined wheat flour, 64
Refractance window (RW), 69, 71, 75, 77, 87
Refrigeration system, 80
Regulatory authorities, 35
Rehydration ability, 81, 84
Residue build-up, 56
Retained sensorial properties, 77
Reverse
- osmosis (RO), 147, 149, 151, 152, 161, 165, 166, 169–172, 178, 180, 181, 185, 189–191, 193, 194, 198, 199, 201–203, 205, 206, 208, 209
- phase evaporation, 110
Rhizopus spp., 269
Ricotta, 190

Root square model, 229
Rosso model, 226, 228

S

Saccharomyces
- *boulardii*, 219
- *cerevisiae*, 38, 113, 226, 309
Salmonella, 12, 264, 267, 268, 282, 288, 289, 291, 307, 309, 310, 312
- species, 12
- *typhimurium*, 265–267, 282, 287, 288, 291
Staphylococcus, 10, 12, 121, 264, 268, 309
- *aureus*, 12, 264, 268, 282, 283, 288, 291
- *epidermis*, 10
Saponification method, 35
Scanning electron microscopy (SEM), 118
Screw configuration, 53, 55–57
Seasonal accessibility, 72
Second
- generation
  - membranes, 146
  - technologies, 75
- order autocatalytic reaction, 249
Sedimentation, 109, 119, 144, 193
- coating, 109
Seed germination, 240, 270
Selectively, 25, 31, 39, 153, 155, 177, 178, 183, 185, 190, 203, 208
- permeable membrane, 154
Self-cleaning membrane, 208
Self-wiping type extruders, 54
Semipermeable
- membrane, 182, 203
- nature, 154
Senescence, 304, 308, 311
Sensitive probiotic microorganism, 122
Sensory
- attributes, 61, 65, 100, 278
- characteristics, 63, 187
- properties, 74, 143, 157, 158, 200, 201, 303
- quality, 198, 278, 311
Sesame oil, 29
Shear stress, 5
Shelf stability, 70, 299, 301, 302, 304, 310, 317
Shellfish, 302
Shigella, 289, 307

Index

Sigmoidal
  equation, 226
  hydration behavior, 237, 243
  model, 239
Simple
  coacervation, 109
  differential equation, 226
Single-screw extruder, 46
Size exclusion chromatography (SEC), 118, 119
Small-angle neutron scattering, 118
Snack foods, 63
Sodium bicarbonate, 100
Solar drying, 71, 74
Solid
  lipid particle emulsions (SLPEs), 107
  powder food material, 51
Solubilization, 29, 95, 237, 239
Solute-transfer rate, 155
Solution diffusion mechanism, 145
Solvating power, 26, 29
Solvent
  evaporation technique, 120
  injection method, 110
  permeability, 145
Sonication
  assisted nanoemulsion, 11
  factors, 13
Sono-electrocoagulation process, 206
Sorption
  diffusion mechanism, 181
  kinetics, 238
Soxhlet extraction, 40
Soy protein isolate, 110
Specific
  bioactive compounds, 25
  mechanical energy, 57
  radiation treated foodstuffs, 312
Spiral wound modules (SWM), 147, 148
Spoilage microorganisms, 255, 282
Spray
  chilling, 103
  cooling, 103
    techniques, 119
  drying, 9, 75, 101–104, 106, 110, 115, 119, 121, 122, 180
  freeze-drying (SFD), 81, 87
    techniques, 81
  nozzle, 105

Sprout, 299, 304, 317
  inhibitors, 307
Square root model, 228
Stability prerequisites, 114
Stabilization, 60, 93, 130, 178, 180, 187, 199–202, 208, 221
  agents, 178
  rice bran, 60
Starch
  extruded products, 52
  gelatinization, 58, 59
  purification, 240
  raw materials, 63
Steady-state metabolic pathways, 225
Steam generation, 205
Sterigmatocystin, 269
Sterilization, 12, 17, 149, 151, 167, 169, 199, 202, 207, 259, 280, 281, 285, 288, 291, 317
Sterols, 36
Stoichiometric matrix, 222, 223
Strawberry, 96, 167, 169, 282
Structural deformation, 45
Substrate utilization, 222, 225
Sugar, 16, 48, 59, 62, 63, 122, 151, 158, 167, 169, 171, 177–181, 185, 191–193, 200, 206–209, 219, 226, 303
  industry, 171, 185, 191, 192, 206
    wastes, 206
  processing, 191
Sulfur dioxide, 158, 192
Supercritical
  fluid (SCF), 25–30, 32, 35, 36, 39, 40, 99, 112, 113
    chromatography (SFC), 25–29, 36, 38–40
    extraction (SFE), 25–38, 40
    technology, 30, 112
  point, 26
Superheated steam, 71, 79, 80, 87
  drying, 79, 80
Surface
  dielectric barrier discharge (SDBD), 287
  modification treatment, 286
  peeling, 8
  tension, 26, 73, 82, 106
  topography, 155
Sustainability, 122, 255–257, 270
  drying, 70

338 Index

processing technologies, 256
release mechanism, 100
Synthetic
  materials, 31
  polymers, 146

T

Tailormade-protected compounds, 129
Tartaric salts, 200
Techniques for encapsulation
  food products, 101
    complex coacervation, 109
    emulsification process, 106
    emulsion technologies, 107
    encapsulation (cyclodextrins), 111
    encapsulation (yeast), 113
    fluidized bed coating, 103
    liposome entrapment, 110
    microgel encapsulation extrusion
      process, 105
    spray cooling-chilling, 103
    spray drying, 101
    supercritical fluid (SCF) encapsulation,
      112
Techno-commercially feasible, 303
Technological evolution, 145, 146
Temperature
  dependent model, 218
  polarization, 155
  profile, 51, 56, 57
Terminology used, 183
  crossflow velocity (CFV), 184
  flux, 183
  fouling (membranes), 184
  molecular weight cut-off (MWCO), 183
  permeance, 183
  retention factor, 183
  transmembrane pressure (TMP), 184
Textural
  attributes, 289, 290
  characteristics, 114
  quality, 75
  softening, 314
  soy protein, 64
  vegetable protein (TVP), 48, 64, 66
Theoretical postulation, 247
Thermal
  degradation, 32, 33, 35

drying, 85
  techniques, 85
energy transfer, 49
evaporation, 165, 180, 189, 193, 194, 198
plasma, 258, 279, 280
power plant, 315
processes, 166
stability, 153, 179
sterilization, 17
technologies, 143, 169, 256, 278
Thermodynamic
  equilibrium, 258, 280
  principle, 80
  stable isotropic dispersions, 109
  temperature equilibrium, 258, 279
Thermolabile
  components, 32, 40
  compounds, 28, 39
Thermo-mechanical process, 47
Thermoplastic polymers, 179
Thermosensitive
  components, 101
  compounds, 194
Thermosonication (TS), 5, 10, 15
Thiamine, 312, 313
Thin-film hydration method, 110
Thiobarbituric acid, 291
Titanium, 6, 178
  oxide (TiO2), 178, 204
Tocopherol, 28, 37, 38, 313
Tomato, 96, 265, 287
  processing, 28
Total soluble solids (TSS), 161, 165, 166, 198
Toxic harmful chemicals, 170
Toxicological hazards, 315
Toxins, 255, 256
Toxoplasmosis, 307
Transglutaminase, 109
Transition zone, 53
Transmembranal
  flux, 166
  pressure (TMP), 155–158, 161, 165, 167,
    183–185, 187, 189, 192, 200, 202, 203
Transmission, 76
  electron microscopy (TEM), 118, 119
Triboelectric sensors, 86
Trichinellosis, 307
Trichothecenes, 34, 269
Triglycerides, 36

Index 339

Tripolyphosphate, 109, 118
Trypsin, 129, 240, 283
Tubular
  membrane, 166
  modules, 148
Tulane virus, 265
Turbid free passion fruit juice, 161
Turmeric, 33, 38
Types of,
  encapsulated food ingredients, 119
    coloring agent encapsulation, 120
    encapsulation (aroma), 119
    encapsulation (carotenoids), 120
    encapsulation (probiotic organisms), 121
    essential oils encapsulation, 121
    phenolic compounds encapsulation, 121
    vitamins encapsulation, 120
  extruders, 52
    single screw extruder, 52
    twin screw extruder, 54
Tyrosine, 313

U

Ultrafiltration (UF), 118, 146, 147, 149,
  151–153, 155, 158, 161, 167, 169, 171,
  172, 177, 178, 180, 181, 185, 187,
  189–194, 198, 199, 201, 202, 204–206,
  208, 209
Ultrasonic, 4–8, 10, 12, 13, 15–18, 120, 171
  generator, 7
  limitations, 4
  process, 9
  technology (food processing), 7
    cooking, 9
    crystallization-freezing, 8
    cutting, 10
    defoaming, 7
    drying, 9
    extraction, 7
    filtration, 8
    microbial inactivation, 10
    nanoemulsion preparation, 11
  treatment, 82
  waves, 82
Ultrasound (US), 3–17, 46, 69, 71, 82, 83,
  87, 156, 158, 256, 278
  assisted
    brining, 16
    conventional cooling, 8

drying technologies, 82
extraction (UAE), 4, 7, 8
extrusion, 15
filtration method, 8
irradiation, 8
washing (fruits), 12
extraction, 18
Ultraviolet (UV), 190, 255, 256, 259,
  262–264, 268, 270, 271, 279, 301
Ungelatinized starch, 59
United Nations risk assessment (UNRA), 315
Unsaturated
  bonds, 94
  fats, 102
Unstructured models, 225

V

Vaccines, 111
Vacuum
  drying, 71, 75, 78, 82
  evaporation, 202
Value-added products, 61
Van zuilichem model, 51
Vanillic acid, 33
Vapor pressure difference, 153, 154
Vegetative microflora, 187
Versatility, 55
Vibrio, 307
  parahaemolyticus, 268
Vigna unguiculata, 35
Vinasse, 170
Viral surrogates, 264
Virgin coconut oil nanoemulsion, 12
Vitamins, 29, 36, 39, 60, 75, 96, 100, 102,
  103, 111, 113, 120, 121, 143, 151, 166,
  221, 312, 313
  minerals, 60
Volatile
  beneficial components, 161
  compounds, 6, 198, 201, 203
  organic aroma compounds, 198
Volume
  minimization, 204
  reduction, 208
Volumetric
  expansion, 249
  heating, 78

340 Index

W

Wall materials, 95, 97, 102, 114, 130
Waste
  generation, 56, 206
  treatment, 204, 205
  water treatment, 149, 171, 178, 206, 207
Water
  absorption index, 58
  activity, 46, 73, 86, 161, 229, 230, 238, 241
  binding capacity, 15
  decontamination, 270
  evaporation, 104
  hydration
    kinetics, 237
    rate, 239
  in-oil (W/O), 11, 12, 106, 107
    in-water (W/O/W), 11, 12
  melon, 96
  migration, 237, 243
  permeability, 241
  solubility index, 58
  sorption kinetics, 238
  transmembrane flux, 166
Weibull distribution model, 246
Wet milling process, 240
Wheat flour treatment, 283
Whey
  processing, 190
  protein, 190
    concentrate (WPC), 97, 116, 189, 190
    isolate (WPI), 110, 116, 190

Wine processing, 199
  clarification (wine), 199
  dealcoholization (wine), 200
  wine tartaric stabilization, 200
World
  Bank, 256
  Health Organization (WHO), 97, 256, 315

X

Xanthan, 116, 122
  gum, 116
Xanthophylls, 96, 313
X-ray, 305, 317
  diffraction (XRD), 118, 119
  emission electron accelerators, 305

Y

Yarrowia lipolytica, 113
Yeast encapsulation, 113

Z

Zearalenone, 269
Zeaxanthin, 96, 120
Zeolites, 201
Zeta potential, 11, 118, 119, 155
Zirconium ceramic materials, 146
Zwietering model, 229
Zygosaccharomyces rouxii, 264, 266